HIGH-ENERGY NON-AQUEOUS BATTERIES

ELLIS HORWOOD SERIES IN PHYSICAL CHEMISTRY

Series Editor: Professor T. J. KEMP, Department of Chemistry, University of Warwick

Atherton, N.	**ELECTRON SPIN RESONANCE SPECTROSCOPY**
Bugaenko, L.T., Kuzmin, M.G. & Polak, L.S.	**HIGH-ENERGY CHEMISTRY**
Burgudjiev, Z.T.	**INFRARED AND RAMAN SPECTROSCOPY OF MOLECULES**
Cisak, A. & Werblan, L.	**HIGH-ENERGY NON-AQUEOUS BATTERIES**
Harriman, A.	**INORGANIC PHOTOCHEMISTRY**
Hildrum, K.I., Isaksson, T. & Tandberg, A.	**NEAR INFRARED SPECTROSCOPY**
Horvath, A.L.	**HANDBOOK OF AQUEOUS ELECTROLYTE SOLUTIONS**
Jankowska, H., Świątkowski, A. & Choma, J.	**ACTIVE CARBON**
Jaycock, M.J. & Parfitt, G.D.	**CHEMISTRY OF INTERFACES**
Keller, C.	**RADIOCHEMISTRY**
Ladd, M.F.C.	**SYMMETRY IN MOLECULES AND CRYSTALS**
Ladd, M.	**INTERATOMIC FORCES IN SOLIDS AND FLUIDS**
Macasek, F. & Navratil, J.D.	**SEPARATION CHEMISTRY**
Mashelkar, R.A., Mujumdar, A.S. & Kamal, R.	**TRANSPORT PHENOMENA IN POLYMERIC SYSTEMS**
Mason, T.J. & Lorimer, P.	**SONOCHEMISTRY: Theory, Applications and Uses of Ultrasound in Chemistry**
Navratil, O., Hala, J., Kopunec, R., Leseticky, L., Macasek, F. & Mikulaij, V.	**NUCLEAR CHEMISTRY**
Paryjczak, T.	**GAS CHROMATOGRAPHY IN ADSORPTION AND CATALYSIS**
Pasiuk-Bronikowska, W., Ziajka, J. & Bronikowski, T.	**AUTOXIDATION OF SULPHUR COMPOUNDS**
Quinchon, J. & Tranchant, J.	**NITROCELLULOSE**
Sadlej, J.	**SEMI-EMPIRICAL METHODS IN QUANTUM CHEMISTRY**
Southampton Electrochemistry Group	**INSTRUMENTAL METHODS IN ELECTROCHEMISTRY**
Tikhomolova, K.P.	**ELECTRO-OSMOSIS**
Timashev, S.F.	**PHYSICAL CHEMISTRY OF MEMBRANE PROCESSES**

BOOKS OF RELATED INTEREST

Balzani, V. & Scandola, F.	**SUPRAMOLECULAR PHOTOCHEMISTRY**
Hearle, J.	**POLYMERS AND THEIR PROPERTIES: VOL. 1**
Horspool, W. & Armesto, D.	**ORGANIC PHOTOCHEMISTRY**
Kennedy, J.F., *et al.*	**CELLULOSE AND ITS DERIVATIVES**
Kennedy, J.F., *et al.*	**WOOD AND CELLULOSE**
Krestov, G.A.	**THERMODYNAMICS OF SOLVATION: Solution and Dissolution; Ions and Solvents; Structure and Energetics**
Lazár, M., *et al.*	**CHEMICAL REACTIONS OF NATURAL AND SYNTHETIC POLYMERS**
Nevell, Z.J.	**CELLULOSE CHEMISTRY AND ITS APPLICATIONS**
Stepek, J. *et al.*	**POLYMERS AS MATERIALS FOR PACKAGING**
Svec, P. *et al.*	**STYRENE-BASED PLASTICS AND THEIR MODIFICATIONS**

HIGH-ENERGY NON-AQUEOUS BATTERIES

A. CISAK
Department of Pharmacy, Medical Academy, Łódź, Poland,

L. WERBLAN
Department of Chemistry, University of Warsaw, Warsaw, Poland

Translation Editor
T.J. KEMP M.A., D.Ph.I., D.Sc.,
Professor of Physical Chemistry, University of Warwick, Coventry

ELLIS HORWOOD
NEW YORK LONDON TORONTO SYDNEY TOKYO SINGAPORE

POLISH SCIENTIFIC PUBLISHERS PWN
WARSAW

English Edition first published in 1993
in coedition between
ELLIS HORWOOD LIMITED
Market Cross House, Cooper Street,
Chichester, West Sussex, PO19 1EB, England
A division of
Simon & Schuster International Group
A Paramount Communications Company
and
POLISH SCIENTIFIC PUBLISHERS PWN, LTD
Warsaw, Poland

Revised and enlarged translation from the Polish original
Wysokoenergetyczne niewodne ogniwa galwaniczne, published in
1986 by Państwowe Wydawnictwo Naukowe, Warszawa

ISBN 83-01-10838-X (PWN, Poland)

Translated by Peter Senn

Printed in Poland

British Library Cataloguing-in-Publication Data

A Catalogue Record for this book is available from The British Library

ISBN 0-13-388596-8

Library of Congress Cataloging-in-Publication Data

available from the Publishers

Table of Contents

Preface

This monograph is devoted to basic and applied studies on the function and construction of high energy galvanic cells. Practical systems embracing a new generation of primary and secondary, mainly lithium, batteries with high-energy electrode materials and electrolyte solutions especially in non-aqueous, organic and inorganic solvents are discussed.

The first two chapters contain a short outline of the historical development of cells, as well as terms, definitions and conventions concerning cells, electrode processes (using a phenomenological approach), and an elementary quantum mechanical interpretation of current-generating electrode processes.

Chapter 3 describes the basis of the kinetics of electrode processes at the interface of the double layer both of metallic (e.g. lithium) and semiconductor electrodes.

Chapter 4 discusses the Onsager theory of the dynamics of irreversible processes and its application in the description of a working, non-aqueous lithium cell. The transport equations balancing the mass, charge, energy and entropy of the components are formulated.

Chapters 5 and 6 concern the properties of anode (metals and alloys) and cathode materials currently applied in lithium cells. Cathode materials (8 subchapters of Chapter 6) include sulphides, chalcogens, halogens and metal oxides. Their properties, especially in the form of their non-stoichiometric compounds, as well as intercalation compounds, organic cathode materials, such as polymers of non-stoichiometric carbon compounds, and liquid inorganic compounds playing the dual role of solvent and cathode material are thoroughly discussed.

Chapter 7 describes the properties of non-aqueous solvents, mainly organic, two-component mixtures and their electrolyte solutions.

Chapter 8 describes the general principles of construction of cells and storage batteries with non-aqueous electrolyte solutions and solid electrolytes. The characteristics of about 60 lithium systems (batteries) with sulphide, oxysulphide, oxide, halogen, chalcogen and polymer cathodes are discussed. For instance, the theoretical values of electromotive force, specific energy, and temperature coefficient of the SEM cell and the cell characteristics under working conditions are given. The work also includes references up to 1991.

This text was written through the close cooperation of both authors, making equal contributions so their names are given in alphabetical order.

Warsaw, May 1991 The Authors

List of Symbols

a	thermodynamic activity of a component in a system
A_{chem}	chemical affinity
A_r, $\overline{A}_r$	chemical or electrochemical affinity of the r-th reaction
A^0	electrical work of a cell at equilibrium
c_i	concentration of the i-th component
c	molar concentration of the i-th component
C_{v}	molar heat capacity at constant volume
C_{p}	molar heat capacity at constant pressure
D, D_i	diffusion coefficient of the i-th component
$\vec{D}$	induced field vectors
$10\,Dq$	crystal field theory parameter
$\mathcal{E}$	energy, internal energy
$\mathcal{E}_{\mathrm{el}}$	electrical energy
$\mathcal{E}_{\mathrm{ec}}$	electrochemical energy
$\mathcal{E}_{\mathrm{s}}$	specific energy of an electrochemical system
$\mathcal{E}^{\neq}$	activation energy
$\mathcal{E}^{\neq}_{\mathrm{eq}}$	activation energy at equilibrium
$\mathcal{E}^{\neq}_{\eta}$	activation energy at overvoltage η
	$\mathcal{E}_{\mathrm{F}}$Fermi level energy
E	electromotive force (e.m.f.); electrode potential; constant in the equation for electrical conductivity Λ
E^0	theoretical e.m.f. under standard conditions; standard potential of an electrode
E_{H}	potential drop across rigid Helmholtz double layer
E_{eq}	equilibrium potential
E_{η}	voltage of a system at overvoltage η
E_r	activation energy of the r-th reaction
e	electron
F	Faraday constant; force; constant in the equation for the electrical conductivity Λ
F_i, F_j	dissipative forces acting on molecules of the i-th or j-th component
f_i	activity coefficient of the i-th component
f_{ij}	partition functions

G	free energy (thermodynamic potential) of a system
G_i	free energy of the i-th component
g	electrical charge density
H_i	enthalpy of the i-th component
i	electrical current
i_0	exchange current
I_{int}	intensive parameters of a system
I_{ext}	intensive parameters of the surroundings
I	current density; current
J	constant in the equation for the electrical conductivity
J_i	flux of the i-th component
J_{el}	total ion flux
J_q	total heat flux
i_a, J_a	anodic current density
i_c, J_c	cathodic current density
K_A, K_{A2}	ion-pair association constant
K_A^T, K_{A3}	ion-triplet association constant
K_D	dissociation constant of an electrolyte; dissociation constant of ion pairs
K_T	ion-triplet dissociation constant
K_i	ionization constant
k	rate constant; Boltzmann's constant
k_A	association rate constant
k_B	Boltzmann's constant
k_D	dissociation rate constant
k_a	anodic process rate constant
k_c	cathodic process rate constant
L, l	distance separating electrodes
l_D	depth of double layer
L_{ij}	Onsager's phenomenological coefficients
N_A	Avogadro's number
n, z	number of electrons participating in an elementary reaction
p	momentum; pressure
P	polarizability; tensor of pressures; energy lost in a cell
q	electrical charge
Q	partition functions
Q_i	molar heat of the i-th component
r_{int}	internal resistance
r_{ext}	external resistance
R	Ohmic resistance
S	entropy; constant in the equation for the electrical conductivity Λ
T	absolute temperature [K]
t	time
t_k	transport number of the k-th component
U	working cell voltage; internal energy of a system
u_i	mobility of the i-th ion
u_+, u_-	mobility of cation and anion
V	volume of a system
ΔV	electrical potential difference
$\underline{V_i}$	partial molar volume of the i-th component
$\vec{v}$	velocity of an ion

υ_{ri}	stoichiometric coefficient
v_i	velocity of the i-th component
v_{di}	diffusion velocity of the i-th component
W	power of a cell or storage battery
W_{max}	maximum cell power
W_S	statistical weight of the state of a system
W_e	solvation energy of an electron
W_{el}	electrical work done by a cell
w	energy efficiency of a cell
X_i	Onsager thermodynamic force acting on molecules of the i-th component
X	external gradient of an electrical field
x_2	mole fraction of the second component in a mixture
z, z_i	valency of the i-th ion
α_a	transition coefficient for the anodic process
α_c	transition coefficient for the cathodic process
$\gamma_1, \gamma_2, \gamma_3$	concentration proportions of free ions, ion pairs and triplets
δ_{ij}	Kronecker delta
ε_0	permittivity of free space
ε_r	relative permittivity of a solvent
ε_{hfr}	high-frequency optical permittivity
ε_s	static permittivity
σ	entropy source
ζ	potential at the boundary between the rigid and diffuse layers in the double layer (electrokinetic or zeta potential)
ζ_l	potential in the diffuse layer as a function of distance from the electrode surface
η	overvoltage
η_a	anodic overvoltage; polarization potential of the anodic process
η_c	cathodic overvoltage; polarization potential of the cathodic process
η	dynamic viscosity coefficient
η_v	volume viscosity coefficient
η_{rot}	rotational viscosity coefficient
η, η^0	cell efficiency coefficient
κ	specific conductivity of an electrolyte solution; reciprocal of ion cloud thickness; coefficient of thermal conductivity
Λ	molar electrical conductivity of an electrolyte solution
Λ^0	molar electrical conductivity at infinite dilution
λ_j	molar conductivity of j-th ions
μ_D	dipole moment
μ, μ_i	chemical potential of the i-th component
$\mu_{el}, \overline{\mu}_i$	electrochemical potential of the i-th component
$\mu_{e(i)}, \mu_e^i$	electrochemical potential of electrons in the i-th phase
v	stoichiometric coefficient of the i-th component in the r-th reaction
v_+, v_-	number of positive and negative ions in a given electrolyte
$\xi_{R,r}$	extent of the r-th reaction
ρ	density of solution; viscous resistance
ρ_l	reaction parameter which is a function of distance
τ	time constant of an electrode reaction; time constant of ion flow from anode to cathode; relaxation time

φ	volume fraction of dissolved electrolyte
Φ	electrical potential at interface; coulombic efficiency of a cell; electron affinity
Φ_a	electrical potential at the anode surface
$\Phi_{c,cs}$	electrical potential at the cathode surface (c) or at the semiconductor cathode surface (cs)
Ψ_{ij}	wave functions
ψ_{int}	internal potential
ψ_{ext}	external potential
ψ_s	surface potential (potential at the surface of an electrode)
ω	reciprocal of viscous resistance

CHAPTER 1

Historical Outline

Electrical energy is the foundation of modern technological development, the early stages of which were difficult. Efficient sources of electric current are a relatively recent discovery, their history stretching back not quite 200 years. Up till the end of the 18th century the only source of electric current was the electrostatic machine, invented by O. Guericke in 1663.

In 1797 Pearson [1] reported that in order to electrolyse water, the electrostatic machine had to be discharged 14 600 times, and that by this process he obtained only ca. 5.5 cm^3 of the gaseous decomposition products – a mixture of hydrogen and oxygen.

Investigations into the nature of electricity had been continuing for a long time – witness the many descriptions of such studies in the scientific literature of the 17th and 18th centuries. In 1678, for example, Swammerdam [2] was investigating the reaction of isolated muscle tissue towards metals, and he observed the contraction of such muscle placed between two pieces of different metals – copper and silver. In 1762, Sulzer [3] noticed that strips of lead and silver placed simultaneously on the tongue "were surprisingly redolent of the taste of iron sulphate solution, although by themselves neither lead nor silver have any taste".

It was Galvani who, in 1786, finally explained the electrical nature of the phenomena observed by Swammerdam and the other workers of that period [4]. Intrigued by Galvani's interpretation, Volta [5] continued these investigations and established that electrical phenomena occur when two different metals are connected by an aqueous solution of a salt. On the basis of this observation he constructed tha first electrochemical battery, which came to be known as the Voltaic pile. Volta first revealed his discovery in a letter to Banks, the President of the Royal Society of London, who organized a demonstration of the new invention at a public lecture. It was subsequently described in a scientific journal [5] and in the daily press. In the same year (1800) the Voltaic pile was used to electrolyse water. The Voltaic pile combined a large number of cells of the same kind and consisted of small pieces of copper or silver sheet interlayered with similar pieces of tin or zinc, and discs of leather saturated with "water, or a fluid on whose nature the success of the entire device depended" [5]. This fluid was the solution of some salt or lye. Somewhat later Volta described a "wet battery" which consisted of a large number of small vessels containing solutions of a salt or lye and pieces of silver and zinc sheet. A wire connected the different metals immersed in the various vessels. Connected in series, they constituted a source of direct current. Using Volta's battery, Davy produced an electric arc in that same year of 1800, and seven years later he electrolysed molten salts or oxides, recovering pure sodium and potassium [6]. He thereby became the discoverer of these two metals. In 1818, he became the first to perform the electrolytic separation of lithium [7] discovered in the previous year by Arfvedson [8]. With the aid of the Voltaic cell, Oersted made his historic discovery of the magnetic effects of an electric current [9]. This paved the way to further knowledge about electromagnetic phenomena and their technological application.

In 1833–1834 Faraday laid the foundations of the quantitative description of electrolytic phenomena when he formulated his two well-known laws of electrolysis. He too coined the terms "electrode, anode, cathode, electrolyte, electrolysis, cation and anion" [10], which are still used in electrochemistry. His perspicacity as an investigator is illustrated by his bold and still correct statement about currents obtained from cells and storage batteries: "Wherever no chemical action takes place, no electric current is generated".

This fundamental work was rapidly followed by an avalanche of information about the application of new cathode and anode materials and oxidizing solutions in cathode spaces. Even at that time, attempts were being made to use manganese dioxide as a cathode material. One of the earliest systems was the Daniell cell (1836) [11], although the possibility of such a construction had been foreseen by Becquerel in 1829 [12]. The zinc-dichromate cell with an unreactive carbon cathode, affording the relatively high voltage of ca. 2.1 V, was invented by Poggendorff in 1842 [13]. The use of graphite or carbon as a current collector was also nothing new in those days as it had already been predicted by Volta *et al.* [13].

New types of cell began to appear, among them zinc-mercury and zinc-silver cells, both constructed by Marié-Davy in 1859 [14].

One of the most important cells was the system built by Leclanché in 1866 [15]. The anode was zinc amalgam and the cathode a porous mixture of manganese dioxide and a graphite current collector, and the electrolyte was a saturated aqueous solution of ammonium chloride. This was the first cell not to contain highly corrosive solutions of acids or bases; it was cheap and shock-resistant, and supplied a voltage of 1.5 V with minimal self-discharging. These properties were sufficient to ensure its rapid and widespread application in many branches of the technology of those times. Of course, the cell also had its disadvantages, e.g. the incomplete consumption of the manganese dioxide, and the low current strength. The Leclanché cell was one of the leading industrial products of its day and was widely used to power telegraph instruments on the railways and in the postal service. In 1876 there were 20 000 Leclanché cells in use in the railway systems of Europe.

The alkali cell with a copper oxide cathode and zinc anode was introduced by Lalande and Chaperon [16].

Already in 1800 – the year in which the Voltaic pile was invented – it had been observed that oxygen could be used as a cathode material; this demonstrated that the construction of a zinc-air cell was feasible. 39 years later there appeared the precursor of the modern fuel cell which made use of two gaseous reactants – hydrogen and oxygen – and platinum current collectors. Grove, its inventor [17], quickly discovered that other gases, such as chlorine, could effectively replace oxygen as a cathode material in this kind of cell and that H_2/Cl_2 systems, like H_2/O_2 systems, could be used in storage batteries [17]. There were obvious advantages in using zinc as an anode material. It had a high negative electrode potential, it could be used in solutions over a wide range of pH from acidic to basic, its salts were readily soluble in water, and the resulting aqueous solutions had a high conductivity. The fact that zinc was cheap was of no less importance.

Amalgamation as a means of reducing corrosion was applied by Wollaston in 1801 and Brugnatelli in 1802, and later by Davy in 1826.

Tin and iron were investigated with a view to their replacing zinc. In 1847, sodium and potassium amalgams were used for the first time in cells [18] and were described in a paper entitled "A new and practical cell of the greatest power, in which potassium is the positive electrode" (the signs of the electrodes were denoted according to the old convention). In the 1880s there were reports of research into the possible use of cadmium, indium and aluminium as cell electrodes [19].

The possible replacement of zinc by magnesium was first noted in 1865 [13], and serious trials were carried out in 1887 [20]. After the rapid development of all kinds of cells and storage batteries in the first half of the 19th century, only a handful of which have been described in the foregoing paragraphs, there followed a long period during which these achievements were put to practical use. The technology and construction of the basic electrode systems which had been developed by that time were improved. The diversity of sizes of cells and storage batteries was broadened enormously to cater for the various requirements of telecommunications, electrical signalling and other uses of electrical energy. To this day the Leclanché cell, introduced in 1866, has retained its leading position in the manufacture of primary cells. Its voltage of 1.5 V and multiples thereof – 3 V, 4.5 V etc. – is still the standard voltage in most devices powered by direct current. It must be remembered, however, that the tremendous amount of effort and technological knowhow involved in improving this cell has achieved an increase in its practical specific energy of several hundred per cent in comparison with that of the original cell. The theoretical specific energy of the Zn/MnO_2 system is 393 $Wh\,kg^{-1}$ at an e.m.f. of 1.54 V [4]. The specific energy attained in practice is of the order of 50 $Wh\,kg^{-1}$, which is near the upper limit of the practical capability of this cell because the ratio of the mass of the case, electrolyte solution, separators and other parts of the cell to that of the electrode materials is very high. A practical specific energy of ca. 50 $Wh\,kg^{-1}$ is too low for today's needs, and that is why the electronics industry is pressing for cells with much greater capabilities. One important line of development has been the construction of cells using the lightest alkali metals and non-aqueous solutions. This has become possible owing to the not very conspicuous but none the less rapid progress accomplished in the electrochemistry of non-aqueous solutions and electrode processes during the last 30 years.

The work done in the mid-19th century on the application of sodium, potassium and magnesium amalgams as anode materials [13, 18] could be regarded as the start of research into high-energy cells. These materials were first tried in systems containing aqueous solutions, but as they self-discharged quickly, their practical application was a failure. A similarly unsuccessful attempt was the application in 1876 of gaseous chlorine as cathode material in a cell with a zinc anode and an aqueous electrolyte solution [19]. In 1948 this work was taken up once more, with magnesium replacing zinc, and a successful Mg/Cl_2 reserve cell supplying high current densities was produced [21].

The high-energy cell containing a magnesium anode, suitable electrolytes – mainly thiocyanates – dissolved in liquid ammonia, and a variety of cathode materials (MnO_2 and PbO_2, or sulphates) turned out to be very efficient. The mercury sulphate cell [22], $Mg|KSCN, NH_3|Hg_2SO_4, Hg$ supplied a high current density of up to 0.25 $A\,cm^{-2}$.

The rapid progress made in cells with highly negative anodes and electrolytes dissolved in non-aqueous solvents began in the 1960s [23, 24], and it is with these problems that this book is concerned.

REFERENCES

[1] G. Pearson, *Phil. Trans. Roy. Soc., London,* **87**, 1, 140 (1797).

[2] J. Swammerdam, *Biblia Naturae*, Vol. 2 (1737), according to the English translation of 1752, after P.P. Mottelay, *Bibliographical History of Electricity and Magnetism*, Chas. Griffin and Co., London (1922), p. 202.

[3] J.G. Sulzer, *Theorie der angenehmen und unangenehmen Empfindungen*, Berlin (1762), according to Benjamin, Park, *The Voltaic Cell*, J. Wiley and Sons, New York (1983).

[4] G.W. Heise and N.C. Cahoon, *The Primary Battery*, Vol. 1, J. Wiley and Sons, New York (1971).

[5] A. Volta, *Phil. Trans. Roy. Soc., London*, **90**, 403 (1800).

[6] H. Davy, *Phil. Trans. Roy. Soc., London*, **98**, 1, 333 (1808).

[7] H. Davy, *Gilb. Ann.,* **59**, 224 (1818), after H.E. Fierz-David, *History of the Development of Chemistry*.

[8] A. Arfvedson, *Ann. Chim. Phys.*, **2**, 10, 82 (1818).

[9] H.C. Oersted, *Ann. Chim. Phys.*, 14, 417 (1820).

[10] M. Faraday, *Experimental Researches in Electricity*, Vol. 1, ed. 2, Richard and J.E. Taylor, London (1849).

[11] J.F. Daniell, *Phil. Mag. III*, **8**, 421 (1836).

[12] A.C. Becquerel, *Ann. Chim. Phys.*, **41**, 5 (1829).

[13] A. Cazin, *Piles électriques*, Gautier-Villars, Paris (1881).

[14] E.H. Marié-Davy, *Compt. rend.*, **49**, 1004 (1859).

[15] G. Leclanché, French patent 71 865 (1966); H.T. Barnett, *The Electrician*, **33**, 564 (1894).

[16] F. de Lalande and G. Chaperon, French patent 143 644 (1881).

[17] W.R. Grove, *Ann. Chim. Phys.*, **58**, 202 (1843).

[18] J. Goodman, *Phil. Mag. III*, **30**, 127 (1847).

[19] D. Tomassi, *Traité des piles électriques*, Paris (1889), pp. 93, 121, 244.

[20] G. Heim, *Electrotechn. Z.*, **8**, 472, 517 (1887).

[21] G.W. Heise, E.A. Schumacher and N.C. Cahoon, *J. Electrochem. Soc.*, **94**, 99 (1948).

[22] O. Aldhart, "High Rate Ammonia Systems with Magnesium Anodes", in: *Proceedings of the 14th Annual Power Sources Conference*, Fort Monmouth, N.J. (1960).

[23] H.F. Bauman, "Development of Lithium Cupric Fluoride Batteries", in: *Proceedings of the 20th Annual Power Sources Conference*, Fort Monmouth, N.J. (1966); K.H. Braeuer, *Organic Electrolyte High Energy Density Batteries*, Status Report; C.M. Arcand and J.R. Tudor, "Nonaqueous Solvent Electrochemical Systems", in: *Proceedings of the 11th Annual Battery Reseach and Development Conference,* Fort Monmouth, N.J. (1957).

[24] A. Brenner, *J. Electrochem. Soc.*, **106**, 148 (1959).

CHAPTER 2

Theoretical Principles of Action of Galvanic Cells

The source of electrical energy in an electrochemical cell is a chemical reaction. The change in free energy involved in such a reaction is the immediate source of electrical energy. All of this free energy change is convertible into electrical energy only under equilibrium conditions, when the rate of the reaction is zero, i.e. when the rates of the forward and reverse reactions are equal. In this limiting case, the maximum quantity of energy obtainable from a system can be found experimentally from voltage measurements when no net electrical current is passing.

A discharging cell is a system in a permanent state of non-equilibrium, whose working parameters vary continuously with time. A cell in a state of equilibrium, i.e. a cell acting only as a source of voltage but not of current, can be described by classical thermodynamics. However, the description of a discharging cell must be extended to include non-equilibrium processes. This is where the thermodynamics of irreversible processes is needed. Then one can assess the energy losses occurring when current is being drawn from a discharging cell and the quantities of energy still available in the later stages of the same cell's operation. These energy losses depend on the rate at which current is delivered – for any given cell a function of the current strength – and their magnitude is given by the polarization loss and the internal resistance of the cell.

Any description of a discharging cell must take into consideration the transfer of mass and electrical charge in cells and the kinetics of cell reactions.

Reversible cells are those in which the changes taking place are characterized by an infinitesimally small difference between the intensive variables of the system I_{int} (such as temperature or pressure) and the intensive variables of the surroundings I_{ext}

$$I_{\text{int}} - I_{\text{ext}} = \pm dI \tag{2.1}$$

The system is in equilibrium when these two variables are equal:

$$I_{\text{int}} - I_{\text{ext}} = 0 \tag{2.2}$$

These two equations are therefore indistinguishable, and any condition guaranteeing reversibility will also guarantee equilibrium.

Fully reversible chemical cells are important in science and technology and include cells which, like the Weston cell, are not practical sources of electrical energy but are used in compensating and reference systems. Such cells usually remain idle, only a very low current passing when measurements are being made. They are therefore in a permanent state of quasi-equilibrium whose duration is limited only by self-discharging – side-reactions which as a rule are minimal. The state of such cells is well described by the chemical thermodynamics of equilibrium states.

The galvanic cells discussed in the opening paragraphs of this chapter can supply useful work in the form of electrical energy. This is done at the expense of the free energy changes occurring during the oxidation and reduction reactions between tle electrode materials contained in such cells. The traditional definition of oxidation as a reaction in which oxygen is gained and reduction as one in which hydrogen is added has been superseded by a more general definition in which oxidation is associated with electron loss and reduction with electron gain. In this book we shall be using the electrochemical definition according to which a chemical species – ion, atom, molecule or functional group of an electrode material – is oxidized when it loses electrons and reduced when it gains them. In a cell, oxidation takes place at the anode, reduction at the cathode. If the direction of current flow in a cell is reversed, the functions of the electrodes change and electrolysis now occurs. When reacting substrates are brought together in a suitable environment, a reaction is set off in which a certain quantity of heat is evolved. Now, although this reaction proceeds rapidly, the heat is not converted into electrical energy; it merely raises the temperature of the system. Altering the temperature or pressure does not usually reverse the reaction; this is an entirely irreversible process. All that happens is that the solution either cools, with some of its heat escaping to the surroundings, or heats up, if the reaction proceeds further, e.g.

$$\text{Zn} + \text{CuSO}_4 \rightarrow \text{ZnSO}_4 + \text{Cu} \ \ \text{(metal)}$$

If this same reaction occurs in an electrochemical cell, one obtains not a certain amount of heat but its equivalent quantity of electrical energy $\mathcal{E}_{\text{el}}$, which, thermodynamically, is equal to the free energy change ΔG:

$$\mathcal{E}_{\text{el}} \text{ (joules)} = nF \text{ (coulombs)} \times E \text{ (volts)} = -\Delta G \tag{2.3}$$

where E is the electromotive force (e.m.f.) of the cell. Relationship (2.3) can be regarded as the electrochemical definition of free energy. The cell operates in such a way that the overall reaction proceeds as a combination of two half-reactions, each in a separate compartment or half-cell. The oxidation reaction occurs in one half-cell in the anode space, i.e. at the interface between the electronegative material and the electrolyte solution, while the reduction reaction occurs in the other half-cell in the cathode space, i.e. at the interface between the electropositive material and the electrolyte solution. For two foregoing reactions we have:

$$Zn^0 \text{ (metal)} - 2e \rightarrow Zn^{2+} \qquad \text{(oxidation)}$$
$$Cu^{2+} + 2e \rightarrow Cu^0 \qquad \text{(reduction)}$$
$$Zn^0 + Cu^{2+} \rightarrow Cu^0 + Zn^{2+} \qquad \text{(overall reaction)}$$

The half-reactions cause equilibria to be set up at their respective interfaces. This results in a potential difference which, when measured under open-circuit conditions, is the e.m.f. of the cell. As soon as the two electrodes are joined by a metal conductor, there is a flow of electric charge which balances the reaction and provides the current which can perform useful work, powering any device connected to the external circuit. The value of the potential difference measured when current is being drawn is called the cell voltage.

For a pair of half-cells in which all the constituents are present in their standard states, the electrical energy is equal to

$$\mathcal{E}_{\text{el}}^0 = nFE^0 \tag{2.4}$$

where E^0 (standard e.m.f.) is the potential difference between the two half-cells in their standard states, i.e. when the activities a of the constituents are unity ($a = 1$).

By convention, the half-cell $H_2 \rightleftharpoons 2H^+ + 2e$ in its standard state is assumed to have zero potential. The standard hydrogen electrode is therefore the generally accepted electrochemical reference system and starting point of the oxidation-reduction (redox) scale of potentials.

The current-generating electrochemical reaction of a cell is a heterogeneous chemical reaction in which the participating chemical species give rise to an electrical potential difference. When the reaction proceeds in separate compartments, it causes electrical charge, i.e. a current, to flow. The direction of a chemical reaction in a cell (written from left to right) is indicated by the sign of the change in free energy ΔG and of the chemical affinity A_{chem}

$$A_{\text{chem}} = -\Delta G \tag{2.5}$$

with respect to one mole of reaction product.

A cell is represented schematically [1] to indicate briefly its composition, e.g.

metal conductor | Zn | Zn^{2+} || Cu^{2+} | Cu | metal conductor

both conductors should be made of the same metal. As far as sign and magnitude are concerned, the electrical potential difference ΔV is equal to the internal potential ψ_{int} of the right-hand metal conductor minus the internal potential of the identical conductor on the left: $\Delta V = \psi_r - \psi_l$. According to an older convention [2] the potential difference ΔV is called the electrical voltage of the cell U, and the internal potentials of the conductor are subtracted in the reverse order, so that $\Delta V = -U$.

If the electrochemical reaction in a cell is written as

$$Zn + Cu^{2+} \rightarrow Zn^{2+} + Cu$$

then according to the above, an electric current flows through the cell from left to right. If that is the direction of the current when the cell is shorted, ΔV is positive, and this will be the case here unless the quotient $[Cu^{2+}]$ / $[Zn^{2+}]$ is extremely small.

If, however, the same reaction is written in the reverse order

$$Cu + Zn^{2+} \rightarrow Cu^{2+} + Zn$$

which corresponds, for example, to the reactions taking place when storage batteries are charged, the cell should be represented as follows:

metal conductor | Cu | Cu^{2+} || Zn^{2+} | Zn | metal conductor

In this case ΔV will be negative, unless the quotient $[Cu^{2+}]/[Zn^{2+}]$ is very small. According to this convention, the electrode written on the left is the anode, that on the right is the cathode.

The e.m.f. [2] of a galvanic cell E is given by the chemical affinity A_{chem} divided by the product of the number n and the Faraday constant F:

$$E = A_{\text{chem}}/nF = -\Delta G/nF \tag{2.6}$$

The electrochemical reaction coefficient n is the number of moles of electrons (faradays) transferred by the cell from left to right during a cell reaction in which one mole of reactants participate. This means that if positive charges are transferred from left to right, n is positive, but if from right to left, n is negative. The limiting value of the electrical potential difference at zero-current flow through the charged cell is called the electromotive force (e.m.f.) [1] and is denoted by E.

The electrode potential or half-cell potential is the e.m.f. of a cell, on the left of whose schematic representation there is a standard hydrogen electrode and on the right the electrode in question. For instance, a zinc electrode or the Zn^{2+}/Zn half cell will be represented by

$$\mathrm{Pt, H_2 \mid H^+ \mid Zn^{2+} \mid Zn} \tag{2.7}$$

The reaction at the zinc electrode will then be

$$\mathrm{Zn^{2+}} + 2e \rightarrow \mathrm{Zn^0} \tag{2.8}$$

which is the half-reaction of the overall reaction

$$\mathrm{Zn^{2+} + H_2 \rightarrow Zn^0 + 2H^+} \tag{2.9}$$

According to the accepted convention, the e.m.f. of this cell in the standard state takes a minus sign (–0.763 V), because hydrogen induces the reduction of zinc ions to metallic zinc; such a reaction does not occur in the standard state (1 atm, $a = 1$). Hence the standard potential of the zinc electrode is also negative: –0.763 V. The symbol Zn/Zn^{2+} indicates the cell $Zn \mid Zn^{2+} \mid H^+ \mid H_2$, Pt in which the spontaneous reaction of zinc dissolving in an acid $Zn + 2H^+ \rightarrow Zn^{2+} + H_2$ proceeds. The e.m.f. of a cell written in this way cannot be called the electrode potential since the hydrogen electrode is not the anode. In accordance with the above definitions it should therefore take the sign opposite to that of the potential, i.e. a plus sign.

If both the hydrogen and investigated electrodes are in their standard states, the potential of the latter is equal to the standard potential of that electrode.

Under standard conditions (activities of ions equal to 1, $p_{H_2} = 1.01325 \times 10^5$ Pa), the electrode potentials defined above can be arranged in the electrochemical series whose signs, if established as indicated above, correspond to the convention now adhered to [1, 3], formerly known as the "European" convention [4].

When a current i flows through a galvanic cell, the corresponding voltage U_i referred to a unit electrode area becomes a function of the algebraic value of the current density $I = i/\mathrm{cm}^2$. The difference $\Delta U = U_i - U_{\mathrm{rev}}$ is called the polarization of the cell. This is positive when a positive current flows from left to right in the schematic representation of the cell and negative if the current flows in the opposite direction.

If an electrode reaction occurs at each electrode, then regardless of whether the cell is supplying electrical energy or taking it up (as, for example, when a storage battery is being charged), the total polarization of the cell is the sum of the overpotentials at the anode η_a and cathode η_c, and the Ohmic loss iR. It must be remembered [2] that for a given state of a cell (phase composition, temperature, pressure) the e.m.f. E remains the same no matter whether the circuit is closed or open ($i = 0$), whereas the cell voltage U_i is a function of the current density. The e.m.f. depends on the kind of materials from which the cell is constructed and not on the rate at which they are consumed when the cell is discharging; the latter is measured by the strength of the current flowing through the cell. This emerges from the previously given definitions of voltage and e.m.f.

The voltage of a cell is an electrical quantity and is the resultant of the conditions under which it works; the e.m.f., however, is a thermodynamic quantity, directly linked with the changes in free energy and chemical affinity [2]. Within and in accordance with the generally accepted principle of electroneutrality, cathode processes proceed independently of anode processes. Cells can therefore be set up from half-cells containing a diversity of electroactive materials; e.g. besides zinc-copper cells, there are zinc-silver or lithium-copper cells. This is why electrode reactions and also the mass and charge transfers associated with the current density at the electrode must be discussed separately for each half-cell.

2.1 THE ENERGY OF A CELL

Generally speaking, changes in any kind of energy are the product of two variables. One must be an intensive quantity, i.e. characteristic of the state of a system but independent of its total mass, the other must be an extensive quantity, i.e. independent of the qualitative properties of the molecules or particles making up the system, but defined by their number, and therefore by the mass of the system. Intensive quantities include pressure, temperature, density, refractive index and all potentials, including chemical and electrochemical potential; mass, volume, heat and entropy are extensive quantities. Chemical (and electrochemical) energy lies within the broader concept of a system's internal energy, since chemical energy is not directly measurable. On the other hand, the changes in this energy due to the breaking and making of bonds during chemical reaction are perfectly measurable, and it is these which are the main object of interest as far as chemical sources of useful energy are concerned. So, in brief, we refer to the free energy of a reaction when we have in mind a change in this quantity, although the potential differences are intensive quantities determining the difference between the states of systems. The electrochemical energy

$\mathcal{E}_{ec}$ – the electrical energy of a current-generating reaction – is therefore the product of the cell e.m.f., a quantity measurable as the maximum cell voltage U_{max} = e.m.f. (intensive quantity), and the electrical charge nF transferred or exchanged during an electrochemical reaction (extensive quantity).

While it is discharging, a cell's electrochemical energy is converted into electrical energy $\mathcal{E}_{el}$, and when the circuit is open the electrochemical energy remains stored inside the cell. Electrochemical energy is thus a kind of potential electrical energy which, given the appropriate external conditions, is directly and entirely convertible into work

$$-\Delta\mathcal{E}_{ec} = EnF, \qquad \Delta\mathcal{E}_{el} = \overline{U}q = \overline{U}\overline{I}t \tag{2.10}$$

($\overline{U}$ and $\overline{I}$ are the mean values of U and I).

This particular kind of energy, which is wholly convertible into work, is called the Gibbs free energy ΔG, or more precisely, the change in Gibbs free energy which can be converted into useful work when p and T are constant, or the change in Helmholtz free energy ΔF when V and T are constant.

2.2 PHASE POTENTIALS AND CELL E.M.F.

The foregoing definitions make it clear that a galvanic cell is a heterogeneous, multiphase system: it must consist of at least three phases – two electrode materials and one electrolyte solution. Except for fuel cells, galvanic cells are usually closed systems, i.e. no matter is exchanged with the surroundings, although at the electrodes, the boundary phases of the system, there is an electrical potential difference characteristic of the cell in question. This potential difference has a strict thermodynamic meaning only for compensatory measurements and for reversible cells, i.e. cells in which no irreversible processes such as corrosion or leak-induced changes in electrolyte composition occur. Such cells are in an equilibrium brought about by the external, compensatory circuit. If these conditions are satisfied, the measured voltage is the cell's electromotive force (e.m.f.) (not really a force at all), which appears under suitable conditions, and is measured and expressed in volts. According to this definition, the e.m.f. is a compound quantity – the sum of the internal potentials ψ_{int} of all the phases making up the cell.

The principle of electroneutrality is always upheld within an electrically conducting phase [6] like a metal or electrolyte solution, i.e. the algebraic sum of all the positive and negative charges is equal to zero: $\sum_i n_i z_i = 0$. At equilibrium there is no excess of one kind of charge within a phase which could be a source of potential; because they repel one another electrostatically they move to the surface of the phase. The excess charge accumulates at the phases surface where it gives rise to an external potential ψ_{ext} differing from the internal potential ψ_{int} defined by the thermodynamic state of this phase. The difference $\psi_{int} - \psi_{ext} = \psi_s$ is called the surface potential. Hence, the internal potential is the sum of the external and surface potentials: $\psi_{int} = \psi_{ext} + \psi_s$, so it is possible that chemically identical phases (with identical ψ_{int}) differ in their external potentials. Two otherwise identical conductors may therefore have different electrical potentials. This difference between the external potentials ψ_{ext} – the Volta potential – is directly measurable. The change in external potential can be variously determined, e.g. electrostatically, by measuring the electrical capacity, but it is impossible to measure experimentally the difference in surface potentials. Hence the difference between the internal potentials of two phases, i.e. the Galvani potential $\Delta\psi_{int}$ (also known as the contact potential), of two electrically conducting phases in contact with each other, is not an experimentally measurable quantity.

The electrical charge causing excess potentials to appear is extremely small, so the chemical composition of the phases remains practically unchanged. This emerges from the well-known relationship between the external electrical potential ψ_{ext} of a sphere of radius r and excess surface charge q; $\psi_{ext} = q/4\pi\varepsilon_0 r$, where ε_0 is the permittivity of free space; $\varepsilon_0 = 8.854 \times 10^{-12}$ C V^{-1}m^{-1}, whereas $4\pi\varepsilon_0 = 1.11 \times 10^{-10}$ C V^{-1}m^{-1}. For a sphere approximately 1 dm^3 in volume of radius $r = 6 \times 10^{-2}$ m the excess electric charge q (in the case of monovalent ions or electrons) producing an electrical potential of 1 V is 7×10^{-12} C; this is equivalent to ca. 10^{-16} moles of electrons or monovalent ions, which is a very insignificant number indeed. Thus we can speak about two phases with the same chemical composition (and activity), but different electrical potentials. The chemical compositions of two such phases do in fact differ, but so slightly that they can be presumed indentical. However, the presence of charged species (ions in particular) in a system or phase means taking into account the fact that their state depends both on the electrical state of the phase (ψ_{int}) and on the latter's chemical composition. To emphasize this, the chemical potential μ_i of these species is called their electrochemical potential μ_{el}. The difference in electrochemical potential μ_{el} between two chemically identical phases (1) and (2) but unequal electrical potentials is proportional to the total electric charge $q = z_i n_i F$ carried by those species and to the difference in electrical potentials $\Delta E = \psi_1$ (phase 1) – ψ_2 (phase 2), between phases:

$$\mu_{el}(1) - \mu_{el}(2) = z_i n_i F(\psi_1 - \psi_2) = nF\Delta E \qquad (2.11)$$

Equation (2.11) thermodynamically defines the electrical potential difference ΔE between two chemically identical phases; the charged component i in these phases is in equilibrium when the electrical potentials are equal, $\psi_1 = \psi_2$; $\Delta E = 0$.

If phases of differing chemical composition are in equilibrium, the separation of the component or the partition between these phases will be such that the electrochemical potentials of this component in the various phases are equal: $\mu_{el}(1) = \mu_{el}(2)$ etc. On the other hand, any division of the difference $\mu_{el}(1) - \mu_{el}(2)$ into a chemical and an electrical part is arbitrary and is physically untenable, because the electrical potential difference between two points in different media can never be measured and has not been defined in real systems. This is why, in the foregoing definitions of a galvanic cell, its voltage U is the electrical potential difference between conductors of the same material. This is due to the fact that in all real electrochemical reactions, including, obviously, those occurring in cells, the principle of electroneutrality is always upheld. In view of this, every transfer of charge of one sign (i) is accompanied by the neutralization or transfer of charge of the opposite sign (j), so that there is no net change in electrical charge:

$$\sum z_i n_i F = -\sum z_j n_j F, \quad \text{i.e.} \quad \sum z_i n_i F + \sum z_j n_j F = 0 \qquad (2.12)$$

So one should consider the chemical potentials of electrically neutral components such as salts, e.g. $\mu_{AgNO_3} = \mu_{Ag^+} + \mu_{NO_3^-}$ and the electrochemical potentials of particular kinds of ions. A more detailed discussion of chemical and electrochemical potentials along with a number of interesting examples will be found in Guggenheim [6]. In a general discussion of cell operation, chemical potentials are of fundamental importance because it is these which define the theoretical cell variables, in particular the theoretical maximum voltage and the theoretical maximum energy efficiency of a current-generating system.

The chemical potentials of a component i are expressed by the partial changes in free energy with respect to that component

$$\mu_i = (\partial G/\partial n_i)_{T,\,p,\,n_j}$$

where

$$dG = -SdT + Vdp + \sum_i \mu_i dn_i \qquad (2.13)$$

A significant characteristic of galvanic cells is that the overall processes in which ions participate take place in such a way that charge transfer does not violate the principle of electroneutrality in the several phases making up the cell. So it is the chemical potentials of substances and not the electrochemical potentials of ions which define these processes. The progress, or advance of an electrochemical (redox) reaction is therefore proportional to the total charge transferred by ions in a solution or by electrons in a metallic conductor, and the electrical work $qE = nFE$ involved is equal to the free energy change $-\Delta G = nFE = \sum z_l n_l FE$, assuming that the process goes on at constant pressure, which is usually the case. The work associated with the unavoidable changes in phase volume during a reaction is taken into account, since the electrical energy is the sum of the Helmholtz free energy ΔF at constant volume and all the phase volume changes in the system $\sum pdV$,

$$nFE = -\Delta F - \sum pdV \tag{2.14}$$

Electrical energy is thus produced at the expense of free energy, and since this decreases, the signs in the above equations are negative.

2.3 REVERSIBLE CELLS

Consider a galvanic cell in which, once the electrical circuit has been closed, current-generating reactions proceed spontaneously because the oxidation processes at the anode spatially separated from the reduction processes at the cathode. At equilibrium, the rates of the electrochemical reactions and the associated flow of current are equal to zero, so there is now no distinction between anode and cathode. However, it can be said that the electrode which is more negative in spontaneous processes becomes the anode, in other words, the source of the electrons supplied to the external electrical circuit by oxidation processes of the type red (1) $\rightarrow$ ox (1) + *ne* taking place at its surface. Simultaneously the more positive electrode becomes the cathode, because stoichiometrically equivalent reductions of the type ox (2) + *ne* $\rightarrow$ red (2) are occurring there. Clearly, at least four chemical species are present during an electrochemical reaction: the oxidized and reduced forms of the electronegative or anode material, i.e. the substance being oxidized at the anode during a spontaneous reaction, and the oxidized and reduced forms of the electropositive or cathode material which is simultaneously reacting spontaneously at the cathode. Each of these reactions taken separately constitues a half-cell whose potential is defined by the ratios of the concentrations of the oxidized and reduced forms, while the standard potential E^0 is defined with respect to a standard half-cell, usually the standard hydrogen electrode. The overall chemical process determining the thermodynamics of the cell is given by the sum of the half-reactions in the half-cells

$$\begin{array}{c} \text{red (1)} \rightarrow \text{ox (1)} + ne \\ \text{ox (2)} + ne \rightarrow \text{red (2)} \\ \hline \text{red (1)} + \text{ox (2)} \rightarrow \text{ox (1)} + \text{red (2)} \end{array}$$

The number of electrons in the first two equations must be identical for the electroneutrality principle to be upheld (after taking the reaction stoichiometry into account).

Suppose a voltage greater than the equilibrium potential (at which $i = 0$) is applied to a reversible cell from an external source (e.g. from a potentiometer), without the polarity being reversed, of course. Even though the signs of the electrodes –(+) for the more positive one and (–) for the more negative one do not change, current now flows in the opposite direction, showing that the functions of the electrodes have changed. When a flow of current is induced, electroreduction occurs at the more negative electrode, which becomes the cathode, and correspondingly, the positive electrode becomes the anode, as always happens in electrolysis. In fact, this case corresponds to the charging of a storage battery, in which the electroactive materials consumed during discharge are regenerated and deposited at the electrodes during the electrolysis induced by an external current source such as a rectifier.

Both half-cells, I and II, i.e. the spaces containing the electrodes and electrode materials, are separated to prevent self-discharging. However, they must be connected by an electrolyte solution if charge transfer is to be possible while electrical neutrality is maintained. This solution usually contains one readily dissociating electrolyte whose ions are present in both electrode spaces. If the electrolyte solution does not actually participate in the electrode reactions, it merely carries current; strictly speaking, though, it does affect the activities of those ionic forms of the electrode materials dissolved in it and electrically neutralized by an appropriate number of ions of the opposite sign. All forms of electrode materials can therefore be treated as systems of electrically neutral substances whose conversions can be described in terms of changes in their chemical potentials.

The concentrations of a common ion in the electrolyte solution, e.g. X^-, may differ in the cathode and anode spaces. So, to take account of the different activities of an ion in the electronegative electrode space (I) and the electropositive electrode space (II) electrochemical potentials μ_{el} or $\overline{\mu}$ have to be introduced, furnished in particular cases with the symbol of the ion or electron, and additionally perhaps with the sign of the charge on the ion or electron, e.g. μ_e or μ_{Ag^+}.

In the general case, the cell can be described according to the accepted conventions as follows:

$$P_I \mid M_I \mid U_{red}, U_{ox}, X_I \parallel X_{II}, D_{red}, D_{ox} \mid M_{II} \mid P_{II}$$

The double lines indicate the interface separating half-cell I from half-cell II; there is a diffusion potential across this interface.

Let P_I and P_{II} be conductors made from the same metal P, taking current from electrodes I and II; the more negative one (M_I) should be written on the left and the more positive one (M_{II}) on the right.

M_I and M_{II} are electrodes, usually of different electron-conducting materials (metal, alloy, semiconductor). Each is in contact with an electrolyte (solid, liquid, or in solution) and suitable electrode materials, each of which is present in its oxidized (ox) and reduced (red) forms due to electron transfer at the electrode. These interfaces are denoted as M_IU_{red} and $M_{II}D_{ox}$. If the cell is discharging, electronegative material U_{red} is reacting at the surface of electrode M_I, releasing electrons during anodic oxidation, while at electrode M_{II} the electropositive material D_{ox} is taking up electrons during cathodic reduction. The form of the reduced electronegative electrode material is usually identical with the material of electrode M_I, e.g. in the Daniell cell it is metallic zinc. In most non-aqueous cells it is metallic lithium, but in the hydrogen cell it is gaseous hydrogen. The same applies to electropositive materials – in the Daniell cell the mass of electrode M_{II} and, at the same time, the form D_{red} of the electropositive material are metallic copper. In other cases, the form of D_{red} can be present in oxidation states other than zero, but these are always lower than the oxidation state of the form of D_{ox}. For example, in some non-aqueous cells, the mass of electrode M_{II} is copper(II) sulphide CuS, which is both a semiconducting electrode and the electropositive material D_{ox} reduced to copper(I) sulphide Cu_2S, i.e. to the D_{red} form, during the cathode reaction. More often, however, the material of D_{ox} or D_{red} is quite different from that of M_{II}, as in the hydrogen-oxygen cell, where D_{ox} is molecular oxygen being reduced at a platinum electrode, or in cells containing organic cathode materials being reduced at a graphite electrode. The variety of possible systems is enormous, and the main types will be described in later chapters. In the general case the partial (half) reactions are

$$\underset{\text{anodic oxidation}}{U_{red} \rightarrow U_{ox} + ne} \qquad \underset{\text{cathodic reduction}}{D_{ox} + ne \rightarrow D_{red}} \tag{2.15}$$

As at least one of the forms (ox) or (red) must be in a formal oxidation state other than zero, the overall reaction must be accompanied by the transfer of an equivalent number of ions of X in order to preserve the stoichiometry

$$U_{red} + D_{ox}X \rightarrow U_{ox}X + D_{red}; \quad \text{e.g. } Li^0 + CuCl \rightarrow LiCl + Cu^0$$

This initiaties ion transfer in electrolyte solutions I and II and across the interface separating half-cells I and II.

In phases in which the conductivity of the conductor P is metallic, the electrons on the one hand and the atoms and positive ions P^+ at the lattice nodes of the metal P on the other are in equilibrium:

$$\mu_{P^+(II)} + \mu_{e(I)} = \mu_{P^+(II)} + \mu_{e(II)} = \mu_P$$

Similar expressions are obtained for phases M_I and M_{II}, even when, to simplify things, $M_I = U_{red}$ and $M_{II} = D_{red}$ (as in the Daniell cell). Then $M_{I,ox} = U_{ox}$ and $M_{II,ox} = D_{ox}$, although it would be better to denote the ions at the lattice nodes of M_I and M_{II} as $M_I^{(+)}$ and $M_{II}^{(+)}$. In this notation $\mu_{e(i)}$ denotes the electrochemical potential of electrons in phase i. This is the situation in which the electrode material is simultaneously the electroactive substance undergoing an electrode reaction. At equilibrium, the equations

$$\mu_{e(I)} = \mu_{e(M_I)} \qquad \text{and} \qquad \mu_{e(II)} = \mu_{e(M_{II})} \tag{2.16}$$

must hold for half-cells I and II, respectively. Hence:

$$\mu_{e(I)} - \mu_{e(II)} = \mu_{e(M_I)} - \mu_{e(M_{II})} = -F(\psi_I - \psi_{II}) = -FE_{eq} \tag{2.17}$$

where $\psi_I - \psi_{II}$ denotes the electrical potential difference, i.e. the equilibrium voltage E_{eq} between the current collectors P_I and P_{II}. This relationship is also true for the more general case where equilibria, defined by the activities of the forms U_{ox} and U_{red} at electrode M_I and the forms D_{ox} and D_{red} at electrode M_{II} in accordance with the Nernst equation, are established at electrodes M_I and M_{II}. It should merely be remembered that the activities of pure phases, both liquid and solid, are equal to unity under standard conditions [5]. The electrical potential difference $\psi_I - \psi_{II}$ is therefore the cell e.m.f. This is independent of the kind of metal the current collectors P are made of, so long as both are made of exactly the same one.

Bearing in mind that there may be at least one common ion X in both solutions, the system for solution I (i.e. half-cell I) is simplified

$$\mu_{M_{I,ox}X(I)} = \mu_{M_{I,ox}(I)} + \mu_{X(I)}$$

and for solution II it is

$$\mu_{M_{II,ox}X(II)} = \mu_{M_{II,ox}(II)} + \mu_{X(II)}$$

Now the full expression for the free energy change ΔG in the cell takes the form of the expanded Nernst equation (for $n = 1$):

$$[\mu_{M_I} - \mu_{M_IX} + RT \ln X_I] + [\mu_{X_I} - \mu_{X_{II}} + RT \ln (X_{II}/X_I)] - \\ - [\mu_{M_{II}} - \mu_{M_{II}X} + RT \ln X_{II}] = -\Delta G \tag{2.18}$$

The expressions under the natural logarithms denote the activity of ion X in solutions I and II, respectively.

This equation defines the magnitude of ΔG when one mole of electric charge flows through the cell: $-\Delta G = FE$ for $n = 1$ and consists of three terms describing the processes occurring in the three parts of the electrochemical system that are essential if the system is to be a source of electric current, i.e. a galvanic cell. The first term refers to the more negative electrode and the solution with which it is in contact – when the cell is discharging this will be the anode space. The second term defines the potential drop across the boundary between solutions I and II. If the solution is the same for I and II then it is the potential associated with the activity gradients of the different ions present in these solutions, i.e. the diffusion potential. The third term is analogous to the first, and refers to the region around the more positive electrode, which becomes the cathode space in a spontaneous current-generating process.

Guggenheim [6] is of the opinion that dividing this equation into these three terms is arbitrary and that breaking down the e.m.f. into three component potentials – two electrode potentials and the diffusion potential – is unwarranted. However, this three-term description does reflect the real situation in practical systems that are chemical sources of electric current. Without a physical division into three parts, enabling the overall chemical reaction to be broken down into the separate electrochemical processes of oxidation, reduction and ion transfer in solution, electrons could not flow in the external circuit, and neither could free energy ΔG be converted into electrical energy. Recently [12] a molecular theory of the liquid junction across which the diffusion potential develops between two different solutions containig the same single binary electrolyte has been derived by a statistical mechanical approach.

Summarizing, one should consider the two extreme situations in which a cell can operate. The first is when a thermostated cell is in a state of quasi-equilibrium. The electrical work then supplied by the cell is nFE; the work done in changing the volume of all the phases in the cell is $\sum pdV$; the heat exchanged during the cell's operation, $T\Delta S$, is $nFT(\partial E/\partial T)$, i.e. the temperature coefficient of the e.m.f. The energy change in the cell is equal to the sum of these three factors

$$\Delta \mathcal{E} = -[nFE + \sum p\mathrm{d}V - nFT(\partial E/\partial T)] \tag{2.19}$$

The change in the heat or enthalpy of the reaction is

$$\Delta H = \Delta \mathcal{E} + \sum p\mathrm{d}V = -nFE + nF\,(\partial E/\partial T) \qquad (2.20)$$

The second extreme situation occurs when the cell is shorted, so that the chemical reactions proceed irreversibly without any electrical work having been performed. In this case the electrical work done $nFE = 0$, the work done in changing the volumes is the same as in the first situation, and the heat evolved is equal to the enthalpy change ΔH. What we are interested in are the changes in free energy occurring at constant pressure. The relationship between these changes and temperature is then connected with changes in the entropy, $\Delta S = -(\partial G/\partial T)_p$. Substituting values of the enthalpy yields the Gibbs–Helmholtz equation

$$\Delta H = \Delta G - T\,(\partial G/\partial T)_p = -nFE + nFT\,(\partial E/\partial T) \qquad (2.21)$$

hence

$$nF\,(\partial E/\partial T) = \Delta S \qquad (2.22)$$

and so the temperature coefficient of the cell is given by changes in entropy.

Clearly, the temperature coefficients of the e.m.f. enable ΔS for a current-generating reaction to be calculated. They are obtained from accurate experimental measurements of the cell e.m.f. at different temperatures. Since ΔG for the same reaction can be calculated from the e.m.f., the enthalpy change ΔH can also be evaluated from these measurements. On the other hand, these same temperature coefficients can be obtained independently by calorimetry. The values obtained by two experimental methods based on totally different physical assumptions can therefore be checked against each other and methodological errors can be eliminated. An example of such a combination of methods are the data in Gibbson and Sudworth [7] who have tabulated the specific energies and theoretical voltages of reactions between elements, e.g. $2Li + F_2 \rightarrow 2LiF$, or of substitution reactions such as $2Li + TiF_4 \rightarrow 2LiF + TiF_2$. The specific energy $\mathcal{E}_s$ is given by $-\Delta G^0$ divided by the sum of the substrate masses. The theoretical voltage of the cell is defined as e.m.f. $= E^0 = -\Delta G^0/nF$, where

$$\Delta G^0 = \Delta G^0_{\mathrm{products}} - \Delta G^0_{\mathrm{substrates}} \qquad (2.23)$$

This monograph [7] gives data not only on known current-generating reactions but also on those reactions that are inhibited for kinetic reasons. In the case of the latter, the variables sought are calculated from other available data such as $T\Delta S$ or $\mathrm{d}E/\mathrm{d}T$. The thermodynamic quantities obtained for various systems are therefore of a general nature and are independent of cell specifications or of the activities of electrolyte solutions because, with reference to potentials and free energies, they are maximum values based on overall reactions obtained by combining half-reactions. We can thus ignore reference electrodes and conventional scales of redox potentials, useful in discussions of half reactions (in half-cells). For example, the E^0 for the Cu/F_2 system and the current-generating reaction $2Cu + F_2 \rightarrow 2CuF$ is almost the same as for the Li/CuS system with the current-generating reaction $2Li + 2CuS \rightarrow Cu_2S + Li_2S$ ($E^0 = 2.42$ V). This means that the free energy changes $-\Delta G$ are practically identical when one mole of electric charge is transferred at the maximum voltage obtainable in these processes (2.42 V). By contrast, the specific energies of the substrates expressed in $\mathrm{W\,h\,kg^{-1}}$ differ markedly because the molar masses of the compounds involved are different. The redox potentials of the above-mentioned cathode and anode reactions as compared to that of a reference electrode (e.g. a hydrogen or lithium electrode) will be quite different too. However, cell construction and the composition of electrolyte solutions are of fundamental technical importance and determine the practical efficiency of an electrochemical system, i.e. the cell power in watts and the current efficiency in watt-hours per kilogram of a cell operating under a load expressed in ohms.

2.4. REFERENCE HALF-CELLS

Some electrode systems which are highly stable and reproducible deserve separate discussion because of the part they play in defining the scale of redox potentials used in electrochemistry, and their application as standards. The basic reference system is the hydrogen half-cell with platinum or platinum black Pt, H_2/H^+. By definition, the redox potential of this system in the standard state ($p_{H_2} = 1.01325 \times 10^5$ Pa, $a_{H^+} = 1$) defines the zero point both on the electrochemical redox potential scale and on acidity scales. These allow one to compare the protonating power of systems, i.e. the ability to transfer protons in Brönsted acid-base reactions, in both aqueous and non-aqueous solutions [8, 9].

The thermodynamic principles of e.m.f. measurements are general and independent of the solvent. Measured potentials should be referred to the hydrogen electrode in the same solvent or to another standard half-cell in that solvent. Cells in which one half-cell contains one solvent, say, water, and the other half-cell contains some other solvent, e.g. a mixed or non-aqueous one, should not be used because the diffusion potentials at the boundary between the solutions are hard to estimate and are subject to considerable error. Moreover, diffusion leads to one solvent becoming contaminated by the other, so that such cells operate unstably [10].

Since the hydrogen half-cell works very well in many non-aqueous solvents, acidity and redox-potential scales characteristic of each of these solvents can be set up, provided the ionizing properties of the solvents with respect to the electrolytes chosen are strong enough to guarantee the necessary ionic conductivity in such solutions. Substances like alkanes cannot therefore be used as pure solvents in solvent systems, although their application as components of mixed solvents containing ionizing solvents is possible.

The hydrogen half-cell's great efficiency and versability is due to the specific mechanism of action of platinum in this system and to the overwhelming tendency of protons to become solvated in almost every solvent. The solvation of protons is limited only by the lack of equally strongly solvated, stable anions, which the principle of electroneutrality in solutions requires. Protons are elementary particles whose dimensions are of the order of 10^{-13} cm; the electrostatic potential is therefore very high in the neighbourhood of such particles which accordingly interact strongly with the electron shells of other atoms and molecules. As a result, free protons cannot exist in condensed phases in physically significant concentrations, but are always present in bound form, as solvates or adsorbates on non-metals such as gels or membranes. In metallic substances there are always free electrons in the form of the Fermi gas, so unsolvated protons adsorbed on the surface of a metal immediately attract electrons and become atomic hydrogen, $H^{\bullet}$, whose adsorption capability depends on the kind of metal. On "soft" metals like mercury and lead, the adsorption energy of hydrogen, ΔG_{ads}, is low, the solvation energy thus prevails, the protons remain in solution and the $H^{\bullet}$ concentration at the surface of these metals is extremely low. By contrast, $H^{\bullet}$ is strongly adsorbed on the platinum metals, particularly on platinum and palladium, since $\Delta G_{ads} << 0$. At the surface of these metals, and especially on the active surface of platinum black, there is a significant concentration of $H^{\bullet}_{ads}$ in equilibrium with solvated protons H^{+}_{solv} . At the same time, adsorption of molecular hydrogen is strong, and equilibria are set up that determine the reversibility and stability of this system:

$$H_{2(gas)} \rightleftharpoons 2H^{\bullet}_{ads}, \qquad H^{\bullet}_{ads} \rightleftharpoons H^{+}_{solv} + Pt(e_F)$$

where e_F denotes free electrons in the platinum at the Fermi level. It is worth noting the fact that it is only at the expense of the adsorption energy that the strong σ-bond in molecular hydrogen is able to dissociate to form atomic hydrogen, $H^{\bullet}_{ads}$. At the same time, the activation energy (activation barrier) of this process is substantially lowered, so equilibrium is quickly established on the surface of the platinum catalyzing this reaction. In electrochemistry, exchange currents i_0 are a measure of the rapidity with which equilibrium is attained, and thus of the reversibility of systems; the larger they are, the more quickly equilibrium is reached. For platinum and palladium they exceed 1×10^{-3} A cm^{-2}. On mercury, however, where such equilibria are unattainable because the adsorption energy of hydrogen is too low, the Hg, H_2/H^+ system is practically irreversible; the exchange current $i_0 = 5 \times 10^{-13}$ A cm^{-2}, that is, ten orders of magnitude lower than in the case of platinum.

In non-aqueous solvents, the lithium reference half-cell Li/Li^+ is frequently used, especially when lithium cells are tested, and the rubidium half-cell Rb/Rb^+ is used for comparing standard potentials in different solvents [11] as Rb^+ ions are equally weakly solvated in most solvents. The mechanism of action of these electrodes relies on the establishment of a low-activation-energy equilibrium, $M^+_{solv} + e \rightleftharpoons M^0$ (electrode). These processes resemble somewhat the equilibrium involved when alkali metals are dissolved and precipitated in liquid ammonia, except that instead of solvated electrons there are solvated salt anions, which remain in solution.

The calomel electrode generally works badly in non-aqueous solutions, but the silver chloride electrode Ag | AgCl is quite satisfactory. The auxiliary potentials of the reference half-cells are determined from the e.m.f. of cells containing a hydrogen electrode as the second half-cell, as for instance in the hydrogen-silver cell: Pt, H_2 | $H^+_{solv} + Cl^-_{solv}$ | AgCl | Ag. The description of this cell implies that it contains only one electrolyte solution common to both electrodes. In fact, however, the regions of the solutions in contact with each electrode differ slightly in composition, and this is essential in order to prevent non-current-generating redox reactions from occurring. In this case, the non-current-generating reaction would be: $Ag^+ + Cl^- + \frac{1}{2}H_2 \rightarrow H^+ + Cl^- + Ag^0$. This principle is a general one and applies to all types of cells and storage batteries. Deviations from it are one of the basic mechanisms by which cells self-discharge. This cell should therefore contain two solutions with the same concentration of HCl, but the solutions around the hydrogen electrode should be saturated with hydrogen, whereas the solution around the silver electrode ought to be saturated with silver chloride (AgCl). The full description of this electrode is

$$\mathrm{Pt} \mid \mathrm{H_2} \mid \mathrm{H^+}\ (a = 1),\ \mathrm{Cl^-} + \mathrm{H_2} \parallel \mathrm{H^+},\ \mathrm{Cl^-}\ (a = 1) + \mathrm{Ag^+} \mid \mathrm{AgCl} \mid \mathrm{Ag} \mid \mathrm{Pt}$$

The e.m.f. of this cell is given by the following equation:

$$nFE = \frac{1}{2}\mu_{H_2} + \mu_{AgCl} - \mu_{Ag} - \mu_{HCl}$$

where the subscripts indicate the substances whose chemical potentials are taken into account in the current-generating reaction. Writing down the cell in such a way that the hydrogen electrode is on the left means that if a spontaneous reaction occurs when the electrical circuit is closed, it becomes the negative electrode, where there is oxidation of hydrogen to hydrogen ions; cathodic reduction $Ag^+ \rightarrow Ag^0$ takes place at the other, positive, electrode. The complete reaction is

$$\frac{1}{2}H_2 + AgCl \rightarrow HCl + Ag$$

when $c_{HCl} < 9\ \mathrm{mol\,dm^{-3}}$. A positive half-cell potential means that the redox electrode system is a stronger oxidizing agent than hydrogen ions are at an activity of $a_{H^+} = 1$ and in equilibrium with gaseous hydrogen. If the potential is negative, the system is a stronger reducing agent than hydrogen at a pressure of 1 atm ($p_{H_2} = 1.01325 \times 10^5$ Pa) and in equilibrium with hydrogen ions under analogous standard conditions.

REFERENCES

[1] "Manual of symbols and terminology for physicochemical quantities and units", *Pure Appl. Chem.*, **51**, 1 (1979).

[2] P. van Rysselberghe, R. Defay, N. Ibl, E. Lange, E. Levart, G. Milazzo and G. Valensi, *Electrochim. Acta,* **12**, 748 (1967).

[3] Electrochemical conventions, *Pure Appl. Chem.,* **37**, 504 (1965).

[4] Proceedings of the 17th Conference of IUPAC, Stockholm (1953), pp. 82–85.

[5] F. Reif, *Statistical Physics*, McGraw-Hill, New York (1967).

[6] E.A. Guggenheim, *Thermodynamics*, 5th ed., North Holland Publ., Co., Amsterdam (1967).

[7] J.G. Gibbson and J.L. Sudworth, *Specific Energies of Galvanic Reactions*, Chapman and Hall, London (1973).

[8] R.F. Cookson, *Chem. Rev.*, **74**, 5 (1974).

[9] R.G. Bates, *Determination of pH*, Wiley, New York (1973).

[10] D.J.G. Ives and G.J. Janz, *Reference Electrodes*, Academic Press, New York–London (1961).

[11] J. Koryta, J. Dvořak and V. Bahačkova, *Lehrbuch der Electrochemie*, Springer-Verlag, Vienna (1975).

[12] F.O. Rainieri and H.L. Friedman, *J. Chem. Phys.*, **94**, 6135 (1991).

CHAPTER 3

Basic Kinetics of Electrode Processes

3.1 STRUCTURE OF THE DOUBLE LAYER

Current-generating reactions in galvanic cells are heterogeneous reactions of electron transfer across the interface between the active substances of the electrode material and the electrolyte solution. In these reactions, electroreduction processes are spatially separated from the conjugated processes of electro-oxidation. When the electrical circuit is closed, the reactions are controlled by the flow of electrons through the external metal conductor and that of solvated ions through the electrolyte solution.

Although the foundations of the theory of electrode processes were laid back in the 1930s [1, 2], the theory is developing continually and is being influenced by research into new types of cell and unconventional current-generating systems. Current-generating processes and reactions take place within the limited space of the electrical double layer at the interface between the electrode material which, strictly speaking, is the outer surface of the phase in which electron conductivity occurs, and the adjacent solution, in which electrolytic conductivity takes place as the result of the transfer of charged particles (also known as ionic conductivity). The structure of the double layer depends on the potential, the electrode material and the composition of the electrolyte solution. The term "double layer" is derived from the fact that there is usually an electrical potential gradient across the interface between the electrically conducting phases as a result of which the molecules and ions in the vicinity become more ordered. Superimposed on this effect are specific surface interactions due to the structure and nature of the electrode material. Furthermore, the structure of the double layer depends on the sign and magnitude of the electrode potential. Electrostatic attraction causes positive ions to accumulate on the surface of a negatively charged electrode (Fig. 3.1). They form the second "plate" of the double layer, which can be compared to a flat condenser. Generally, all

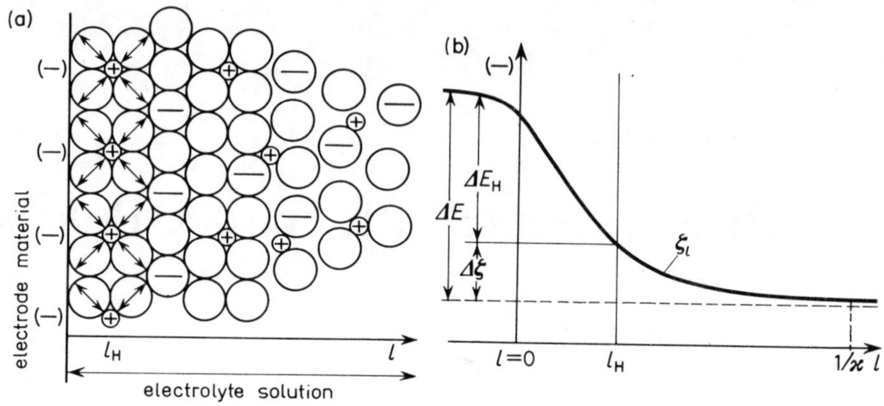

Fig 3.1 (a) Structure of the double layer at the interface between a negatively-charged electrode material and an electrolyte solution; (b) schematic distribution of potential in the double layer; l – distance from electrode, l_H – minimum distance between centres of ions in the solution and the electrode surface, $1/\varkappa$ – conventional depth of diffuse layer in the electrolyte solution measured from l_H. ⊕ cations, ⊖ anions, ↔ solvent molecules with permanent or induced dipole moments, ΔE_H – potential drop across the rigid Helmholtz plane, $\Delta\zeta$ – electrokinetic potential at the interface between the rigid and diffuse layers, ΔE – electrode potential equal to $\Delta E_H + \Delta\zeta$, ζ_l – values of the potential in the diffuse layer as a function of distance l.

positive metal ions except the largest monovalent ions are strongly solvated, so between the metal surface and the cations there are still a number of solvent molecules, some associated with the cation and some not belonging to the immediate solvation spheres. If the adsorption capability of solvent molecules at the electrode surface is sufficiently great, solvated cations will accumulate at a certain distance from the electrode. Small lithium cations are particularly strongly solvated owing to their small radius. Their charge density is therefore high in comparison with that of other cations, and it is this which defines above all the force of interaction between ions on the one hand and polar and polarizable solvent molecules on the other. A plane called the Helmholtz plane can be drawn through the layer of ions nearest the electrode. The distance of the Helmholtz plane from the electrode surface corresponds to that between the plates of a condenser. As charges of one sign accumulate at the electrode surface, ions of the opposite sign are attracted, and the concentration of the electrolyte solution is therefore much greater in the double layer than in the bulk of the solution. The strong attraction between the ions and molecules located between the electrode surface and the Helmholtz plane results in the latter acquiring a highly ordered structure – hence the term "rigid double layer", even though particular ions, atoms and molecules are subject to continual translation and exchange due to thermal motion and vibrations. Further away from the electrode the intensity of the electric field lessens, interactions are weaker and thermal movements loosen up the structure, thereby reducing the ordering of the ions. That is why this part of the double layer is called diffuse. Double layers vary in depth, depending on the electrode potential and the composition of the solution. Usually, however, the rigid double layer is 0.5–2 nm deep, whereas the diffuse layers may extend into the bulk of the solution over a distance of several nm. A somewhat different situation exists when anions accumulate at a positively charged electrode. Anions are generally larger than cations and, with the exception of fluorides, are mostly much less strongly solvated than cations. Moreover, many simple anions such as chlorides, bromides, iodides and sulphides are strongly adsorbed at the surfaces of uncharged or positively charged metal electrodes. As a result, anions can become attached directly to the electrode surface, displacing from it even strongly polar solvent molecules. Solvated cations accumulate on this adsorbed anion layer. The diffuse double layer stretches away from here and gradually passes into the statistically disordered electrolyte solution typical of its bulk, where the intensity of the electrical field is already negligibly small. The distribution of potentials is quite similar even in very dilute solutions, because the presence of an electrical field causes ions to migrate instantaneously to the electrode surface where they form a double layer in which the ion concentration is fairly high in comparison with the solution's stoichiometric concentration. The excess concentration in the double layer is large enough to neutralize the excess electrical charge on the electrode, so that practically the whole potential drop occurs within the double layer. Since the condition for electron transfer reactions to occur and the driving force behind them are potential gradients, electrode processes take place only within the double layer, and then chiefly within the Helmholtz plane (the rigid layer) adjacent to the electrode, where the electrical field intensity is greatest. If we take the potential drop across a 1 nm-deep double layer to be 1 V, an electrical field intensity of 10^7 V cm^{-1} results, which then illustrates the magnitude of the electrical forces acting on the dipoles and electrical charges of the molecules and ions in the double layer. It must, however, be stressed that despite the presence of such large electrical forces, no part of the double layer is completely rigid and impervious. Besides thermal movements due to increased ionic concentration, it is the elevated osmotic pressure of the electrolyte which tends to equalize concentrations throughout the electrolyte solution and gives rise to the diffuse double layer. Mutual interaction and the influence of opposing forces bring about a state of dynamic equilibrium in which mass and charge can move back and forth across the double layer; this is a necessary condition for electrode reactions to occur [3].

The potential (with respect to the bulk of the solution) in the diffuse layer ζ_l, for distances between l_H and $(1/\kappa)$ (Fig. 3.1), can be calculated from the electrical charge density g [C cm^{-3}], which is

$$g = F \sum c_i z_i \exp(-z_i F \zeta_l / RT) \tag{3.1}$$

where c_i and z_i are the respective concentrations and charges of the ions in the double layer. The spatial (geometric) variables are included in Poisson's equation which, in the one-dimensional case where the potential depends on the distance l for a flat electrode, takes the form

$$\frac{\partial^2 \zeta_l}{\partial l^2} = -4\pi g/\varepsilon \tag{3.2}$$

where ε is the relative electrical permittivity of the liquid medium. Since ζ_l is large, the expression $\exp(-z_i F \zeta_l / RT)$ in equation (3.1) is much smaller than unity and, after expansion, can be written down as a Taylor series

$$\frac{\partial^2 \zeta_l}{\partial l^2} = -\frac{4\pi g}{\varepsilon} = \frac{4\pi F}{\varepsilon}\left(\sum c_i z_i - \sum c_i z_i^2\right)\frac{F\zeta_l}{RT} \tag{3.3}$$

Since the solution must remain electrically neutral, $\sum c_i z_i$ must be equal to zero. All the constants can be combined into a single constant κ, i.e. $4\pi F^2 \sum c_i z_i^2 / \varepsilon RT = \kappa^2$. Then $\partial^2 \zeta_l / \partial l^2 = \kappa^2 \zeta_l$. The solution with respect to ζ_l is

$$\zeta_l = \zeta e^{-\kappa l} \tag{3.4}$$

where ζ is the potential at the interface between the rigid and diffuse layers, i.e. the potential of the Helmholtz plane with respect to the bulk of the solution. The parameter κ has the dimension of reciprocal distance, so the difference $[(1/\kappa) - l_H]$ corresponds to the depth of the diffuse layer. It is worth noting that an expanded form of this equation was applied by Debye and Hückel to calculate activities in electrolyte solutions.

The potential drop across the rigid Helmholtz plane ΔE_H is defined by the distance l_H between the surface of the metal electrode and the Helmholtz plane in which the ions responsible for the surface charge ρ have accumulated. This charge is derived from the excess ions of one sign that have concentrated in the Helmholtz plane. An identical charge q of the opposite sign [C cm^{-2}] is electrostatically bound to the metal surface and is a part of the overall charge that accumulate on the metal to balance out the resultant charge of all the ions present in the double layer adjacent to the electrolyte solution. The rigid layer can be treated as a flat condenser, and its atomic structure can be disregarded in simplified calculations although it explains the properties of the double layer. The electrical capacity C_H of the rigid layer is given by

$$C_H = \varepsilon_0 \varepsilon_r s l_H^{-1} = \varepsilon s l_H^{-1}$$

and the potential drop ΔE_H by

$$\Delta E_H = 4\pi \rho l_H \tag{3.5}$$

The total potential drop across the double layer ΔE is equal to the sum of both terms

$$\Delta E = \Delta E_H + \Delta\zeta \tag{3.5a}$$

Where the electrode charge with respect to the bulk of the solution is zero, i.e. at the electrocapillary zero, the structure of the double layer changes. As electrostatic interactions weaken, the influence of adsorption gains ground. In the absence of strongly adsorbing ions and organic molecules, the position of the electrocapillary zero is defined by the most readily adsorbing inorganic anions; their adsorption capability usually increases in the series $ClO_4^- < SO_4^{2-} < Cl^- < Br^- < I^- < S^{2-}$ and with polarizability and anionic charge.

In the case of the mixed solvents used in non-aqueous cells, the composition of one of the components, usually the more polar one, is enriched in the double layer [4, 5]. This specific adsorption due to stronger interactions such as chemisorption is rather rare in organic solvents, but can play an important part in some inorganic solvents. The nature and structure of the electrode material are also significant factors in adsorption processes – but that is another problem. The double layer structure in the three cases discussed is illustrated schematically in Figs. 3.1–3.3. Figure 3.1 shows the structure of the double layer and the schematic distribution of potential at the surface of a negatively charged electrode material and an electrolyte solution containing small, strongly solvated cations, as well as larger, weakly solvated anions in a non-aqueous solvent. Lithium perchlorate in dimethyl sulphoxide is an example. The double layer structure at the surface of an uncharged electrode in an electrolyte solution is illustrated in Fig. 3.2. The adsorption of polar solvent molecules at the less-clearly-indicated rigid part of the double layer is characteristic.

At an interface, however, particularly that between the electrolyte and the electron-conducting phase (electrode material), there will always be a double layer as a result of diverse specific surface interactions which cause the electrolyte's anions and cations to separate. Depending on the case in question, there is preferential adsorption of one kind of ion associated with

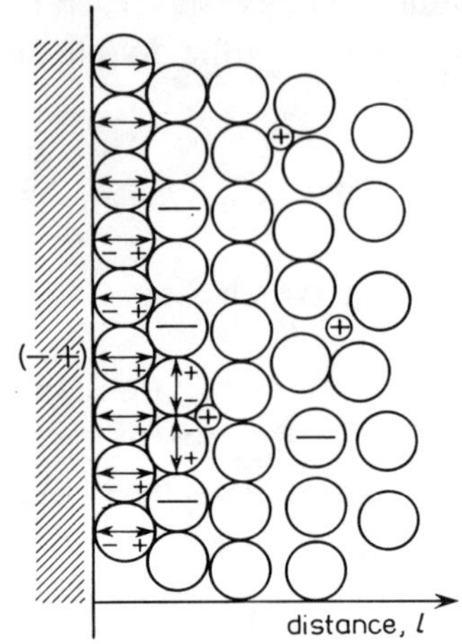

Fig 3.2 Double-layer structure at the surface of an uncharged electrode in an electrolyte solution: (—+) – charges induces on the metal after dipole adsorption.

the formation (due to adsorption) of layers of oriented dipoles of neutral molecules. For instance, a layer of dipoles whose negative ends (containing electronegative atoms such as oxygen) are oriented towards the electrode surface induces a partial charge of the opposite sign on the metal surface, while the other end of the dipole, carrying a partial positive charge, orients itself in the opposite direction, i.e. towards the solution. Anions preferentially accumulate on the solution side of such an ordered layer of dipoles, and beyond them, a less ordered layer of as many cations as are necessary to keep both the double layer and the electrolyte solution electrically neutral.

The double layer structure at the surface of a positively charged electrode material is shown in Fig. 3.3 The layer of adsorbed solvent is displaced by weakly solvated anions which become adsorbed directly onto the electrode surface. In a fairly concentrated, weakly-solvated electrolyte such as tetramethylammonium perchlorate, the double layer may become deprived of solvent because a large number of its molecules are displaced from the rigid double layer into the bulk of the solution [5]. The outermost layer of atoms on the surface of the electrode can also be included in the double layer because a potential gradient is set up across it and the electrical charge density in it is no longer constant.

The potential distribution in and the electrical capacity of the double layer can be estimated on the basis of a simple model in which the metal electrode is treated as a thin layer devoid of atomic structure. This is the quantum-mechanical jellium model, in which the electrolyte solution near the electrode is represented by a charged plane a certain distance away from the electrode, corresponding to the rigid Helmholtz layer [6].

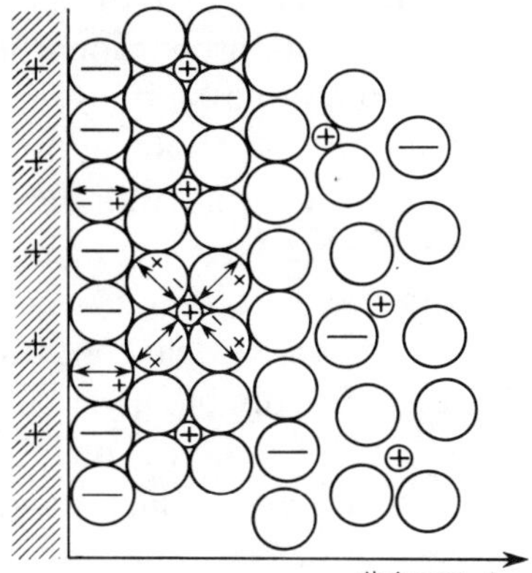

Fig. 3.3 Double-layer structure at a positively charged surface of an electrode in an electrolyte solution.

A more complex model, in which the electrolyte solution consists of polarized dipoles, has been investigated by Torrie *et al.* [26], but it requires very complicated calculations. A detailed completely molecular description of the electrical double layer based on reference hypernetted chain (RHNC) theory has been now developed for molecular fluids near planar surfaces [27]. It is an extension of the image-dipole, classical continuum theory with which it agrees to the first approximation.

The double layer structure in semiconducting electrodes is different because semiconducting materials are much poorer conductors of electric current [8]. There are two kinds of electrical charge carrier in semiconductors: electrons in the conduction band and "holes" in the valence band. The holes are gaps (free orbitals) left by electrons transferred to the conduction band and thus have an apparent positive charge, so that the electroneutrality principle within the mass of the semiconductor is not violated. The holes have a certain mobility: as electrons move in to occupy them, they leave behind new holes. The overall effect is that the holes appear to move in the direction opposite to that of valence electrons. A hole therefore behaves like a virtual, positively-charged particle which has the mass of an electron but is usually less mobile. The concentration of charge carriers in semiconductors is several orders of magnitude less than in metals, and may be far less than the concentration of ions in dilute solutions of electrolytes. For example, the concentration of charge carriers, i.e. of free electrons, in metallic lithium is ca. 4×10^{22} cm^{-3}; in other metals it is generally of the order of 10^{22} cm^{-3}. In a ca. 0.1 M solution of an electrolyte the concentration of ions is ca. 10^{20} cm^{-3}, whereas in many semiconductors the concentration of mobile holes and electrons is 10^{14}–10^{17} cm^{-3}. According to Gouy's theory, the size (extent) of the double layer is proportional to the reciprocal of the square root of the charge carrier concentration, whereas the conductivity is proportional to the charge they carry, their effective concentration and mobility. Near the interface between a metal and an electrolyte solution a metal can therefore be treated to a good approximation as a phase with an infinitely large electrical conductivity, whereas at the interface between a semiconductor and an electrolyte solution the situation may be the opposite – an electrolyte solution can sometimes be regarded as an almost ideal conductor of electricity in comparison with a semiconductor. The double layer structure at a semiconductor surface is thus the reverse of that at a metal surface. The diffuse potential gradient is present in the semiconductor and not in the electrolyte solution (Fig. 3.4). The main influence on the kinetics of electrode processes is therefore exerted by the concentration of charge carriers in the semiconductor, which can change by many orders of magnitude, depending on the cell voltage at which such an electrode is operating. So, the voltage at which a semiconductor electrode operates chiefly influences the statistical parameter of the reaction, i.e. the entropy, and not the energy parameter of the reaction, as happens with metal electrodes.

Figure 3.4 shows the electrostatic potential E at the interface between a semiconductor and an electrolyte solution. There are two kinds of charge carrier in semiconductors. Since, however, holes are fictitious particles,

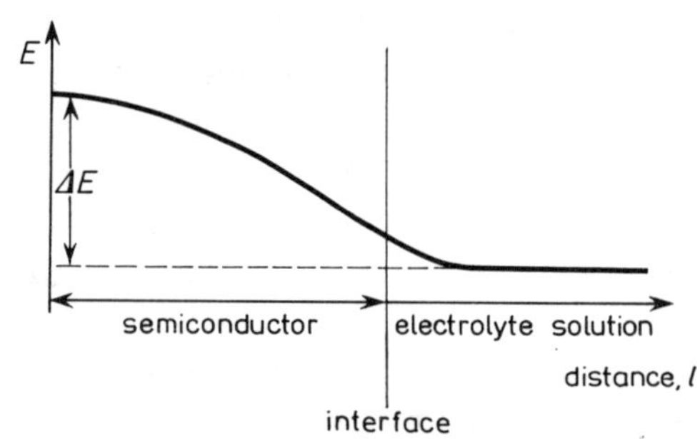

Fig. 3.4 Potential gradient ΔE at the interface between a semiconductor and an electrolyte solution.

the actual carriers are the electrons in the conduction band and the electrons in the valence band – which are excited electronic states in the semiconductor crystal. Hence the nature and concentration of contaminants in a semiconductor affect the concentration of charge carriers significantly and the double layer structure indirectly. Mechanical treatment also affects the electrochemical properties of such materials, altering as it does the grain (crystal) dimensions and the potentials at the contact points between crystallographically different crystal faces.

3.2 ELECTRON TRANSFER PROCESSES

As far as current-generating reactions are concerned, the basic process is the elementary act of electron transfer across the interface between the electrode material and the electrolyte solution, because it is indispensable to the occurrence of oxidation and reduction reactions in the separate anode and cathode spaces. These reactions are conjugated because the two spaces are electrically connected by the external circuit. However, electrons can cross this interface without any redox (and therefore current-generating) reactions occurring. The photoelectric effect is one example, but the simplest case of interest here is the ionization of alkali metals in aprotic solvents such as ethers and amines, and especially ammonia. At temperatures below –33°C (240 K) all alkali metals readily dissolve in liquid ammonia producing solutions containing from 15 to 20% mol. of the metal, except caesium, whose solutions can contain any amount, i.e. from 0 to 100% mol., of the metal. Pure caesium is precipitated at the eutectic point (ca. 16% mol. Cs in solution), which indicates that no stable ammoniate complexes of caesium can exist. On the other hand, lithium does form solvates of this type [9], the precipitation of an eutectic of composition $Li(NH_3)_4$ being one indication of this.

The case with which alkali metals dissolve in liquid ammonia is due to the low activation energy and to the small change in the free energy of this reaction, which proceeds because the metal is solvated and ionized. For example, the free energy change ΔG when potassium is solvated $K(solid) \rightarrow K^+(solv) + e(solv)$ is $-9\ kJ\,mol^{-1}$. As is well known, a solvated electron in liquid ammonia is a stable ion and such solutions are also stable. At metal concentrations less than $10^{-2}\ mol\,dm^{-3}$ these solutions behave like strong electrolytes. If, however, the solvent has a low relative electrical permittivity [10, 11], a solvated electron is still stable but metal cations are weakly solvated. Metals dissolve in such solvents to yield solutions having the properties of weak electrolytes, following the establishment of left-shifted equilibria [7] such as

$$M\ (metal) \rightleftharpoons M\ (solv), \qquad 2M \rightleftharpoons M^-\ (solv) + M^+\ (solv)$$

with minimal participation of other equilibria like

$$M\ (solv) \rightleftharpoons M^+\ (solv) + e\ (solv)$$

Hence the solubility of alkali metals in solvents with a low relative electrical permittivity is generally poor. By contrast, the well-known phenomenon of metallic or electronic conductivity is observed at larger concentrations of metals in liquid ammonia. When the metal concentration exceeds 3–4% mol., electronically-conducting aggregates suspended in the liquid ammonia solution are formed (in the intermediate region the solutions are microscopically non-homogeneous). The electronic conductivity of concentrated solutions of alkali metals in liquid ammonia is due to the formation in such solutions of conduction bands analogous to the conduction bands in metals. The fact that such bands are able to form in the solution of a dielectric is attributed to the tunnelling of electrons from the site where they were solvated to another solvation site which may be several nanometres from the first one. The following example will illustrate the point. In liquid ammonia the concentration of NH_3 is ca. 44 mol dm^{-3} containing 4–5% mol. of metal, the concentration of solvated electrons and ions is practically the same, i.e. ca. 2 mol dm^{-3}, or ca. 1.2×10^{24} of each in 1 dm^3 of solution. For each electron there are thus 1×10^{-24} dm^3 or 1×10^{-27} m^3 of solution, i.e. each electron is contained in a cube of side 10^{-9} m or 10^{-8} dm = 1 nm. So this is the order of magnitude of the average distance between solvated electrons, at which they behave as if they were in the crystal lattice of a metal. The dielectric properties of the solvent are now of secondary importance – the ammonia molecules seem only to cause the metal's crystal lattice to dilate and loosen up, and they do not impede the motion of the electrons. The conductivity of such solutions is therefore proportional to the concentrations of electrons and in solutions containing 20% mol. of lithium in ammonia (specific conductivity of the order of $1.5 \times 10^4\ \Omega^{-1}\,cm^{-1}$) it reaches that of liquid mercury. Electrons can easily cover distances of the order of 1 nm regardless of the presence of a dielectric such as liquid ammonia because of the wave-like, quantum-mechanical properties of matter, which are especially striking in the case of electrons because of their small mass. The wave properties of electrons allow them to cross energy barriers without having to attain the full activation energy equal to the height of this barrier, that is, in a manner forbidden by classical mechanics. This is the tunnel effect, which quantum mechanics takes account of.

When an alkali metal is immersed in liquid ammonia it is wetted and the outermost layer of metal atoms at the interface becomes partially solvated by ammonia molecules. The tunnel effect also allows electrons to penetrate the solution for short distances of the order of nm, where they are solvated if they remain there long enough for the polar molecules to reorient themselves around the negatively-charged electrons. The work involved in such electron transfer depends strongly on how far an electron moves away from the metal; this dissolves if the energy gained when the electron is solvated exceeds the energy required to remove an electron to an infinite distance from the metal, i.e. the work function of the electron

[12]. The metal at the surface now becomes positively charged, so the repulsion between the positive ions on the metal surface increases, facilitating the transfer of partially solvated metal ions from the surface to the solution where they become wholly solvated. The process continues until a saturated state equilibrium is attained, when mobile electrons returning to the metal charge its surface negatively and facilitate the return of metal cations to the crystal lattice of the metal. A quasi-stationary double layer characteristic of the system thereby comes into existence. In liquid ammonia at –70°C this double layer is ca. 0.5 nm deep. The dissolution of a metal in this type of solvent therefore depends principally on the work function of the electron from the metal φ, and on the solvation power of the aprotic solvent which in turn depends, *inter alia*, on the dipole moment μ_D and the polarizability of its molecules. Dissolution ensues if $\varphi < W_e + 3k_BTel_D/\mu_D$, where W_e is the solvation energy of the electron, l_D is the depth of the double layer, k_B is Boltzmann's constant, T is the absolute temperature and e is the charge on the electron. Since the limiting value of φ is around 2.5 eV for ammonia, only alkali metals dissolve in liquid ammonia. A double layer also forms when other metals ($\varphi > 2.5$) are immersed in liquid ammonia, but the electrons that leave the metal to enter the double layer stay there, electrostatically bound to the metal, which does not dissolve. The double layer is therefore a region in which the two phases – metal and solution – penetrate each other, since the electron cloud of the metal penetrates the electrolyte solution to a depth approximately the same as that of the rigid part of the double layer. If a neutral solvent such as ammonia or hexamethylphosphoramide (HMPA) contains traces of substances reacting with the solvated ions, the metal starts to corrode. The most frequent such contaminant is water; the electron is hydrated and molecular hydrogen and hydroxide ions are evolved:

$$e(H_2O) + e(H_2O) \rightarrow H_2 + 2OH^-$$

The metal hydroxide is formed at the same time:

$$M - ne \rightarrow M^{n+}, \qquad M^{n+} + n\,OH^- \rightarrow M(OH)_n$$

A more or less permanent layer of hydroxide is formed on the surface of the corroding metal. If the layer of products is coherent and adheres tightly to the metal, corrosion becomes kinetically inhibited, and the metal is said to be passivated. However, this apparently undesirable process often turns out to be extremely useful in cells, because it reduces the self-discharging of the cell and permits the use of solvent systems in which a given electrode material (e.g. lithium) is thermodynamically unstable. The passivating film provides sufficient kinetic stability to secure the durability of the electrochemical cell. Systems comprising an alkali metal immersed in liquid ammonia are examples of the theoretically simplest electrochemical half-cell that can function in cells as an anode due to the presence of a strongly electronegative alkali metal. By joining this half-cell via an electrolytic bridge to a second half-cell having a lower electrochemical potential (a cathode), a cell is formed. If the electrical circuit is closed, the cell produces an electric current because the alkali metal is oxidized and the cathode material reduced by electrons moving from anode to cathode. This movement of electrons is the effective electric current, and is followed by the further dissolution of the alkali metal, and by the transfer of anions from the inter-electrode space to the anode space and of cations to the cathode space in order to balance electrostatically the ions formed during the current-generating reaction. So begins the movement of solvated ions and electrons. The further both half-cells move away from thermodynamic equilibrium, the stronger is the electric current. The state of the whole cell is defined by the irreversible processes associated with mass and charge transfer.

The displacement of an electrochemical system from equilibrium is given by the overvoltage η. However, the magnitude of the electric current i – a direct measure of the rate of the electrochemical reaction – depends not only on the overvoltage but also on the speed with which the system can return to equilibrium when the disturbance causing the displacement from equilibrium ceases, i.e. when the electrochemical reaction is stopped. The ability of a system to return to equilibrium depends on its degree of reversibility, measured by the reaction rate constant k. Accordingly, systems with high values of k are highly reversible, i.e. they have short relaxation times. Systems with not very high values of k are said to be quasi-reversible. A low k is characteristic of irreversible systems, and if k is extremely low, the system is absolutely irreversible: its relaxation time is immeasurably long, which means that it can never attain thermodynamic equilibrium.

As far as galvanic cells are concerned, each half-cell is a system to be considered separately, because diverse electrode processes take place in half-cells and their mechanisms may be completely different.

The greater the reversibility of an electrode system, the larger the current obtainable from a cell, since the working efficacy of the whole cell is restricted by its least efficient part.

The reversibility of electrochemical systems is measured by the exchange current i_0, which is directly proportional to the rate constant k of the electrochemical reaction. So, one of the principal tasks of the theory of electrode processes is to calculate k for selected electrochemical systems. The Butler–Volmer relationship [13–15] links the current i and the exchange current i_0

$$i = i_0[\exp(\overleftarrow{\alpha}F\eta/RT) - \exp(-\overrightarrow{\alpha}F\eta/RT)] \quad (3.6)$$

where $\overrightarrow{\alpha}$ is the transfer coefficient of the product-forming reaction, while $\overleftarrow{\alpha}$ is the transfer coefficient of the reverse, substrate-forming reaction, and η is the overvoltage. This last quantity is the difference between the equilibrium potential of the system E and the voltage V measured when a current i is flowing with respect to the reference electrode, i.e. the electrode whose potential is constant with respect to that of the hydrogen electrode. The transfer coefficients α define the rate of change of current strength as a function of the overvoltage

$$\mathrm{d}\eta/\mathrm{d}\ln i = RT/\alpha F$$

where $\eta = E - V$, hence

$$\alpha = (RT/F)/(\mathrm{d}\eta/\mathrm{d}\ln i) \quad (3.7)$$

Obviously, the reactions proceeding in either direction, → and ←, are conjugated: $\overrightarrow{\alpha} + \overleftarrow{\alpha} = n/\nu$, where n is the number of electrons transferred during the multistep reaction $A + ne \rightarrow B$, and ν is the stoichiometric coefficient of the reaction. For example,

$$2Hg \rightarrow Hg_2^{2+} + 2e$$

where $\nu = 2$.

If $n = \nu$, $\overrightarrow{\alpha} = 1 - \overleftarrow{\alpha}$. In a basic one-step electron-transfer reaction, $n = \nu = 1$. In this particular case, the Butler–Volmer relationship simplifies, and a single transfer coefficient, often denoted by $\beta = \overrightarrow{\alpha}$ or simply by α, is used instead of $\overrightarrow{\alpha}$ and $\overleftarrow{\alpha}$. Then

$$i = |i_a| - |i_c| = i_0\{\exp[\alpha zF\eta/RT] - \exp[-(1-\alpha)zF\eta/RT]\} \quad (3.8)$$

If no further complications arise in connection with the reaction mechanism, α approaches 0.5 and the plot of $i = f(\eta)$ is symmetrical with respect to the sign of the potential and direction of current flow. Hence α (or β) is also known as the symmetry factor.

Equation (3.8) indicates that the current i in the circuit is the difference between the currents $\overrightarrow{i}$ and $\overleftarrow{i}$, which are proportional to the reaction rates in the two directions. If $i_a = i_c = i_0$, $i = 0$, even though i_0 can take high values. This is a condition of equilibrium which is fulfilled if $E = V$. Then $\eta = 0$, and

$$\exp(\alpha F\eta/RT) = 1 \quad (3.9)$$

Useful current thus always flows when the thermodynamic equilibrium is disturbed.

For an elementary electron transfer reaction

$$i = i_0[\exp(\alpha F\eta/RT) - \exp(-\beta F\eta/RT)] \qquad (3.10)$$

where $z = 1$ and $1 - \alpha = \beta$.

Two limiting cases arise, depending on the magnitude of the overvoltage. If this is low, i.e. of the order of millivolts, the exponent can be considered very small, so on expanding the equation in a Taylor–McLaurin series only the first two terms of the expansion need be retained [13]:

$$\exp(\alpha F\eta/RT) = 1 + \alpha F\eta/RT + (\alpha F\eta/RT)^2/2! + (\alpha F\eta/RT)^3/3! + \ldots \exp(-\beta F\eta/RT) = 1 - \beta F\eta/RT + \ldots$$

then

$$i = i_0(1 + \alpha F\eta/RT - 1 + \beta F\eta/RT) = i_0(\alpha F\eta/RT + \beta F\eta/RT) \qquad (3.11)$$

If the polarization curve $i = f(\eta)$ is symmetrical, $\alpha = \beta = 0.5$, and relationship (3.11) becomes

$$i = i_0 F\eta/RT \qquad (3.12)$$

if $\eta << RT/zF$ and $i = i_a = -i_c$. In this case $n = z$.

A similar expression can be obtained for the more general case, so long as the condition $\alpha + \beta = 1$, i.e. $\beta = 1 - \alpha$ is satisfied. Let $F/RT = q$; then

$$i = i_0[\alpha q\eta + (1 - \alpha)q\eta] = i_0 q\eta \qquad (3.13)$$

Under certain experimental conditions the exchange current i_0 is constant and can be combined with the constant q to yield $i_0 q = B$. Equation (3.13) can then be written as $i = B\eta$, which is analogous to Ohm's law; the current is proportional to the overvoltage. The coefficient B has the dimensions of conductivity. Within the range of applicability of this linear approximation, the current is low because the overvoltage must be low (in principle under 10 mV). Then the ohmic losses in metallic conductors and not very dilute solutions of electrolytes can be disregarded, while the magnitude of B will be defined by the characteristics of the double layer and the electrode reaction. B can be treated as the conductivity of the double layer, and its reciprocal $1/B = R_w$ as the effective resistance of this part of the system. A similar expression is obtained for $z \neq 1$ = const (integers only), which enables i_0 to be determined if z is known

$$i_0 = i/q\eta = iRT/zF\eta \qquad (3.14)$$

Since the current in heterogeneous electrochemical reactions depends on the surface area of the electrodes, the normalization condition requires the exchange current i_0 to be expressed in amperes per unit electrode surface area, i.e. in $A\,m^{-2}$.

The second limiting case occurs at high overvoltages, when $\eta > RT/zF$, i.e. when $\eta > 0.06/z$ [V] at room temperature. Then the rate of the reaction proceeding in the direction opposite to that imposed by the sign of the applied voltage is so low that its contribution to the resultant current can be ignored and a logarithmic form of the simplified Butler–Volmer relationship is obtained

$$i = i_0 \exp(|\alpha| zF\eta/RT) \tag{3.15}$$

The form of equation (3.15) now allows α (or $\overleftarrow{\alpha} = \beta$ if the polarization of the electrodes is reversed) to be determined if a known value of i_0, obtainable from measurements at low overvoltages, is substituted. The equation for resultant currents $i_\pm$ can be written as

$$\ln i_\pm = \ln i_0 + |\alpha| zF\eta/RT$$

or

$$\log i_\pm = \log i_0 + |\alpha| zF\eta/2.303RT \tag{3.16}$$

The equation of the polarization curve $\eta = f(i)$ is obtained after rearrangement:

$$\eta = \frac{2.303RT}{|\alpha| zF} \log i_\pm - \frac{2.303RT}{|\alpha| zF} \log i_0 \tag{3.17}$$

The second term on the right-hand side contains nothing but constants and can be replaced by the combined constant a. The fraction on the first term also contains constants which can be combined into b. Equation (3.17) then takes the well-known form of the Tafel equation for anode currents i_a and cathode currents i_c [13]

$$\eta_a = a + b \log i_a, \qquad \eta_c = n - b \log i_c \tag{3.18}$$

The polarization curve $\eta = f(i)$ described by the Tafel equation for anode and cathode currents and the means of determining the exchange current i_0 are shown on Fig. 3.5. The coefficients α and β as well as the exchange current i_0 can be determined from the plot of the polarization curves, but only on condition that they are independent of the overvoltage η, i.e. provided that the mechanism and rate of the electrode reaction do not depend on the overvoltage. The lower the value of η, the more readily is this condition fulfilled, which in turn makes it harder to evaluate α or β. Since, however, the values of these coefficients are determined independently,

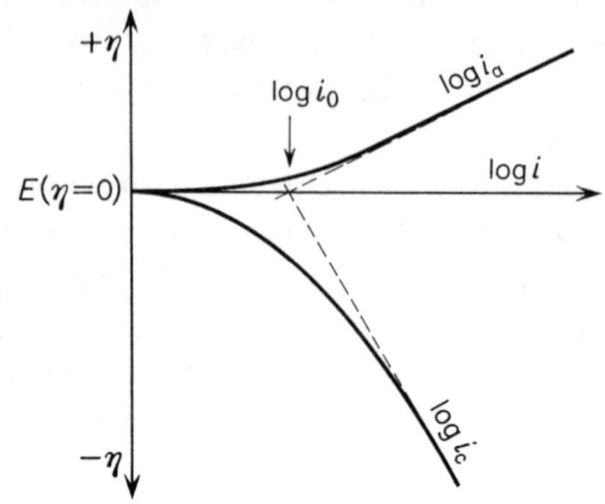

Fig. 3.5 Polarization curves $\eta = f(i)$ described by the Tafel equation for anode i_a and cathode currents i_c, and the evaluation of the exchange current i_0.

each can be established within a different range of η values. Agreement of the exchange currents determined at low and high η from anode and cathode polarization curves confirms the results obtained and proves that the electrode processes are not being seriously disturbed, and that the mechanism of the electroreaction has not changed either. On the other hand, exchange current values are usually inaccurate owing to the difficulty of calculating the effective surface area of the electrodes and to the marked influence of all secondary processes, e.g. adsorption of contaminants and changes at the electrode surfaces due to the electrode processes. Therefore, the order of magnitude and a comparative scale of exchange currents are usually reliable. Measurable exchange currents may range from ca. 1 A cm^{-2} to ca. 1 pA cm^{-2} – 12 orders of magnitude. It is considered that if $i_0 > 10^{-4}$ A cm^{-2} electrochemical systems are reversible, if $i_0 < 10^{-8}$ A cm^{-2} they are irreversible, and if i_0 falls between these values they are quasi-reversible.

At low overvoltages, the transfer coefficient α can also be determined from the relationship between the logarithm of the exchange current and the logarithms of the concentration of one form of the redox system when the concentration of the other form is constant, however, these values of α are not very accurate. The physical meaning of the transfer coefficient is not always unequivocal because the double layer structure affects the experimental values of α or β.

3.3 ACTIVATION PROCESSES IN REDOX REACTIONS

The transfer coefficient α also has a thermodynamic interpretation: like other chemical reactions, electrode processes require a certain activation energy $\mathcal{E}^{\neq}$ in order to pass from the initial to the final stage, i.e. from substrate to product. In electrochemical reactions the magnitude of the activation energy $\mathcal{E}^{\neq}$ depends on the activation energy at equilibrium $\mathcal{E}^{\neq}_{\text{eq}}$ when the overvoltage $\eta = 0$, and on the electrical energy input $F\eta$ for non-equilibrium state when $\eta \neq 0$. If the empirical magnitude of α is also of thermodynamic significance, then

$$\Delta\mathcal{E}^{\neq} = \alpha F\eta \tag{3.19}$$

where $\Delta\mathcal{E}^{\neq} = \mathcal{E}^{\neq}_{\text{eq}} - \mathcal{E}^{\neq}_{\eta}$.

If $\mathcal{E}^{\neq}_{\eta}$ is the activation energy of a process in systems displaced from thermodynamic equilibrium by an overvoltage η, the transfer coefficient α, considered without the direction of the reaction being specified, is the ratio of the change in the activation energy to the change in the electrical energy of the system:

$$zF\eta = zF(E_{\eta} - E_{\text{eq}}), \qquad \alpha = \Delta\mathcal{E}^{\neq}/(zF\eta) \tag{3.20}$$

As usual, E_{eq} is the equilibrium potential, E_{η} is the voltage of the system displaced from equilibrium by η at which the resultant current is of strength i [A] and density i/cm^2. These relationships can be presented graphically by means of simplified Morse curves showing the changes in the system's energy vs. the changes in the interatomic distances between the substrates. Basically, these are planes in multidimensional space, but for electrode reactions in the double layer one can assume that the substrates take up certain average positions with respect to the electrode surface, determined by the value of the potential gradient and chemical, intermolecular and adsorption interactions. The plot then illustrates the potential energy changes in the substrate system as a function of the parameter ρ which has a distance dimension perpendicular to the electrode surface. Usually, the energy is denoted on the ordinate axis only by a comparative scale, but an electron-volt scale (eV) is also used sometimes. The distance parameter ρ on the abscissa is either a comparative quantity only or, less often, units such as Å are given (see Fig. 3.6).

Let the electrode potential with respect to the solution, or more precisely the potential gradient in that part of the double layer where charge transfer reactions occur, be ΔE. The finite dimensions of atoms allow only a certain maximum approach of substrates to electrodes, but values of ΔE can vary within broad limits and they determine in large measure the direction and progress of the electrode reactions.

At equilibrium the electrochemical potential $\Delta\mu_{\text{el}}$ does not change, i.e. $\Delta\mu_{\text{el}} = \Delta\mu + zF\Delta E = 0$. This is a dynamic state, brought about because the rates of the reactions proceeding in opposite directions ($\rightarrow$) and ($\leftarrow$) are equal. These rates v are determined by the magnitude of the charges transferred via 1 cm^2 of electrode surface and depend on the reaction rate constants $\overrightarrow{k}$ and $\overleftarrow{k}$, the concentrations of the oxidized and reduced forms of the substrates $c_{i,j}$, and the activation energies of the processes $\mathcal{E}^{\neq}$:

$$\overrightarrow{v} = \overrightarrow{i}/zF = \overrightarrow{k}c_i \exp(-\mathcal{E}^{\neq}/RT)$$
$$\overleftarrow{v} = \overleftarrow{i}/zF = \overleftarrow{k}c_j \exp(-\mathcal{E}^{\neq}/RT) \quad (3.21)$$

The activation energy of electrochemical processes $\mathcal{E}^{\neq}$ is the sum of the chemical activation energy of the oxidized or reduced form $\mathcal{E}^{\neq}_{ij}$, regardless of the electrical fields in the double layer and the electrical energy $zF\Delta E$:

$$\mathcal{E}^{\neq} = \mathcal{E}^{\neq}_{ij} + (-\alpha zF\Delta E) \quad (3.22)$$

The coefficient α includes that part of the electrical energy which directly affects the changes in the activation energy of charge transfer processes – hence the name transfer coefficient.

At electrochemical equilibrium $\overrightarrow{v} = \overleftarrow{v}$, so $\overrightarrow{i} = \overleftarrow{i}$. Then $\Delta E = E_{eq}$ if the whole electrical potential gradient lies in that part of the double layer where the electrode reactions take place, usually its rigid part. This condition is approximately satisfied for not very dilute solutions of electrolyte. Then for other values of the applied voltage the electrode potential $E_\eta = |\eta| + E_{eq}$.

If $\overrightarrow{i} = \overleftarrow{i} = i_0$, *the exchange current* i_0 *is defined as follows:*

$$i_0 = \overrightarrow{k}zFc_i \exp[-(\mathcal{E}^{\neq}_i - \alpha zFE_{eq})/RT]$$
$$= \overleftarrow{k}zFc_j \exp[-(\mathcal{E}^{\neq}_i - \beta zFE_{eq})/RT] \quad (3.23)$$

If, in addition, $c_i = c_j$, the equilibrium potential corresponds to the formal redox potential $E^{0\prime}$ of the system c_i/c_j.

An overvoltage η appears at voltages E_η different from the equilibrium potential. Then the current

$$\overrightarrow{i} = \overrightarrow{k}zFc_i \exp\{-[\mathcal{E}^{\neq}_i - \alpha zF(E_{eq} + \eta)]/RT\}$$

Clearly, the overvoltage is the factor exponentially changing the magnitude of the resultant currents $\overrightarrow{i}$ and, in like manner, $\overleftarrow{i}$. The exchange current i_0 can be found from the above expression

$$\overrightarrow{i} = \overrightarrow{k}zFc_i \exp[-(\mathcal{E}^{\neq}_i - \alpha zFE_{eq})/RT] \exp[\alpha zF\eta/RT] \quad (3.24)$$
$$= i_0 \exp(\alpha zF\eta/RT)$$

and also $\overleftarrow{i} = i_0 \exp(-\beta zF\eta/RT)$ (see equation (3.15)).

In this way, three electrical parameters defining the magnitude of the resultant current – the charge or discharge current of the cell – have been found. These parameters are the exchange current i_0, the transfer coefficient α and the overvoltage η. The electrode system determines the magnitudes of i_0 and α, but the overvoltage also depends on the electrical circuit with its loads or other sources of current. Hence i_0 and α characterize the electrochemical system (the half-cell) and depend on its thermodynamic parameters, in particular on the activation energy $\mathcal{E}^{\neq}_{ij}$ and the reaction rate constants. The evaluation of these parameters is at present one of the main research problems in theoretical electrochemistry.

At equilibrium, the activation energies of the anode and cathode processes are equal to each other

$$\mathcal{E}^{\neq}_{eq} = \mathcal{E}^{\neq}_{i} + \alpha zFE_{eq} = \mathcal{E}^{\neq}_{j} - \beta zFE_{eq} \tag{3.25}$$

On substituting i_0 in the equation for the exchange current (see equation (3.23)) and combining constants, we get $i_0 = \theta\exp(-\mathcal{E}^{\neq}_{eq}/RT)$, or on taking logarithms $\ln i_0 = \ln\theta\,\mathcal{E}^{\neq}_{eq}/RT$. Consequently, the exchange current increases exponentially as the activation energy of the electrode process decreases.

The various energy parameters and the connections between them are shown on the diagram in Fig. 3.6. This is the usual presentation of the mechanism of a chemical reaction [16]. It is based on the convention which assumes that the dynamics of reacting systems can be illustrated in a system of Cartesian coordinates, together with the effects resulting from the presence of energy barriers and the electronic and oscillatory states of subsystems such as molecules. The energies corresponding to the states of the substrates and products are illustrated with the aid of curves similar to Morse curves, and that is why they are often approximately parabolic in shape. The points at which these curves intersect denote situations when substrates and products have configurations of identical energy – that corresponding to the energy $\mathcal{E}^{\neq}$ of the active complex. This is, of course, a highly simplified presentation, but one which is readily comprehensible. An exact description of polyatomic systems requires the introduction of a multidimensional phase space which is crossed by the trajectories of the reaction. In systems of rectangular coordinates, the abscissa acts as the reaction coordinate ρ_l to which the dimension of length (l) can be ascribed for the geometrical cross-section of the system in question, e.g. an electrode-substrate reaction in the double layer. This kind of analysis leads to equations similar to equations of motion, which can be solved either as Schrödinger or Heisenberg equations in order to obtain information on the temporal and spatial changes which this system undergoes during the reaction. The stationary states of such a system are characterized principally by

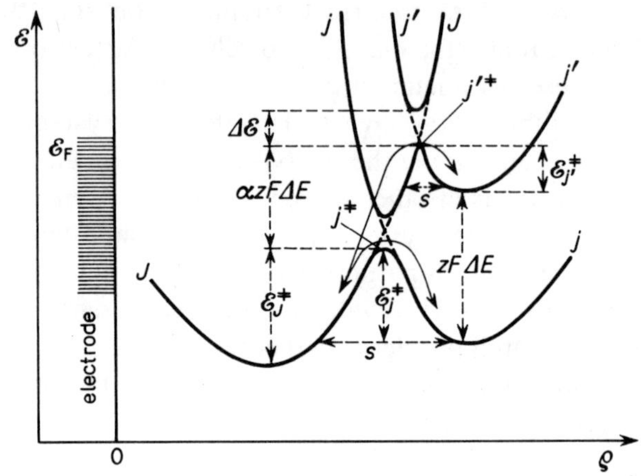

Fig. 3.6 Potential energy curves for electron transfer between any solvated ion j and its atom J which has an incomplete solvation shell, if it lies within the double layer. Curve JJ – energy states of atom J in the double layer; curves jj and $j'j'$– energy states of ion j in the double layer at lower (E) and higher (E') electrode potentials; $\mathcal{E}_F$ – Fermi potential, $j^{\neq}$, $j'^{\neq}$ – ion in excited state, close to the activated complex state, s – width of the potential barrier.

their energy levels. They are illustrated by means of the energy parameter $\mathcal{E}$ on the ordinate of a potential energy plot. On this figure (Fig. 3.6), the zero on the abscissa corresponds to the energy states in the electrode at the interface between the electrode and the electrolyte solution. The energy scale on the ordinate axis is purely conventional; in other cases the direction of this axis indicates increasing energy, from negative values to the zero corresponding to the work function of the electron from a given phase into a vacuum. If a system's energy increases, e.g. by the addition of electrical energy $\Delta\mathcal{E}_{el} = zF\Delta E$, the potential energy curve of a subsystem – a substrate or a product of the electrode reaction – rises. Figure 3.6 shows the transfer of an electron from any solvated ion j to its partially solvated atom J. Curve JJ represents the various states of atom J in the double layer. The analogous curves jj and $j'j'$ represent the respective energy states of ion j in the double layer for two electrical potentials separated by a differences of ΔE. $\mathcal{E}_F$ denotes the Fermi potential. The broken lines indicate the regions where the potential energy curves intersect. They are important regions, where the energy of the substrates is equal to that of the products. This precondition must be fulfilled if a chemical reaction, and electron transfer in particular, is to proceed. Substrates are energetically indistinguishable from products in these regions, which are therefore in a state intermediate between substrate and product. Moreover, they are unstable, a metastable state whose energy corresponds to that of excited states.

This intermediate state is called the active or activated complex, because its energy is higher than the energy of the substrates by the activation energy, which enables the potential barrier to be crossed. The activated complex is treated as an unstable though real chemical species which may decompose to give either the products or the original substrates of the reaction. The latter course clearly inhibits the reaction, but it must occur if a dynamic equilibrium between the reacting systems is to be set up, and this in turn must happen if electrode processes and chemical reactions in general are to be reversible. The reaction rate v is proportional to the concentration of the activated complex $c^{\neq}$ and the coefficient $k_B T/h$, in which k_B is Boltzmann's constant, T the absolute temperature and h is Planck's constant. This coefficient has the dimensions of reciprocal time t^{-1}, hence its name – the frequency coefficient $[s^{-1}]$: $v = c^{\neq} k_B T/h$. Independently of this, the law of mass action also holds, as a result of which the concentrations of substrates c_i and the concentration of the activated complex (products c_j) are in equilibrium

$$c^{\neq} \rightleftharpoons c_1 + c_2 + \ldots + c_i$$

The rate of reaction v is proportional to the concentration of substrates $v = kc_1c_2 \ldots c_i$. The coefficient of proportionality k is the rate constant of the reaction, hence $kc_1c_2 \ldots c_i = c^{\neq} k_B T/h$. Combining the rate constants yields the equilibrium constant of the activated-complex-forming reaction $K^{\neq} = c^{\neq}/(c_1c_2 \ldots c_i) = kh/(k_B T)$.

The free energy change $\Delta G^0 = -RT \ln K$; the equilibrium constant K includes the quotient of the partition functions Q, which include the participation of every kind of internal and translational energy

$$\Delta G = \Delta G^0 - RT \ln(Q_j/Q_i) = -RT \ln K - RT \ln(Q_j/Q_i)$$

$$\Delta G/RT = -\ln K - \ln(Q_j/Q_i) = -\ln(KQ_i/Q_j)$$

After delogging,

$$KQ_i/Q_j = \exp(-\Delta G/RT)$$

hence

$$K = (Q_j/Q_i)\exp(-\Delta G/RT) \tag{3.26}$$

For the activated-complex-forming reaction

$$K = (Q^{\neq}/Q_1Q_2Q_i)\exp(-\mathcal{E}^{\neq}/RT) \tag{3.27}$$

and

$$k = (k_B T/h)(Q^{\neq}/Q_1Q_2Q_i)\exp(-\mathcal{E}^{\neq}/RT) = A\exp(-\mathcal{E}^{\neq}/RT) \tag{3.28}$$

This is the well-known Arrhenius equation, in which A is the frequency factor in the statistical theory of reaction rates, and $\mathcal{E}^{\neq}$ is the activation energy, i.e. the difference between the initial state of the substrates and the energy state of the activated complex. When an electric field is applied the potential energy curves shift; this is shown on Fig. 3.6 only for curve jj, as a result of which the activation energies also change. The reaction is thus accelerated, its activation energy reduced, and the reverse reaction slowed down because the activation energy of the reverse process rises, e.g. $\mathcal{E}_j^{\neq} + \Delta zF\Delta E = \mathcal{E}^{\neq\prime}$.

When the equilibrium is disturbed, i.e. when the rates of the two processes differ, $v_1 \neq v_2$, an electric current flows through the circuit. The reaction kinetics, and hence the strength of the resultant current, are determined by two fundamental statistical parameters – the frequency factor and the activation energy. The electrochemical work in this case is therefore linked with the free energy change occurring as the system passes from substrates to products or vice versa. The arrows on Fig. 3.6 $\leftrightarrow$ joining the two subsystems by means of a continuous line through the active complex state show this. Transfer becomes possible when these systems interact because the wave functions ψ describing the state of subsystems J and j overlap

$$q_J \psi_J + q_j \psi_j = \psi \tag{3.29}$$

At the same time the Pauli exclusion principle applies, so that during interactions the same state cannot belong to two different systems. This is why the potential energy curves move apart in the region of intersection, although the region in which the lower sections of the curves (of lower potential energy) join at the point corresponding to the activation energy is equivalent to that in which activated complexes are formed (Fig. 3.6). The stronger the interaction between the subsystem, the larger the gap between the lower and upper parts of the potential curves, and the greater the probability that the system will cross the activation energy peak, move down the lower potential energy curve, and pass from substrates to products (or vice versa), i.e. react. The probability that transfer will occur here is nearly unity; this is an adiabatic transfer or reaction. Conversely, if the interaction between the subsystems is weak, the energy gap in the region of the active state is small and very probably the subsystem goes up or down the same potential energy curve, e.g. curve JJ, without crossing over to curve jj or $j'j'$. In this situation no chemical reaction takes place; the substrates are merely excited to higher energy levels from which they return to their initial states while releasing their energy to the surroundings. These are non-adiabatic reactions, in which the probability of transfer may be very small indeed. This probability is determined by the frequency factors in the equation for the reaction rate constant, so, according to Ehrenfest [15] it is important to divide electron transfer reactions into adiabatic and non-adiabatic. Though their activation energies may be the same, an adiabatic reaction may proceed rapidly but a non-adiabatic one slowly [16]. There is a wide gap $\Delta\mathcal{E}$ between the upper and lower potential energy curves of adiabatic reactions. Obviously, the magnitude of the activation energy $\mathcal{E}^{\neq}$ also has an influence, and an exponential one at that, on the reaction kinetics through the exponential factor in this equation: $\exp(-\mathcal{E}^{\neq}/RT)$. The minus sign indicates that the higher the value of $\mathcal{E}^{\neq}$, the slower the reaction. In adiabatic reactions, the way in which the system overcome the potential barrier set up by the energies close to the activation energy $\mathcal{E}^{\neq}$ (see Fig. 3.6) is significant in establishing the mechanism or path of the reaction.

Many paths are possible, but in electrochemical processes radiationless transfer is important. In view of the law of conservation of energy, this means that energy gained or lost by a subsystem must change the energy states of other subsystems within the reacting system, e.g. the oscillational excitation energy of the products, or the energy of electrons collected by an electrode of a suitably high potential. Only after relaxation processes leading to an exchange of energy with the surroundings have occurred, e.g. when a current has flowed, can subsystems return to their initial states; the free energy change in the system is expended and converted, say, to the work of the electric current, or heat. There are two principal mechanisms in radiationless electron transfer; the activation mechanism, in which the electron crosses the potential barrier after having gained the requisite activation energy $\mathcal{E}^{\neq}$, or the tunnelling mechanism, when the electron tunnels through the potential barrier associated with the activation energy. The first mechanism is preferred when the potential barrier is wide (the distance J–j on the abscissa between potential energy curve minima is long) but not very high (low activation energy). This is the classical mechanism by which the conversion A→B is achieved through an activated complex $(A + B)^{\neq}$: A→$(A + B)^{\neq}$→B. The second mechanism is a quantum-mechanical effect and can occur because of the wave properties of matter. According to classical mechanics, an electron or any other particle must, on arrival at a potential barrier, either rebound from it, or cross it after having obtained a quantity of energy equal at least to the activation energy; in other words, it must climb up to the full height of the energy barrier. This is the classical activation mechanism discussed earlier. However, quantum mechanics has shown that particles of matter behave like wave packets capable of penetrating or tunnelling through the potential barrier. If the barrier is narrow, the particle can tunnel through it without having to acquire all the required activation energy, and the height of the barrier is then of less importance. So the tunnelling mechanism is preferred if a potential energy barrier is tall and narrow. The state of a system just prior to the barrier being penetrated could also be said to be a kind of activated complex, although resembling more nearly the substrate than the product because of the former's lower energy. By means of the tunnel effect, an electron is transferred to a second system having an identical energy but a configuration approximating to that of the product. This transfer is denoted by the lower-case *s* on Fig. 3.6. This mechanism can be presented schematically as $A \rightarrow A^* \xrightarrow{s} B^* \rightarrow B$ (the asterisk indicates an excited state). The tunnel effect is also important in electrochemistry because it explains the appearance of electrons in the double layer at the electrode surface. Accordingly, an electrode process would merely be a disturbance of the equilibrium between electrons moving from the electrode material to the double layer and those moving in the opposite direction. An almost identical mechanism explains the electronic conductivity of concentrated solutions of alkali metals in liquid ammonia. It is therefore worth having a closer look at the tunnel effect.

Classical mechanics ascribes wave properties to those particles which have zero rest mass and which move at the velocity of light (photons). However, since the discovery of the photoelectric effect and the introduction of relativistic quantum mechanics it has been an accepted fact that photons and any other particles without a rest mass must have a momentum $p = h\overline{\nu}/c = h\lambda$, where h is Planck's constant (which has the dimensions of action – energy × time), $\overline{\nu}$ is the radiation frequency (in Hz), c is the velocity of light, and λ is the wavelength. On the other hand, the momentum p of a corpuscular particle of rest mass m and velocity υ is $m\upsilon$. Wave mechanics combines these two equations by means of Planck's constant to give the de Broglie equation, $m\upsilon = h/\lambda$, and assumes dual corpuscular and wave properties for all particles whose wavelengths are inversely proportional to their relativistic masses: $\lambda_p = h/m\upsilon$. Because Planck's constant has a small value, the wavelengths λ_p of particles heavier than hydrogen atoms at ordinary temperatures are much smaller than the dimensions of an atom and can therefore be disregarded. The electron holds a special position here because of its very small rest mass, which is barely 0.068% of the rest mass of a proton. The wavelength of such a small mass is quite considerable and depends on the kinetic energy of the particle $\mathcal{E}_{kin} = m\upsilon^2/2$ which is linked to the velocity υ.

After substituting

$$\lambda_p = h/(2m\mathcal{E}_{kin})^{1/2} \tag{3.30}$$

By definition, the energy gained by electrons accelerated by a potential difference of 1 V is 1 eV; this corresponds to a wavelength λ_p of ca. 1.2 nm (12 Å). So at the energies which electrons acquire during electrochemical reactions, i.e. < 6 eV , the wave properties of electrons are very strongly marked, and it is these, apart from the charge, which mainly determine the behaviour of electrons in electrochemical processes, and in particular facilitate transitions through spaces considered by classical mechanics to be inaccessible. If a moving electron comes up against a potential barrier of height $\mathcal{E}_x$, it can penetrate this barrier to a distance x dependent on $\mathcal{E}_x$ and on the energy of the electron $\mathcal{E}_e$; the probability of penetration associated with the amplitude of the wave function Ψ_e of the electron at position x decreases exponentially with increase in x from the barrier threshold. For an electron leaving the electrode to a distance x

$$\Psi_e = q \exp(-\kappa x) \tag{3.31}$$

where $\kappa = 2\pi\,[2(\mathcal{E}_x - \mathcal{E}_e)m_e]^{1/2}/\,h$; q is a constant and m_e is the mass of the electron. The ability of electrons to penetrate the potential barrier falls to zero only when the height of the barrier has risen to infinity. When $\mathcal{E}_x \neq \infty$, $\Psi_e \neq 0$. If the potential barrier now has a finite width l, the probability of finding the electron at a distance l from the barrier threshold is finite, which means that it has crossed the potential barrier, even though $\mathcal{E}_e < \mathcal{E}_x$. In this case, $\mathcal{E}_x$ corresponds to the activation energy, and the process described is called the tunnel effect. At given values of $\mathcal{E}^{\neq}$ or $\mathcal{E}_x$, the probability of electrons tunnelling through the potential barrier rises rapidly as this decreases in width; very narrow barriers are almost "transparent" to electrons, even though they may be very high on the scale of magnitudes normal in electrochemistry. This is significant because just such a very narrow but high potential barrier exists at the surface of metallic electrode materials, due to the entire excess charge of a charged electrode accumulating exclusively at its surface. Wider potential barriers do occur in the double layer of the electrolyte, and here too, the electron transfer mechanism depends on the shape of the activation energy barrier. This is defined by the redox system and the double layer structure, the latter depending on the composition of the solution and the electrode potential.

The transfer of an electron changes the electronic configurations of the substrates. Oxidation and reduction reactions can therefore be treated as dynamic systems [18]. One example of this approach are the generalized dynamics in which the interactions between electrons and atomic nuclei (nuclides) treated as wave packets are considered as functions of time. The general quantum-mechanical function then reads:

$$S = \int_{t_1}^{t_2} < \Psi(\alpha) | \, i\partial/\partial t - H \, | \, \Psi(\alpha) > \mathrm{d}t$$

Thus one can introduce the model wave function $\Psi(\alpha)$ whose changes in time can be effectively studied. Here, α is a set of parameters describing the motion of real, classical particles. In this way the motion of a system along the reaction coordinate ρ_l can be calculated for a selected model of the system using suitable programs and computers [19].

When cells and storage batteries (but not compensating systems) are operating, the currents flowing through the external circuits are usually large during charging and discharging. So electrochemical current-generating systems are mostly far from equilibrium when electrode over-voltages are significant. It is in these cases that the quantum-mechanical mechanisms of electron transfer come into their own, because the potential gradients also extend further from the electrode and the substrates can exchange electrons even at a distance of the order of 1 nm from the electrode surface. As they are fermions, the free electrons in a metal cannot occupy one, common, precisely defined energy level, but occupy an enormous number of energy levels lying exceedingly close to one another, which together make up the conduction band, well known from the theory of the metallic state. Not all free electrons participate in redox reactions (electron transfer). Only those of the highest energy $\mathcal{E}_F$ do so, i.e. those in the top level of the conduction band in the metal (the Fermi level), and in that part of the band $\pm kT$ wide with respect to the Fermi level, i.e. in the region $\mathcal{E}_F \pm kT$. The set of activated complexes forming as a result of the radiationless transfer of these electrons is equivalent to one particular activated complex in the solution, whose state corresponds to the Fermi level. This level determines the surface potential ψ_s of the electrode material and influences the ratio of the rate constants due to the cathode current i_c and the anode current i_a of the reacting electrochemical system. The following relationship exists between these constants

$$k_a/k_c = \exp\,[e_0(\psi_s - E_l) - \mathcal{E}_F - \mathcal{E}_R - \mathcal{E}_M]/(k_B T) \tag{3.32}$$

in which e_0 is the elementary electrical charge and $\psi_s - E_l$ is the potential drop across the part of the double layer in which electrode reactions occur. In many cases, especially in concentrated electrolyte solutions, this region is restricted to the rigid part of the double layer. Then, $E_l = E_H$. $\mathcal{E}_F$ denotes the energy of the electrons in the Fermi level of the electrode material, $\mathcal{E}_R$, the solvation energy and other intermolecular interaction energies in the solution, and $\mathcal{E}_M$ the lattice energy of the electrode material. At equilibrium, $i_c = i_a$, so there is no net current (cell or storage battery in a compensating system). The current i flowing through the system as a result of non-equilibrium electrode reactions is given by the general theoretical formula [20]

$$i \doteq C \int_{-\infty}^{+\infty} F(\mathcal{E}')N(\mathcal{E}')P(\mathcal{E}')\,\mathrm{d}\mathcal{E}' \tag{3.33}$$

in which $F(\mathcal{E}')$ is the distribution function of electronic energy levels in the metal in accordance with Fermi–Dirac statistics, $N(\mathcal{E}')$ is the distribution function of the free energy levels of the dissolved substrates in that region of the double layer where electron transfer occurs, $P(\mathcal{E}')$ is a statistical coefficient expressing the probability of electron transfer, sometimes called the "transparency" coefficient, and C is an integration constant accounting for the concentration of the substrates.

The principle of energy conservation requires that radiationless transfers take place between equipotential levels of equal energy. However, if there are electrons in the metal at energy levels slightly above the lowest free levels of the dissolved substrates, electrons may be transferred with the formation of activated complexes at higher oscillation levels; after the system has relaxed, this may induce a temperature effect.

The current density i/cm^2 at instant t is given by the number of transferred charges per second per unit surface area of electrode. The size of this charge depends on the stream of substrate molecules or ions arriving at the double layer, i.e. on the concentration gradient and the coefficient of diffusion. For concentration gradients c_j,

$$i_j = c_j \int_{-\infty}^{+\infty} N(\mathcal{E}')\,\mathrm{d}\mathcal{E}'$$

This quantity is equal to the number of electrons crossing the interface per second:

$$i_e = c_x \int_{-\infty}^{+\infty} F(\mathcal{E}')P(\mathcal{E}')\,\mathrm{d}\mathcal{E}' \qquad (3.34)$$

where i_j is the net current due to a reaction in a non-equilibrium process, $P(\mathcal{E}')$ is the tunnelling probability coefficient for electrons leaving the metal when $\mathcal{E}_x = \mathcal{E}_F$, and $F(\mathcal{E}')$ for levels filled by electrons of energies $\mathcal{E}' \neq \mathcal{E}_F$ is also given by Fermi–Dirac statistics

$$F(\mathcal{E}') = (\mathcal{E}')^{1/2} / \,[1 + \exp(\mathcal{E}' - \mathcal{E}_F)\,/(kT)] \qquad (3.35)$$

Unfilled energy levels also exist in a redox system in solution; their distribution can be described by the analogous Fermi function, so they correspond to the Fermi level in a metal. By altering the electrode potential ΔE, the Fermi level in the metal is shifted with respect to the energy levels of the substrates in the electrolyte solution. As the energy of the Fermi level in the metal rises, an increasing number of available energy levels in the double layer of the redox system can be filled by electrons from the metal. The current then increases until the electrode reaches a potential at which all the available vacant energy levels in the redox system's double layer have been filled by electrons. The progress of the electrode reaction then depends on the magnitude of the flux of substrates arriving at the electrode, determined at an instant t. Raising the electrode potential higher does not increase the current any more, unless the conditions of substrate transport (e.g. the depth of the diffusion layer changes), or unless the potential is attained at which a different redox system starts to react, such as a neutral electrolyte added to improve the electrical conductivity of the solution or if the solvent decomposes. The maximum current obtained under these conditions is known as the limiting current.

The consecutive states of the substrates are represented as harmonic oscillators of different energies, so that a set of potential energy curves can be used. To obtain an expression for the rate constant k, a power series expansion is applied in the quantum-mechanical approach [17]

$$k = \sum_{n=-1}^{\infty} c^{n+1} T^{n/2} \exp(-\mathcal{E}^{\neq}/k_{\mathrm{B}}T) \tag{3.36}$$

in which c^{n+1} are the virial coefficients of the expansion. The problem is then to evaluate these coefficients and the activation energy. They can be derived for selected model systems using the method of successive approximations on a computer [14]. If one starts with the model of a harmonic oscillator, the basic Hamiltonian operator

$$H = H_{\mathrm{osc}} + H_{\mathrm{met}} + H_{\mathrm{solv}} + H_{\mathrm{int}} \tag{3.37}$$

can be introduced. The mass centres of the reacting molecules or solvated ions are assumed to remain stationary in their equilibrium positions in the double layer. The Hamiltonian, i.e. the molecular energy, thus reduces to the oscillation Hamiltonian H_{osc}, which includes the energy states equivalent to the vibrations of the reacting subsystem, e.g. of a solvated ion. H_{met} is the energy operator of the reacting electrons in the metal electrode

$$H_{\mathrm{met}} = (p^2/2m_{\mathrm{e}}) - \mathcal{E}_{\mathrm{L}} - e\mathcal{E}_{rj} \tag{3.38}$$

where p is the momentum, m_{e} the mass of the electron, $\mathcal{E}_{\mathrm{L}}$ the crystal lattice energy of the metal, and $e\mathcal{E}_{rj}$ the proportion of potential energy derived from other ions in the double layer, i.e. those not participating in the redox process. H_{solv} is a function of the solvent treated as a uniform polarizable dielectric which can be completely characterized by the optical electrical permittivity $\mathcal{E}_{\mathrm{hfr}}$ and the static electrical permittivity $\mathcal{E}_{\mathrm{s}}$, the frequency ω_P associated with the polarizability P and vectors D, the field induced by the ionic forms of the redox system

$$H_{\mathrm{solv}} = \frac{2\pi}{\varepsilon_{\mathrm{hfr}}^{-1} - \varepsilon_{\mathrm{s}}^{-1}} \int \mathrm{d}V(P^2 + \omega_P^{-2}\dot{P}^2) + \frac{\varepsilon_{\mathrm{hfr}}^{-1} - 1}{8\pi\varepsilon_{\mathrm{hfr}}} \int \mathrm{d}V(D_1^2 - D_2^2) \tag{3.39}$$

The intermolecular interaction term H_{int} includes the solvation energy of the electron $H_e = 2\pi \int dV \overline{P}\,\overline{P}_{ke}$, where $\overline{P}_{ke}$ is the interaction of a solvated electron from the Fermi level in the form of a polarized anion complexed with the surrounding solvent, V has the dimensions of volume, $\omega = 2\pi\overline{\nu}$, and $\overline{P}$ is the polarization field intensity. This Hamiltonian also includes the vibration of the centre of mass of the solvated substrates. All these terms make up the basic Hamiltonian of the system, H. The wave function of the system Ψ contains three component functions corresponding to different kinds of energy associated with adiabatic electron transfer Ψ_{el}, oscillation Ψ_{osc} and rotational energy Ψ_{rot}; the latter can also include interactions with solvent molecules.

The foregoing discussion has mentioned most of the energies and types of interactions currently taken into account by theory. An apparently straightforward electron transfer, even in the simplest redox system of the type $j^{n+}/j^{(n-1)+}$ turns out on closer examination to be a complex process requiring many interactions to be considered, including interactions with the electrically neutral solvent molecules. When ions are reduced to neutral atoms, or conversely, when uncharged atoms or particles are converted into electrically charged forms of products, the orientation of the solvating molecules and the solvation energies of the substrates undergo a marked change. The activation energy involved in regrouping solvation shells may make up the largest portion of the activation energy of the entire electron transfer process. The real activation energy thus depends on the repolarization energy, which reflects the changes in the interactions of the solvated substrates as a result of electrode reactions and on specific interactions occurring in the double layer at the interface. When the latter interactions are very strong, the influence of the adsorption energy, reflecting further changes in chemical bond length and the geometry of the reacting subsystems [14, 21–24], on the magnitude of $\mathcal{E}^{\neq}$ may be dominant.

Thus recent theoretical research has been attempting to construct new models of the double layer and the electrolyte solution which represent more faithfully their complex structures. Such a model treats the electrolyte solution as a mixture of polarizable dipoles (representing solvent molecules) and electrical charges (ions).

Electrode A^-, an electron source, is located at a distance x from atom B, an electron acceptor (B may be a solvated ion or the reactive part of a molecule). Consider the reversible electron transfer reaction between them

$$A^- + B \rightleftharpoons B^- + A$$

The probability $\kappa(E, \varepsilon, x)$ of this reaction occurring depends on the magnitude of the external field E applied at point x, $W(E,x)$. The overall average rate of this reaction k is then given by

$$k = \int d\varepsilon \int dx \int dE\, W(E, x)\kappa(E, \varepsilon, x)\rho_B(x)N(\varepsilon)f(\varepsilon)$$

where ρ_B is the average density of ionic charges at point x.

The theory of adiabatic electron transfer at metal electrodes is still being developed by Kuznetsov *et al.* [25].

A considerable difficulty inhibiting further progress in this field is that all too little is known of the double layer structure when an electric current is flowing, especially at the high current densities usual in operating cells.

Nevertheless, the general theory of electrode processes has enabled detailed though complicated relationships describing the exchange current and rate constants of electrode processes to be derived coherently and consistently from the basic statements of statistical physics and quantum mechanics.

3.4. PROCESSES AT SEMICONDUCTOR ELECTRODES

Electron transfer at a semiconductor electrode differs markedly from that at a metal electrode. The main reason for this is the low concentration of charge carriers in semiconductors, so the double layer structure is quite different, and consequently, the mechanisms of the electrode processes are also different. On the other hand, the larger the concentration of charge carriers in the semiconductor, i.e. the more completely the conduction band is filled, the more the semiconductor resembles a metal; then the double layer structure and electrode mechanism are very similar to those characteristic of metal electrodes. This results from the usual definition of metals and semiconductors. In the latter it is only the wider energy gap between bands and the lack of suitable components capable of acting as electron donors and acceptors which inhibits electrical conductivity. The conductivity of semiconductors is therefore extremely sensitive to the presence of contaminants even in trace quantities, and to lattice defects, and so to the physical form of such materials. Electrode mechanism and processes may then be specific to particular semiconductor materials and even to different samples of the same material if its production has not been standardized or not strictly controlled. The remainder of this discussion will refer to a semiconductor whose conductivity is good enough for the system to be practicable, but in which the concentration of the carriers is small enough and that of the electrolyte solution high enough for almost all the potential drop to occur within the semiconducting phase. Then the high concentration of charge carriers in the electrolyte ensures that the solution's electrical conductivity is far higher than the specific conductivity of the semiconductor. But in an electrochemical system in which this semiconductor is used as an electrode, any alteration in the voltage causes a large and instantaneous change in the carrier concentration; the Fermi level in the semiconductor electrode changes too. Rectification may occur, depending on the width of the energy gap between the conduction and valence bands, and on the level of contaminants, and the resulting polarization curves will be strikingly asymmetrical. When potentials are high and negative, the conduction band fills with electrons, and the semiconductor electrode assumes the properties of a metal electrode [6, 8]. Conversely, when potentials are positive, the reduction in electron concentration in the conduction band may bring about such a large increase in the electrode's resistance that the semiconductor electrode now functions as a virtual insulator.

The electrochemical potential of electrons in semiconductors is established by the position of the Fermi level, and in electrolyte solutions by the redox potential of a system capable of exchanging electrons, which can be regarded as the Fermi potential of the electrolyte solution. Electrical contact occurs at the interface between the two phases (i.e. electrolyte and semiconductor), and the Fermi levels $\mathcal{E}_F$ are equalized in both. Electrons are transferred from one phase to the other and an inflection appears in the energy levels in the semiconductor. The region in which these potentials change defines the region of the double layer (Fig. 3.7). $\mathcal{E}_n$ and $\mathcal{E}_M$ are the conduction bands in the semiconductor and metal respectively, the broken lines indicate the prevalent energy levels, $\mathcal{E}_{ij}$ for the oxidized and reduced forms of the redox subsystem J/j reacting at the semiconductor electrode under non-equilibrium conditions when the external circuit between the electrodes is shorted. The levels of the second, conjugated redox subsystem M/M^+ at the metal electrode (which must react at the same time) are also indicated. The situation illustrated schematically on Fig. 3.7 corresponds to the cathode process at the semiconductor electrode and the anode process at the other electrode (dissolution of the metal anode). The balancing out of Fermi levels between the two phases is equivalent to the equalization of the electrochemical potentials. A potential difference of $\mathcal{E}_F - \mathcal{E}_M$, equivalent to the voltage ΔU must be set up when a current is flowing between the electrodes. The unit energy change in the system is $\mathcal{E} = e\Delta E$ or $N_A e\Delta E$ when one mole of electrical charges (i.e. 96 500 C) is transferred across the interface; as always, e is the elementary negative electrical charge.

The depth of the double layer (l_D), i.e. the region in a semiconductor electrode where the energy bands are inclined, can be determined from its electrical capacity.

The value of l_D is approximately comparable with the reciprocal of the Debye–Huckel constant κ; thus

$$l_D \approx \frac{1}{\kappa} = \left(\frac{\varepsilon k_B T}{8\pi n_i e^2}\right)^{1/2} \tag{3.40}$$

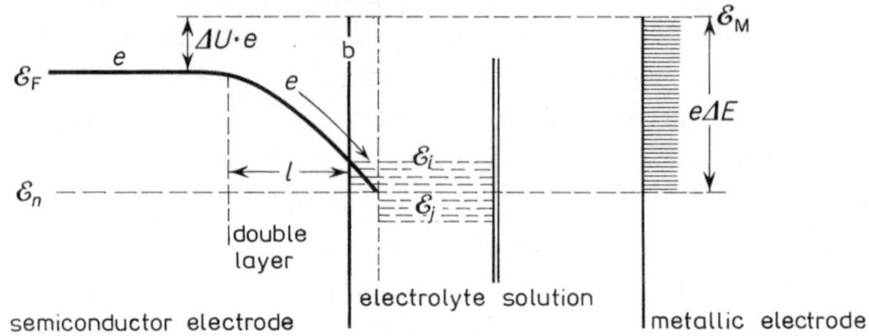

Fig. 3.7 Region of electrical potential changes in the double layer between the semiconductor and electrolyte solution. $\mathcal{E}_M$ and $\mathcal{E}_n$ – conduction bands in the metal and semiconductor, $\mathcal{E}_F$ – Fermi level, l – depth of the double layer in the semiconductor, b – boundary between the semiconductor (solid phase) and the electrolyte (liquid phase), $\mathcal{E}_i$, $\mathcal{E}_j$ – energy states of the reduced and oxidized ion or molecule of the electroactive substance in the solution; ΔUe – energy released in the external circuit.

In semiconductor theory, the value $l_D = (\varepsilon\varepsilon_0 k_B T/n_i e^2)^{1/2}$ is that depth from the semiconductor surface at which the spatial charge density falls to the value $1/e = 1/2.718$, i.e. to ca. 37% of the maximum surface value (see Fig. 3.7). The electrical double layer can be assumed to extend this distance into the bulk of the semiconductor. The electrical permittivity of semiconductors ε usually lies within the range $10 < \varepsilon < 20$ [24] (if the electrical permittivity of a vacuum ε_0 is assumed equal to 1). Depending on the values of ε and the concentration n_i of charge carriers in the semiconductor, l_D may range widely, from ca. 10^{-6} to ca. 10^{-3} cm at room temperature . For example, in a semiconductor in which $n_i = 10^{16}$ cm^{-3} and $\varepsilon = 12$, $l_D = 5 \times 10^{-7}$ cm [24], but in pure germanium, where $n_i = 10^{14}$ cm^{-3} and $\varepsilon = 16$, the electrical field, and hence the electrical double layer, extends into the semiconductor to a depth of $l_D = 4 \times 10^{-4}$ cm. As a result, almost all of the potential drop across the double layer occurs within the semiconducting material. In metals, where n_i is of the order of 10^{22} cm^{-3} and $\varepsilon = 1$, $l_D = 1 \times 10^{-8}$ cm; this means that the charge accumulates on the surface of the metal. The depth of the Helmholtz double layer in concentrated electrolytes is ca. 5×10^{-8} cm; it is in this part of the electrolyte where practically all the potential drop between the interface (electrode surface) and the bulk of the electrolyte solution occurs when metal electrodes are used. The double layer structure of semiconductor electrodes therefore differs from that of metal electrodes. The fact that almost all the potential drop occurs within a semiconductor electrode means that the activation energy of a process taking place at the surface of such an electrode, but still in the solution, i.e. in the Helmholtz layer, is largely independent of the potential. So it is this potential which primarily affects the value of the frequency factor, and it influences the kinetics of electrode processes on semiconductors mainly via changes in the frequency factor [6].

In cells, semiconductor electrodes usually act as cathodes; the substrate, i.e. the substance undergoing reduction, is the cathode material, whose chemical bonds are largely covalent. Breaking covalent bonds, even polarized ones, usually requires considerable activation energy, which is why such reactions are only slightly reversible, and why the use of these materials is limited to primary cells.

The flow of electrons from, say, a lithium anode, where they are released during the reaction $Li^0 \rightarrow Li^+ + e$, causes electrons to appear in the antibonding orbitals of atoms both on the surface and in the bulk of the cathode. This causes a similar kind of weakening of chemical bonds in the electrode material as an electron deficit, and makes feasible the reduction reaction whose products pass into solution or are precipitated, e.g. by cations in solution, which are needed to balance out the negative charge transferred to the products of electroreduction. Having accepted electrons, metal cations form a new crystal lattice in which the atoms occur at lower or zero oxidation states, e.g.

$$Cu^+ - Cl^- + Li^+ + e \rightarrow Cu^0 + LiCl$$

This kind of process will occur in n-type semiconductors at a rate υ_l proportional to the frequency factor A, the effective density of electron states in the conduction band, N_n at the electrode surface, and the flux J_{M^+} of positive anti-ions M^+ binding the reduction products at the electrode. Hence

$$v_1 = AN_n J_{M^+} \exp(-\mathcal{E}_1^{\neq}/k_B T) \qquad (3.41)$$

where $\mathcal{E}_1^{\neq}$ is the activation energy of a process of rate v_1.

The participation of valence band charge carriers in the electroreduction process of rate υ_2 is small because of the low mobility of holes and the presence of an energy gap $\Delta\mathcal{E}_{n,p}$ between the valence and conduction bands. The rate ratio of υ_1 and υ_2 is

$$\frac{\upsilon_1}{\upsilon_2} = \frac{N_n}{N_p} \exp \frac{\mathcal{E}_F - \mathcal{E}_b + \gamma\Delta\mathcal{E}_{n,p}}{k_B T} \qquad (3.42)$$

where N_n and N_p are the effective densities of the carrier states in the conduction and valence bands respectively, $\mathcal{E}_F$ is the Fermi level in the semiconductor, $\mathcal{E}_b$ is the energy of the boundary of the conduction band, γ is a numerical coefficient greater than 0.5 dependent on the difference in activation energies of the processes occurring within the conduction and valence bands.

Electroreduction is inhibited on p-type semiconductors because there it proceeds in the valence band [6]. The maximum rate of the process v_2 is low and saturation currents appear at an early stage, so p-type semiconductors are unsuitable as cathode materials in galvanic cells.

The increasing spatial charge and number of carriers means that the semiconductor will become more and more metallic in nature. At the same time, the depth l_D decreases and the potential gradient in the rigid Helmholtz layer steepens; the double layer structure approaches that existing at a metal-electrolyte solution interface. There is thus a continuous transition between the properties of semiconductor and metal electrodes, even though in the extreme cases (a metal and a pure semiconductor with a fairly broad energy gap between the valence and conduction bands) the double layer structures are totally different.

Recent experimental and theoretical developments concerning the structure of the double layer are reviewed by Parsons [28]. Advanced theory of charge-transfer in electrochemical systems is discussed by Lorenz [29].

REFERENCES

[1] R.W. Gurney, *Proc. Roy. Soc.*, **A134**, 137 (1931).

[2] A.N. Frumkin, V.S. Bagotski, S.A. Yofa and B.N. Kabanov, *The Kinetics of Electrode Processes* (in Russian), Izd. Mosk. Univ., Moskva (1952).

[3] K.J. Vetter, *Elektrochemische Kinetik*, Springer-Verlag, Berlin (1961).

[4] A. Cisak, *Roczniki Chem.*, **36**, 1895 (1962).

[5] A. Cisak, *Roczniki Chem.*, **37**, 1025 (1963).

[6] H. Gerischer, "Semiconductor Electrochemistry", Ch. 5 in: *Physical Chemistry. An Advances Treatise* (H. Eyring, ed.), Vol. 9A, Academic Press, New York–London (1970).

[7] J. Jortner and N.R. Kestner, *Electrons in Fluids*, Springer-Verlag, Berlin (1973).

[8] B.K. Makarenko, Y.M. Makarenko, P.A. Sereda and N.Y. Bobrova, *Elektrochimia*, **9**, 705 (1973).

[9] L. Werblan and A. Cisak, *Wiadomości Chem.*, **27**, 843 (1973).

[10] T. Pyle and C. Roberts, *J. Electrochem. Soc.*, **115**, 247 (1968).

[11] J. O'M. Bockris and A.K.N. Reddy, *Modern Electrochemistry*, Vol. 2, Plenum Press, New York (1970).

[12] J.S. Newman, *Electrochemical Systems*, Ch. 8, Prentice-Hall Inc., Englewood Cliffs, NJ (1973).

[13] T. Erdey-Gruz, *Kinectics of Electrode Processes*, Akadémiai Kiadó, Budapest (1972).

[14] P.P. Schmidt, in: *Specialist Periodical Reports, Electrochemistry* (H.R. Thirsk, ed.), Vol. 5, The Chemical Society London (1975), p. 21.

[15] P. Ehrenfest, *Ann. Phys.*, **51**, 327 (1916).

[16] N. Calusaru, *Electrochim. Acta*, **12**, 1507 (1967).

[17] P.P. Schmidt and H.B. Mark, *J. Chem. Phys.*, **58**, 4290 (1973).

[18] R.A. Marcus, *J. Chem. Phys.*, **41**, 2614, 2624 (1964).

[19] R.R. Dogonadze and A.M. Kuznetsov, *Physical Chemical Kinetics* (in Russian), Vol. 2, Viniti, Moskva (1973).

[20] J.J. Bikerman, *Phil. Mag.*, **7**, 33, 384 (1942).

[21] R.R. Dogonadze and Y.A. Chizmadzhev, *Dokl. Akad. Nauk SSSR*, **144**, 1077 (1962).

[22] R.R. Dogonadze and Y.A. Chizmadzhev, *Dokl. Akad. Nauk SSSR*, **145**, 848 (1962).

[23] M. Green, in: *Modern Aspects of Electrochemistry* (J. O'M. Bockris, ed.), Vol. 2, Butterworth, London (1959); M. Green, *J. Chem. Phys.*, **31**, 200 (1959).

[24] Ch. Weissmantel and C. Hamann, *Grundlagen der Festkörperphysik*, VEB Deutscher Verlag der Wissenschaften, Berlin (1981).

[25] A. Kuznetsov, R.R. Nazmutdinov and M.S. Shapnik, *Elektrokhimiya*, **22**, 776 (1986).

[26] G.M. Torrie, P.G. Kusalik and G.N. Patey, *J. Chem. Phys.*, **90**, 4513 (1989).

[27] D.R. Bérard and G.N. Patey, *J. Chem. Phys.*, **95**, 5281 (1991).

[28] R. Parsons, *Chem. Rev.*, **90**, 813 (1990).

[29] W. Lorenz, *J. Phys. Chem.*, **95**, 10566 (1991).

CHAPTER 4

Transport Processes in Cells

The thermodynamics of irreversible processes as applied to electrochemical cells are the subject of this chapter. The mechanisms of the electrode reactions and ion transfer in a cell working under stationary conditions and supplying a moderate current are such that a minimum of energy is dissipated. According to this principle, formulated by Onsager, Prigogine and Glansdorff [1, 6], the thermodynamic migration, diffusion and heat fluxes of ions and solvent molecules depend linearly on thermodynamic forces, namely, electrical field strength and gradients of temperature, pressure and concentration of the solution's components. The point of the following discussion should be obvious: the structural and operational design of a cell that is to work stably over a broad range of temperatures and external loads must be supported by an accurate mathematical description of all important cell processes, including non-linear ones like electrode reactions.

4.1. ONSAGER EQUATIONS AS APPLIED TO ELECTROCHEMICAL CELLS

The Prigogine–Glansdorff variational principle [6] is a general statement of evolution processes and includes the thermodynamic principles of irreversible processes formulated earlier by Onsager [1]. Since the stationary processes considered are close to the equilibrium state we confine the discussion to the Onsager scheme, according to which fluxes and thermodynamic forces are fundamental quantities. It is assumed after Onsager that any flow of tensor of order n depends on all the forces of the same order acting in the system. Scalar flows depend on the scalar forces of chemical or electrochemical affinities. Vector flows like ionic flux, diffusion or heat flow depend linearly on the gradients of electrical potential, concentration, temperature and pressure. If fluxes are denoted by $\vec{J}$ and thermodynamic forces by $\vec{X}$, the Onsager equation for the i-th component is

$$\vec{J}_i = \sum_j L_{ij}\vec{X}_i \tag{4.1}$$

The phenomenological coefficients L_{ij} are functions of the state parameters $L_{ij} = L_{ij}(T, p, \phi)$ [2–4]. These coefficients must be determined experimentally, though not necessarily all of them. Onsager showed that the matrix of coefficients L_{ij} was symmetrical, i.e. $L_{ij} = L_{ji}$. One must therefore determine the diagonal coefficients L_{ii} and those non-diagonal ones L_{ij} which are easiest to measure. Since the entropy source σ is given by

$$\sigma = \sum_i \vec{X}_i\vec{J}_i = \sum_{ij} L_{ij}\vec{X}_i\vec{X}_j \tag{4.2}$$

and according to the second law of thermodynamics this must be positive for an irreversible process and zero for a state of equilibrium, it should be assumed that the matrix L_{ij} is positively defined, i.e. all the diagonal coefficients are positive $L_{ii} > 0$. All minors $L_{ii}L_{jj} - \frac{1}{4}L_{ij} + L_{ji}^2 > 0$ etc. are positive as far as the determinant $\det|L_{ij}| > 0$. Hence, cross processes are less intensive than the straight ones. This is important: to a first approximation, when little is known about the system, the effect of cross processes on the mass, charge and energy balances can be ignored. The above relationships show that there are two complementary aspects to the optimization of electrode processes and cell design. One is the need to describe the cell by a set of linear equations; these are based on equations of ion transport in solution and on the kinetic equations of electrode reactions that emerge from Onsager's minimal energy-dissipation principle. The other, material aspect is the choice of appropriate electrode materials, pure or mixed solvents and ions; these, however, can only be found by experiment.

Let us examine the Onsager equations for electrolyte solutions in greater detail and look for the simplest possible way of expressing the phenomenological coefficients L_{ij} by means of the coefficients of ionic mobility, diffusion and thermal conductivity. The migration, mass and electrical fluxes as well as the thermodynamic forces appearing in the Onsager equations must be defined. For electrolyte solutions these fluxes are most conveniently found by multiplying the concentrations of components c by their velocities $\vec{\upsilon}$. The migration flux of the i-th component is then

$$\vec{J}_i = c_i\vec{\upsilon}_i. \tag{4.3}$$

Consider the relations between L_{ij}, in order to obtain the maximum information for a minimum experimental outlay. In electrolyte solution these components are ions of both signs, electrically neutral ion pairs, ion triplets of both signs (if they form), and larger clusters of ions (appearing in concentrated solutions or in solutions of low polarity), all of which are solvated. Solvent molecules must also be considered. The mobilities of the ions depend on the degree of solvation. Solvation shells of ions in regions adjacent to the electrodes may differ from those in the bulk of the solution. The diffusion (mass) flux [5–7] in a "stationary" electrolyte solution of a cell is given by the products of the component concentrations and their diffusion velocities

$$\vec{J}_{\mathrm{d}i} = c_i \, \vec{v}_{\mathrm{d}i} \tag{4.4}$$

In such a case, the sum of the diffusion fluxes of all molecules is zero [2]

$$\sum_i \vec{J}_{\mathrm{d}i} = 0 \tag{4.5}$$

The electrical flux of the i-th component is calculated from (4.3) by multiplying $\vec{J}_i$ by $z_i F$

$$\vec{J}_i = z_i F (c_i \vec{v}_i) \tag{4.5a}$$

where $z_i F$ is the electrical charge referred to 1 mole of component of charge number z_i. The third important vector flux is the heat flow in the cell $\vec{J}_q$. This will be defined later by equation (4.19).

If chemical reactions occur, their products form scalar fluxes. The magnitude of such a flux J_j^{ch} is given by the rate of formation of component j in time dt

$$J_j^{\mathrm{ch}} = \frac{\mathrm{d}\xi_j}{\mathrm{d}t} \tag{4.5b}$$

where ξ is the extent of the reaction. It is assumed following Onsager that the scalar flux J_j^{ch} of the rate of formation of the j-th component depends linearly on all the chemical affinities of the reactions occurring in the system [2]

$$J_j^{\mathrm{ch}} = \frac{\mathrm{d}\xi_j}{\mathrm{d}t} = \sum_{r=1} L_{jr} \bar{A}_r \tag{4.6}$$

where $\bar{A}_r$ is the chemical affinity of the r-th reaction. The respective equations for the vector fluxes of migration $\vec{J}_i$ and heat $\vec{J}_q$ are

$$\vec{J}_i = \sum_j L_{ij} \vec{X}_j + L_{iq} \vec{X}_q \tag{4.7}$$

$$\vec{J}_q = \sum_j L_{qj}\vec{X}_j + L_{qq}\vec{X}_q \tag{4.8}$$

Onsager proved that the matrices of the phenomenological coefficients L_{jr}, L_{ij} and L_{qi} are symmetrical

$$\begin{aligned} L_{jr} &= L_{rj} \\ L_{ij} &= L_{ji} \\ L_{qi} &= L_{iq} \end{aligned} \tag{4.9}$$

The forces inducing these flows are the following: the basic thermodynamic forces involved in electrode processes are the electrochemical affinities of the electrode reactions $\overline{A}_j$

$$\overline{A}_j = \mu_j^{\overline{\mathrm{M}}} - \mu_j^{\overline{\mathrm{s}}} \tag{4.10}$$

where $\mu_j^{\overline{\mathrm{s}}}$ is the electrochemical potential of the j-th ion in the solution given by

$$\mu_j^{\overline{\mathrm{s}}} = z_j F\Phi^{\mathrm{s}} + \mu_{j(p,T)}^0 + RT \ln c_j f_j \tag{4.11}$$

and z_j its valency; F is Faraday's constant, Φ^{s} the internal (Galvani) potential at any point in the solution, $\mu_{j(p,T)}^0$ the standard chemical potential of the ion in solution, and f_j the activity coefficient of the j-th ion

Correspondingly, $\mu_j^{\overline{\mathrm{M}}}$ is the electrochemical potential of the j-th ion passing from a metal, e.g. a lithium anode, into the solution

$$\mu_j^{\overline{\mathrm{M}}} = z_j F\Phi^{\mathrm{M}} + \mu_{j(p,T)}^0 \tag{4.12}$$

The thermodynamic force $\vec{X}_j$ inducing the j-th ion to migrate in the solution is the electrochemical potential gradient

$$\vec{X}_j = -\mathrm{grad}\, \overline{\mu}_j^{\mathrm{s}} \tag{4.13}$$

The electrochemical potential $\overline{\mu}_j = f(\Phi, p, T, c)$ and its gradient are generally given by the partial derivatives of the parameters Φ, p, T, c:

$$\begin{aligned} \mathrm{grad}\, \overline{\mu}_j = {} & \left(\frac{\partial \overline{\mu}_j}{\partial \Phi}\right)_{p,T,c} \cdot \mathrm{grad}\, \Phi + \left(\frac{\partial \overline{\mu}_j}{\partial p}\right)_{T,c,\Phi} \cdot \mathrm{grad}\, p + \left(\frac{\partial \overline{\mu}_j}{\partial T}\right)_{p,c,\Phi} \cdot \mathrm{grad}\, T + \\ & + \sum_i \left(\frac{\partial \overline{\mu}_j}{\partial c_i}\right)_{p,T,\Phi} \cdot \mathrm{grad}\, c_i \end{aligned} \tag{4.14}$$

where

$$\left(\frac{\partial \overline{\mu}_i}{\partial \Phi}\right)_{p,T,c} = z_j F$$

$$\left(\frac{\partial \overline{\mu}_j}{\partial p}\right)_{T,c,\Phi} = V_j \tag{4.15}$$

$$\left(\frac{\partial \overline{\mu}_j}{\partial T}\right)_{p,c,\Phi} = -S_j$$

$$\left(\frac{\partial \overline{\mu}_j}{\partial c_i}\right)_{p,T,\Phi} = \frac{RT}{c_i}\left[\delta_{ij} + \left(\frac{\partial \ln f_j}{\partial \ln c_i}\right)_{p,T,\Phi}\right]$$

Summarizing, the electrochemical potential gradient is

$$\begin{aligned} \operatorname{grad} \overline{\mu}_j = {} & z_j F \operatorname{grad} \Phi + \overline{V}_j \operatorname{grad} p - \overline{S}_j \operatorname{grad} T + \\ & + \sum_i \frac{RT}{c_i}\left[\delta_{ij} + \left(\frac{\partial \ln f_j}{\partial \ln c_i}\right)_{(p,T,\Phi)}\right] \operatorname{grad} c_i \end{aligned} \tag{4.16}$$

In the above equations $z_j F$ denotes the partial molar charge, $\overline{V}_j$ the partial molar volume, $\overline{S}_j$ the partial molar entropy, δ_{ij} the Kronecker delta, f_j the activity coefficient and c the electrolyte concentration. The contribution to the thermodynamic forces of the pressure gradient as $pV_j \operatorname{grad} \ln p$ is ca. 1/1000 of the other forces [5], and will therefore be disregarded in subsequent discussion. To omit this force is to assume that the cell electrolyte solution is virtually incompressible. In further analysis of the important relationship (4.7), the coefficients L_{ij} and L_{iq} are interpreted after a simple rearrangement of the Onsager equations. It is assumed that coefficients L_{iq} can be expressed by L_{ij} as follows:

$$L_{iq} = \sum_j L_{ij} Q^* \tag{4.17}$$

The Onsager equations take the form

$$\vec{J}_i = \sum_j L_{ij}(\vec{X}_j + Q^* \vec{X}_q) \tag{4.18}$$

The physical sense of Q^* can be understood if the flux $\vec{J}_i$ is multiplied by Q^* and the sum $\sum_i \vec{J}_i Q_i^*$ subtracted from $\vec{J}_q$

$$\vec{J}_q - \sum_i \vec{J}_i Q_i^* = \sum_j (L_{qj} - \sum_i L_{ij} Q_i^*) \vec{X}_j + (L_{qq} - \sum_i L_{iq} Q_i^*) \vec{X}_q$$

Taking the Onsager equations (4.17) into account, the first term disappears, leaving an equation defining the heat flux [2]

$$\vec{J}_q = \sum_i \vec{J}_i Q_i^* + (L_{qq} - \sum_i L_{iq} Q_i^*) \vec{X}_q \tag{4.19}$$

It is clear from this equation that Q_i^* is the heat transferred by 1 mole of the i-th component. If there were a temperature difference within the cell, the thermodynamic force inducing a flow of heat would be given by

$$\vec{X}_q = -\text{grad} \ln T \tag{4.20}$$

At constant temperature, $\vec{X}_q = 0$, so Q_i^* is a purely kinetic quantity due to cross effects linking the fluxes with the forces coupled with them. Prigogine's principle applies to heats Q_i^*; if there is mechanical equilibrium and the fluxes in the solution tend to zero, there is no overall heat transfer [6]. Under the influence of the electrochemical potential gradient $\vec{X}_j$, the migration flux $\vec{J}_i$ is

$$\vec{J}_i = -\sum_j L_{ij} \left[z_j F \,\text{grad}\, \Phi - \left(S_j + \frac{Q_j^*}{T} \right) \text{grad}\, T \right] + \\ + \sum_i \frac{RT}{c_i} \left[\delta_{ji} + \left(\frac{\partial \ln f_j}{\partial \ln c_i} \right)_{p,T,\Phi} \right] \text{grad}\, c_i \tag{4.21}$$

Equation (4.21) is the starting point for interpreting the phenomenological coefficients L_{ij}. Let us now return to equation (4.5a). The migration flux of ions in an electrical field at constant temperature, when $\text{grad}\, T = 0$, and in a solution of homogeneous concentration, when grad $c_i = 0$ for every c_j, can be defined as the product $\vec{J}_{\text{el}.j} = z_j F c_j \vec{v}_j$, because $\vec{v} = u\vec{E}$, hence $\vec{v}_j = u_{ij}\,\text{grad}\, \Phi$, and

$$\vec{J}_{\text{el}.j} = z_j u_{ij} c_j F \,\text{grad}\, \Phi \tag{4.22}$$

Therefore

$$L_{ij} = u_{ij} c_j \tag{4.22a}$$

where u_{ij} is the mobility of the j-th ions in the presence of i-th ions, and c_j is the concentration of the j-th ions. The molar ionic conductivity λ_j is directly linked to the mobility u_{ij} which, as is well-known from electrical conductivity theory [3, 8], is a function of temperature, the dynamic viscosity coefficient, the electrical permittivity of the solvent, and the ionic concentration

$$\lambda_j = u_{ij} F \tag{4.23}$$

Knowing the electrical conductivity, one can define the first term in equation (4.21): $-\sum_i u_{ij} c_j z_j F \text{ grad } \Phi$. The next term in this equation can be rewritten as $\sum_j u_{ij} c_j (S_j + Q_j^*/T) \,\text{grad}\, T$, where $(S_j + Q_j^*/T)$ is the entropy transferred by 1 mole of the j-th component. This quantity can be found experimentally from measurements of the electrochemical potential gradient with respect to the temperature gradient

$$S_j + \frac{Q_j^*}{T} = -\frac{\text{grad}\, \bar{\mu}_j}{\text{grad}\, T} \tag{4.24}$$

The quantity $\alpha_j = \left(S_j + \frac{Q_j^*}{T} \right) / F$ is the ionic equivalent of the Seebeck coefficient.

Under given conditions, the diffusion flux in a multicomponent solution is similar in form to the last term in equation (4.21)

$$(\vec{J}_i)_{p,T,\Phi} = -\sum_j \sum_k L_{ij} \frac{RT}{c_k} \left[\delta_{jk} + \left(\frac{\partial \ln f_j}{\partial \ln c_k} \right)_{p,T,\Phi} \right] \text{grad}\, c_k \tag{4.25}$$

This equation is comparable to Fick's generalized law of isothermal and isopotential diffusion in a multicomponent solution

$$(\vec{J}_i)_{p,T,\Phi} = -\sum_k D_{ik} \,\text{grad}\, c_k \tag{4.26}$$

and enables the phenomenological coefficients $L_{ij} = u_{ij} c_j$ to be expressed by means of the mutual diffusion coefficients D_{ik}

$$D_{ik} = \sum_j RT u_{ij} \frac{c_j}{c_k} \left[\delta_{jk} + \left(\frac{\partial \ln f_j}{\partial \ln c_k} \right)_{p,T,\Phi} \right] \tag{4.27}$$

If the values of D_{ik} are known from experiment, the ionic mobility u_{ij} can be calculated from equation (4.27). Equation (4.27) enables one to calculate the ion mobilities u_{ij}, provided the experimental mutual diffusion coefficients D_{ik} are known. The relationship between the mobilities and the mutual diffusion coefficients holds for dilute electrolyte solutions. The concentration dependence of the mobility is the result of relaxation and electrophoretic effects and of their mutual interaction. The relaxation effect is stimulated by the outer electric field which is superimposed on the Brownian motion of ions. In more concentrated solutions, the latter is modified first of all by the electrophoretic effect. Thus, if the experimental mutual diffusion coefficients are unknown, they can be evaluated from the molar conductivity of the electrolyte modified by removing the relaxation effect.

If the current passing through a solution is defined by

$$\vec{J}_{\mathrm{el}} = \sum_j \sum_i L_{ij} z_j z_i F \operatorname{grad} \Phi \tag{4.28}$$

and the numbers of positive and negative charges is the same

$$\sum_i z_i c_i = 0 \tag{4.29}$$

then the expression for the number of j-th ions transferred is [5]:

$$t_j = \frac{\vec{J}_{\mathrm{el}.j}}{\vec{J}_{\mathrm{el}}} = \frac{\sum_j l_{ij} z_j}{\sum_j \sum_i L_{ij} z_i z_j} \tag{4.30}$$

Values of t_j are found experimentally. A relationship describing the potential gradient at any point in the solution may be found from equation (4.21), (4.29) and (4.30):

$$\operatorname{grad} \Phi = -\sum_i \frac{t_i}{z_i F} (\operatorname{grad} \mu_i + Q_i^* \operatorname{grad} \ln T) - \frac{\vec{J}_{\mathrm{el}}}{\kappa} \tag{4.31}$$

where κ is the specific conductivity of the electrolyte solution

$$\kappa = F^2 \sum_i \sum_j z_i z_j L_{ij} \tag{4.32}$$

The potential gradient expressed by equation (4.31) and directed opposite to the e.m.f. is present in every working galvanic cell. The first term in equation (4.31) refers to ionic diffusion, the second to thermodiffusional flow: the potential drop is therefore defined by these two effects which go by the combined name of the "thermodiffusional gradient". The third term in this equation is the ohmic loss $\vec{J}_{\mathrm{el}}/\kappa$ induced by the electrode potential gradient.

In accordance with equation (4.6), the scalar flows of the reactants of electrochemical reactions are given by

$$J_i = \sum_r L_{ir} \overline{A}_r \tag{4.33}$$

where $\overline{A}_r$ is the electrochemical affinity of the reactions. In the electrode layers $\overline{A}_r$ depends on the spatial coordinates. It may therefore be assumed that an electrochemical potential gradient does exist. This gradient is the source of the vector flux of ionic species constituting the current in the solution and of the electron current in the metal. The electrochemical affinity gradient is

$$\operatorname{grad} \overline{A}_r = \operatorname{grad} \sum_i \nu_{ri} \overline{\mu}_i = \sum_i \nu_{ri} (z_i F \operatorname{grad} \Phi + \operatorname{grad} \mu_i) \tag{4.34}$$

Substituting the right-hand side of equation (4.31) for grad Φ yields

$$\begin{aligned} \operatorname{grad} \overline{A}_r = \sum_i \nu_{ri} (\operatorname{grad} \mu_i)_T - \sum_i \sum_k z_i \nu_{ri} \frac{t_k}{z_k} (\operatorname{grad} \mu_k + \\ + Q_k^* \operatorname{grad} \ln T) - \frac{\vec{J}_{\mathrm{el}}}{\kappa} \end{aligned} \tag{4.35}$$

The linear relationship between the cathode or anode currents J_c or J_a and the electrochemical affinity gradient of the electrode reaction is satisfied when the excess of affinity difference under conditions $\overline{A}_r$ and at equilibrium $\overline{A}_{0r}$ is less than the kinetic energy RT

$$\overline{A}_r - \overline{A}_{0r} < RT \tag{4.36}$$

Experiments have shown that this relationship is exponential. Marcelin and De Donder [10, 11] have proposed the following relationship for the anode and cathode densities:

$$\begin{aligned} J_\mathrm{a} &= \frac{zF}{s} \cdot \frac{\mathrm{d}\xi_\mathrm{a}}{\mathrm{d}t} = \frac{zF}{s\tau} \exp\left(\frac{\overline{A}_\mathrm{a} - \overline{A}_{0\mathrm{a}}}{RT}\right) \\ J_\mathrm{c} &= \frac{zF}{s} \cdot \frac{\mathrm{d}\xi_\mathrm{c}}{\mathrm{d}t} = \frac{zF}{s\tau} \exp\left(\frac{\overline{A}_\mathrm{c} - \overline{A}_{0\mathrm{c}}}{RT}\right) \end{aligned} \tag{4.37}$$

where s is the surface area of the electrode and τ is the time constant of the electrode reaction [7, 12].

As the kinetics of electrode reactions have been dealt with in Chapter 3, no further discussion of these equations will be undertaken here.

4.2 EQUATIONS BALANCING MASS, CHARGE, ENERGY AND ENTROPY

The task of the thermodynamics of irreversible processes is to describe systems in non-equilibrium states and to formulate differential or integro-differential equations linking thermodynamic quantities. It is clear that all the quantities involved are functions of time and spatial coordinates. Descriptions of the spatial and temporal changes in these quantities make use of averaged quantities referring to the medium as a continuum. If the number of independent differential equations is equal to the number of functions sought and the initial and boundary conditions are correctly formulated, solutions are usually found by computational methods. For a cell, the desired functions concern the distribution of concentration $c_i(x, y, z, t)$, temperature $T(x, y, z, t)$, velocity $\overline{v}(x, y, z, t)$, pressure $p(x, y, z, t)$ and potential $\Phi(x, y, z, t)$. These equations take the form of local balances of mass, energy, momentum, angular momentum and charge. In cells in which the electrolyte is at rest relative to the electrodes, $\overline{v}(x, y, z, t) = 0$, and the equation for the conservation of momentum can be disregarded. High-energy cells are so designed that the electrode and the electrolyte are sealed off in steel casings. The pressure may then be assumed to be equalized and $p = \text{const}\ (x, y, z, t)$. Therefore, equations balancing mass, energy and charge are sufficient to balance the concentrations of components, temperature distribution and potential.

4.2.1 *Mass balance*

The continuity equation balancing the mass $M = \sum c_i M_i = \text{const}$, where M_i is the molar mass of the i-th component, for a point $n(x, y, z, t)$ and its surroundings within the electrolyte takes the form

$$\frac{\partial c_i}{\partial t} = -\mathrm{div}\vec{J}_i + \sum_r \nu_{ir} J_r \tag{4.38}$$

This equation defines the change in concentration of the i-th component of the electrolyte solution in a time $\mathrm{d}t$ around a point $x + \mathrm{d}x$, $y + \mathrm{d}y$, $z + \mathrm{d}z$ due to its migration-diffusion flow and to the chemical reactions in which the i-th component participates; $\vec{J}_i$ is the migration-diffusion flux, and J_r is the chemical flux describing the rate of formation or decomposition of a quantity ν_{ir} of the i-th component in the r-th reaction. Using the equations of chemical kinetics, J_r can be expressed by

$$J_r = \frac{\mathrm{d}\xi_r}{\mathrm{d}t} = \nu_{ir} k_{0r}(T) \exp(-\mathcal{E}_r^{\neq}/RT)\, c_1^{\nu_{1r}} c_2^{\nu_{2r}} \ldots \tag{4.39}$$

where ξ_r is the extent of the r-th reaction, ν_{ir} is the stoichiometric coefficient of substance i forming or decomposing in the r-th reaction, $k_{0r}(T)$ is the rate constant of the r-th reaction, and $\mathcal{E}_r^{\neq}$ is the activation energy of this reaction. After expanding and taking equations (4.21), (4.22a) and (4.27) into account, this equation can be written as

$$\frac{\mathrm{d}c_i}{\mathrm{d}t} = \mathrm{div}\left\{\sum_f [z_j u_{ij} c_j F \,\mathrm{grad}\, \Phi - (S_j T + Q_j^*)\,\mathrm{grad}\ln T + \sum_k D_{jk}\,\mathrm{grad}\, c_k]\right\} +$$

$$+ \sum_r \nu_{ir} k_{0r}(T) \exp(-\mathcal{E}_r^{\neq}/RT)\, c_1^{\nu_{1r}} c_2^{\nu_{2r}} \ldots \tag{4.40}$$

For stationary processes it is assumed that $\partial c_i/\partial t = 0$.

4.2.2 *Charge balance*

Multiplying both sides of equation (4.38) by $z_i F$ yields the continuity equation for the charge of i-th ions

$$\frac{\partial(z_i c_i F)}{\partial t} = -\mathrm{div}(z_i F \vec{J}_i) + F \sum_r z_i \nu_{ir} J_r \tag{4.41}$$

Summing over all components and denoting

$$q = \sum_i z_i c_i F, \qquad \vec{I} = \sum_i z_i F \vec{J}_i \tag{4.42}$$

and remembering that $F \sum_i \sum_r z_i \nu_{ir} J_r = 0$, the law of charge conservation is obtained

$$\frac{\partial q}{\partial t} + \mathrm{div}\,\vec{I} = 0 \tag{4.43}$$

where q is the charge density, and $\vec{I}$ is the current passing through the electrolyte solution in a galvanic cell.

4.2.3 *Energy balance*

The energy-balance continuity equation can be expressed by changes in either internal energy or enthalpy. A large number of data on molar enthalpies, thermal capacities at constant pressure, and free energies (Gibbs energies) for aqueous and non-equeous solutions and for solid phases have been tabulated in the electrochemical literature. It is therefore more convenient to derive the energy equation by balancing enthalpy.

Consider the enthalpy change in a small volume element of electrolyte solution. The change may be due to a variety of causes: heat inflow from the exterior; components migrating into the volume element, bringing with them their partial enthalpies; the heat of chemical reactions occurring there, including the solvation and desolvation of ions during the association processes, and, finally, internal friction. Let H_i be the partial molar enthalpy of the i-th component and c_i its concentration. The enthalpy can now be balanced by the continuity equation

$$\frac{\partial}{\partial t}\sum_i c_i H_i = -\mathrm{div}\left(\sum_i c_i H_i \vec{v} + \vec{J}_q + \sum_i F_i \vec{J}_i + 2\eta(\mathrm{grad}\,\vec{v})^2 + \right. \\ \left. + \eta_v(\mathrm{div}\,\vec{v})^2 + \frac{\partial p}{\partial t} + \vec{v}\,\mathrm{grad}\,p\right) \tag{4.44}$$

where $\vec{J}_q$ is the heat flux, $\vec{J}_i$ is the migration-diffusion flux, $-\mathrm{div}\sum_i c_i H_i \vec{v}$ the convection enthalpy, $(\partial p/\partial t + \vec{v}\,\mathrm{grad}\,p)$ a pressure-dependent term, $2\eta(\mathrm{grad}\,\vec{v})^2 + \eta_v(\mathrm{div}\,\vec{v})^2$ the dissipated energy brought about by the internal friction processes generated by the shear (dynamic) and volume viscosities η and η_v respectively, and $\sum_i F_i \vec{J}_i$ is the Joule heat.

For an electrolyte solution in a closed cell, this equation can be simplified if it is assumed that the cell solution is incompressible, and that the pressure within the cell is constant, and does not depend on the spatial coordinates. Then $\partial p/\partial t = 0$, $\mathrm{div}\,\vec{v} = 0$ and $\mathrm{grad}\,p = 0$. Note also that two terms appear when the left-hand side of equation (4.44) is differentiated

$$\frac{\partial}{\partial t}\sum_i c_i H_i = \sum_i H_i \frac{\partial c_i}{\partial t} + \sum_i c_i \frac{\partial H_i}{\partial t} \tag{4.45}$$

After carrying over the first term of the differentiated left-hand side of equation (4.44) to the right, taking into account the mass-balance continuity equation (4.38), the fact that $(\partial H_i/\partial T)_p = C_{p_i}(T)$ and $\mathrm{grad}\,H_i = C_{p_i}(T)\,\mathrm{grad}\,T$, and that the sum

$$\sum_r \left(\sum_i H_i \nu_{ir}\right) = -\Delta H_r \tag{4.46}$$

defines the enthalpy of the r-th process or chemical reaction, and rearranging, one obtains a somewhat simplified expression for the enthalpy of the cell electrolyte

$$\sum_i c_i C_{p_i}(T)\left(\frac{\partial T}{\partial t}+\frac{\vec{J}_i}{c_i}\operatorname{grad} T\right)$$
$$=-\operatorname{div}\vec{J}_q+\sum_r \Delta H_r J_r-\sum_i \vec{J}_i \operatorname{grad}\overline{\mu}_i \tag{4.47}$$

Let us analyse the importance of the separate terms: $\vec{J}_q$ is the heat flux exchanged with the surroundings. The main thermodynamic force inducing this flux is the local temperature gradient

$$\vec{J}_q=-\lambda\overline{C}_p \operatorname{grad} T \tag{4.48}$$

where λ is the coefficient of thermal conductivity and $\overline{C}_p$ the molar thermal capacity of the solution at constant pressure. The third term is the sum of the migration-diffusion fluxes of neutral $\sum_i \vec{J}_i \operatorname{grad}\mu_i$ and charged components. For the latter,

$$-\sum_i \vec{J}_i \operatorname{grad}(z_i F\Phi+\mu_i^0+RT\ln c_i)$$
$$=-\vec{I}\sum_i \operatorname{grad}\left[(\Phi-\Phi_i^0)+\frac{RT}{z_i F}\ln c_i\right] \tag{4.49}$$

where $\vec{I}$ is the electric current and $\vec{X}_{\text{el}}=-\operatorname{grad}\left[(\Phi-\Phi_i^0)+\frac{RT}{z_i F}\ln c_i\right]$; Φ is the local electrical potential, and Φ_i^0 is the local standard potential of cations, anions or ion triplets. By denoting the average molar thermal capacity as $\overline{C}_p=\sum_i C_{p_i}$ and the average molar thermal velocity as $\overline{\upsilon}=\sum_i \frac{\vec{J}_i}{c_i}$, one can obtain another form of the energy equation for the electrolyte solution in a cell

$$c_i\overline{C}_p\left(\frac{\partial T}{\partial t}+\overline{\upsilon}\operatorname{grad} T\right)=\lambda \operatorname{div}\overline{C}_p \operatorname{grad} T+$$
$$+\vec{I}\sum_i \vec{X}_{\text{el}}+\sum_r \Delta H_r J_r \tag{4.50}$$

4.2.4 *Entropy balance*

The irreversible processes in electrolyte solutions giving rise to transport of mass, momentum and energy cause an increase in entropy. The entropy increase in a closed, convection-free cell is [1, 12–16]

$$\frac{\partial S}{\partial t} = -\mathrm{div}\vec{J}_S + \sigma \tag{4.51}$$

where $\vec{J}_S$ is the entropy flux and σ the entropy source.

$$\vec{J}_S = \frac{1}{T}\left(\vec{J}_q - \sum_i \bar{\mu}_i \vec{J}_i\right) \tag{4.52}$$

where $\vec{J}_q$ is the heat flux given by $\vec{J}_q = -\lambda C_p \,\mathrm{grad}\, T$ (see equation (4.48)), $\vec{J}_i$ and $\bar{\mu}_i$ are the flux and electrochemical potential of the i-th component respectively. According to Onsager [1], the entropy source

$$\sigma = \sum_r J_r \bar{A}_r + \vec{J}_q X_q + \sum_i \vec{J}_i X_i + \eta \sum_i (\mathrm{grad}\, \vec{v}_i)^2 \tag{4.53}$$

the symbols being the same as before: $\bar{A}_r$ – the chemical affinity of the r-th reaction is defined by equation (4.6), J_r by equation (4.39), $\vec{J}_q$ by equation (4.48) and $\vec{X}_q$ by equation (4.20); $\vec{X} = \vec{F}/T - \mathrm{grad}\,(\mu_i/T)$, where $\vec{F}$ (if non-zero) is the resultant external force acting on the ions and molecules. The term $\eta \sum_i (\mathrm{grad}\, \vec{v})^2$ is equal to the entropy production due to friction. If one assumes that the solution is incompressible and that it is at rest, the last terms in equation (4.53) is equal to zero.

From the Clausius–Prigogine equation

$$\frac{\mathrm{d}S}{\mathrm{d}t} = \frac{\mathrm{d}_e S}{\mathrm{d}t} + \frac{\mathrm{d}_i S}{\mathrm{d}t} \tag{4.54}$$

the first term in equation (4.51) is

$$\frac{\partial_e S}{\partial t} = -\mathrm{div}\, \vec{J}_S \tag{4.55}$$

and the second term in equation (4.51) is

$$\frac{\partial_i S}{\partial t} = \sigma \tag{4.56}$$

where $\mathrm{d}_e S/\mathrm{d}t$ is the entropy exchanged with the surroundings and $\mathrm{d}_i S/\mathrm{d}t$ is the entropy produced by irreversible processes. The dissipation function P is important: it is the product of temperature and the entropy source, integrated over the volume

$$P = T\sigma = \int_V T(r,t)\,\sigma(r,t)\,\mathrm{d}V \tag{4.57}$$

This product is the quantity of heat produced in the cell per second. This heat arises from the dissipation of electrochemical energy, flux momentum and flow energy (Joule heat). This energy is lost for electrical work. In the general case, dissipation functions can be defined by

$$P = T\sigma = \sum_i J_i^* X_i^* \tag{4.58}$$

The units of the product $J_i^* X_i^*$ are joule s^{-1}, i.e. watts. Every quantity $J_i^* X_i^*$ is transformed from the unstarred quantities J_i and X_i respectively ($J_i X_i = \mathrm{watt\,K^{-1}}$). The smaller the value of P, the more efficient the cell. One could say that the so-called energy crisis is caused by entropy sources.

As far as cells are concerned, the researcher's task should be to choose materials, components and technologies which minimize entropy sources. Equation (4.53) for the entropy source will be used in Section 4.4 to determine cell efficiency.

4.3 OPEN-CELL POTENTIAL DISTRIBUTION; CLOSED-CELL VOLTAGE

The most important force in a galvanic cell is the electrical potential gradient. So a physically justified equation relating the internal (Galvanic) potential Φ to the x-coordinate directed perpendicularly to the anode or cathode surface must be found. A relationship like this can be obtained using statistical mechanics, and for an open cell it is virtually in its simplest form (Gouy–Chapman theory) [19]

$$\Phi(x) = \frac{2RT}{F} \ln \frac{1 + \mathrm{e}^{-\kappa x} \tanh\left(\dfrac{F\Phi^0}{4RT}\right)}{1 - \mathrm{e}^{-\kappa x} \tanh\left(\dfrac{F\Phi^0}{4RT}\right)} \tag{4.59}$$

in which

$$\kappa = \left[\frac{2000 F^2 c(\gamma_1 + \gamma_3)}{RT\varepsilon_0\varepsilon}\right]^{1/2}$$

is the reciprocal of the Debye length l_D (the depth of the double layer), c is the electrolyte concentration in mol dm^{-3}, γ_1 and γ_3 are fractions expressing the relative contents of solvated free ions and charged ion triplets, and tanh is the hyperbolic tangent. Equation (4.59) is applied to the range $0 \leq x \leq L/2$ where L is the distance between the surface of the two cell electrodes, and x is a value chosen from the set belonging to L. For the above range of values of x (anode space) the anode potential is negative and Φ^0 is denoted as Φ_a^0 ($\Phi_a^0 < 0$). For the cathode space $L/2 \leq x \leq L$ one can apply the same equation to define the potential in the diffuse layer of a semiconductor electrode, where $\Phi^0 = \Phi_{cs}^0$ is the surface potential of the semiconductor; it is positive. Obviously, the concept of a semiconductor cathode surface, and especially a porous one, is here conventional; it depends on the method by which the semiconductor has been prepared. The reader will remember that within a semiconductor there is a double layer with quite a complex structure [17, 18] as set out in Chapter 3. The dependence of Φ_c^0 on x within the semiconductor should be treated as a model. The values of Φ_a^0, Φ_c^0 and Φ_{cs}^0 refer to the case when a cell is in quasi-equilibrium. The changes in potential Φ with the electrode distances x is shown in Fig. 4.1. The plot of potential change Φ vs. interelectrode distance, the x-coordinate perpendicular to both electrodes imagined as two parallel planes, is shown in Fig. 4.1 for a Li/CuS system [44, 45] (see also Chapter 8).

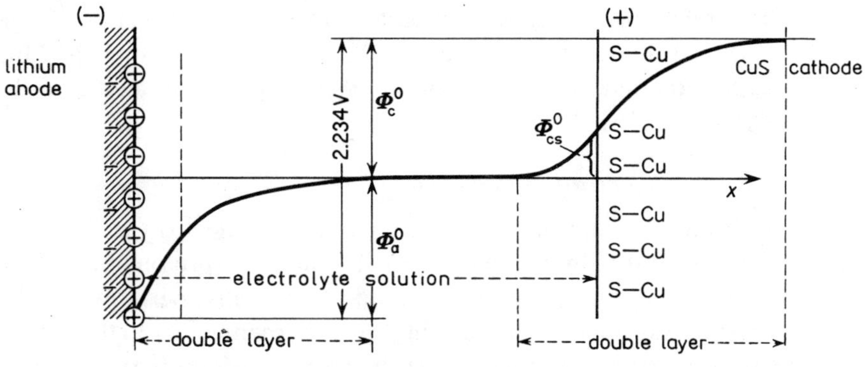

Fig. 4.1 Potential changes Φ vs. interelectrode distance x in a stationary Li/CuS cell [44, 45]: Φ_a^0 – anode potential, Φ_{cs}^0 – surface cathode potential.

According to the Gouy–Chapman theory, the open-cell thermodynamic force dΦ/dx is given by

$$\frac{d\Phi}{dx} = \left[\frac{8000RTc(\gamma_1 + \gamma_3)}{\varepsilon_0 \varepsilon}\right]^{1/2} \sinh\left(\frac{F\Phi}{2RT}\right) \tag{4.60}$$

where sinh denotes the hyperbolic sine. After taking account of the property of sinh: sinh ln $f(x) = \frac{1}{2}f(x) - 1/f(x)$, equation (4.60) is rearranged to give

$$\frac{\mathrm{d}\Phi}{\mathrm{d}x} = \left[\frac{8000RTc(\gamma_1 + \gamma_3)}{\varepsilon_0 \varepsilon}\right]^{1/2} \frac{2\mathrm{e}^{-\kappa x}\tanh[F\Phi^0/(4RT)]}{1 - \mathrm{e}^{-2\kappa x}\tanh^2[F\Phi^0/(4RT)]} \tag{4.60a}$$

As soon as the cell is closed by an external resistance, electrons begin to flow from anode to cathode. Equilibrium distributions are disturbed throughout the system. The reader will have found a description of such a reaction and its associated activation overvoltages in Chapter 3. The flow of electrons in the external circuit is balanced by the flow of an equal number of cations (e.g. Li^+) in the electrolyte solution. The ensuing disturbance of charge distribution can be described by concentration overvoltages. If equation (4.60a) is to define the potential gradient in the diffuse layer of a semiconductor electrode, Φ_c^0 or Φ_{cs}^0 must be substituted for Φ^0. The relationship between Φ and x in the bulk of the semiconductor is very complicated, even in pure, crystalline semiconductors like silicon or germanium. It is more difficult to write the Poisson–Boltzmann equation for, say, a porous CuS semiconductor with microcrystals of Li_2S precipitated on its surface and within the pores (a Li/CuS cell; electrode reaction: $2Li^+ + CuS \rightarrow Li_2S + Cu^0$) [44, 45], because it is not clear how the spatial charge density ρ in the surface layer is to be represented [19]. This relationship can be represented arbitrarily by a section of a saturation curve so that its initial value is Φ_{cs}^0 and the asymptote Φ_c^0. In such an approximation, the magnitude of Φ_{cs}^0 should be treated as a theoretical parameter [18].

4.3.1 *Concentration overvoltages*

Consider the current passing through the cell electrolyte. Suppose this is a solution of lithium perchlorate $LiClO_4$ in a non-protonating solvent with a fairly low electrical permittivity (and chemically inert towards the lithium anode). Assume that the current flow is perpendicular to the surfaces of the two electrodes, and further that the current is stationary $\partial q/\partial t = 0$. Taking the migration-diffusion flux and equations (4.22), (4.23) and (4.26) into consideration, the current $\vec{I}$ is most simply given by [39, 40]

$$\vec{I} = \underbrace{-\sum_i z_i^2 \Lambda_i c_i \frac{\mathrm{d}\Phi}{\mathrm{d}x}}_{\text{migration current}} - \underbrace{\frac{RT}{F}\sum_i \frac{z_i}{|z_i|}\Lambda_i \frac{\mathrm{d}c_i}{\mathrm{d}x}}_{\text{diffusion current}} \tag{4.61}$$

where Λ_i is the molar conductivity. In the cell solution it is the Li^+ ions which carry the current; they are released by the lithium anode and react with the cathode ions, usually producing insoluble compounds. Perchlorate anions ClO_4^-, pair up with lithium cations, thus ensuring electrical neutrality away from the electrode layers. Equation (4.61) sometimes called the Faraday or Nernst equation, can be written separately for cations (Li^+) and anions (ClO_4^-)

$$I_+ = -\Lambda_+ c\frac{\mathrm{d}\Phi}{\mathrm{d}x} - \frac{RT}{F}\Lambda_+\frac{\mathrm{d}c}{\mathrm{d}x} \tag{4.62}$$

$$I_- = \Lambda_- c\frac{\mathrm{d}\Phi}{\mathrm{d}x} - \frac{RT}{F}\Lambda_-\frac{\mathrm{d}c}{\mathrm{d}x} \tag{4.63}$$

Taking into account that $I_- = 0$ equation (4.63) yields

$$\frac{\mathrm{d}\Phi}{\mathrm{d}x} = \frac{RT}{F}\left(\frac{\mathrm{d}\ln c}{\mathrm{d}x}\right) \tag{4.64}$$

Substituting (4.64) in (4.62) one obtains

$$I = -2\frac{RT}{F}\Lambda_+\frac{\mathrm{d}c}{\mathrm{d}x}; \qquad \frac{\mathrm{d}c}{\mathrm{d}x} = -\frac{IF}{2RT\Lambda_+} \tag{4.65}$$

and using (4.65) one can obtain an expression for $\mathrm{d}\Phi/\mathrm{d}x$

$$\frac{\mathrm{d}\Phi}{\mathrm{d}x} = -\frac{I}{\Lambda_+}\frac{1}{c(x)} - \frac{RT}{F}\left(\frac{\mathrm{d}\ln c}{\mathrm{d}x}\right) \tag{4.66}$$

Integrating (4.66) gives the gradient of the potential $\mathrm{d}\Phi$ as a function of the distance x from the interface. Assuming that Λ_+ is experimentally determined as a function of the electrolyte solution concentration (it is a number), one can write

$$\int_{(0)}^{(x)} \mathrm{d}\Phi = -\frac{I}{\Lambda_+}\int_{c_s}^{c_x}\frac{\mathrm{d}x}{c(x)} - \frac{RT}{F}\int_{c_s}^{c_b}\mathrm{d}\ln c(x) \tag{4.67}$$

The electrolyte concentration depends on the x coordinate: two diffusion layers are distinguishable near the electrodes separated by the bulk of the solution. For diffusion layers, the concentration c is assumed to change linearly within the interval from $x = 0$ to $x = d_a$ for the anode and from $x = L - d_c$ to $x = L$ for the cathode (d_c, d_a – diffusion layer depths at the cathode and anode respectively). For the bulk of the solution the integral

$$I\int_0^L \frac{\mathrm{d}x}{\Lambda_+ c(x)} = r_{\mathrm{int}} I \tag{4.68}$$

is assumed equal to the ohmic loss of voltages as $I \to 0$, where r_{int} is the internal resistance of the electrolyte, c_s and c_b are the electrolyte concentration at the anode surface and in the bulk of the solution respectively, and c_x is the concentration at a distance x from the interface. For the anode diffusion layer it is assumed that

$$c(x) \stackrel{\text{def}}{=} c_s + (c_b - c_s)\frac{x}{d_a}; \qquad \frac{c_b - c_s}{d_a} = \frac{dc}{dx} = -\frac{IF}{2RT\Lambda_+} \tag{4.69}$$

and

$$c(x) = c_b - \frac{IF}{2RT\Lambda_+}(x - d_a) \tag{4.70}$$

whereas

$$c_s \stackrel{\text{def}}{=} c_b + \frac{IFd_a}{2RT\Lambda_+} \tag{4.71}$$

Let the concentration overvoltage at the anode be

$$\eta_{c,a} \stackrel{\text{def}}{=} \Phi(d_a) - \Phi(0) \tag{4.72}$$

and $I_{L,a}$, the limiting value of the anode current (limiting current), be

$$I_{L,a} \stackrel{\text{def}}{=} \frac{2RT\Lambda_+ c_b}{Fd_a} \tag{4.73}$$

Then

$$\eta_{c,a} = \frac{2RT}{F}\ln\left(1 - \frac{I}{I_{L,a}}\right) + \frac{RT}{F}\ln\frac{c_b}{c_s} \tag{4.74}$$

Rearrangement yields a relatively simple equation for the anode concentration overvoltage

$$\eta_{c,a} = \frac{RT}{F}\ln\left(1 - \frac{I}{I_{L,a}}\right) \tag{4.75}$$

and a similar equation for the cathode concentration overvoltage

$$\eta_{c,\,cat} = \frac{RT}{F}\ln\left(1 - \frac{I}{I_{L,c}}\right) \tag{4.76}$$

Using the derived relationships (4.75) and (4.76), and the expressions for the anodic and cathodic activation overvoltages (see Chapter 3), the cell terminal voltage U can be estimated for a 1:1 electrolyte [39, 40]

$$U = E^0 - \left[\frac{RT}{\alpha_a F}\ln\frac{I}{i_{0,a}} + \frac{RT}{F}\ln\left(1 - \frac{I}{I_{L,a}}\right)\right] -$$
$$-\left[\frac{RT}{\alpha_c F}\ln\frac{I}{i_{0,c}} + \frac{RT}{F}\ln\left(1 - \frac{I}{I_{L,c}}\right)\right] - r_{int}I \tag{4.77}$$

where α_a and α_c are the anodic and cathodic charge transfer coefficients, $i_{0,a}$ and $i_{0,c}$ the anodic and cathodic exchange currents, $I_{L,a}$ and $I_{L,c}$ the anodic and cathodic limiting currents, and E^0 the cell e.m.f. The dependence of the cell voltage U on the current I as expressed by equation (4.77) is in good agreement with experiment, at least for a one-step electrode reaction, and when passivation layers do not play a predominant part during cell operation.

4.4 ASSESSMENT OF CELL EFFICIENCY

The limit of cell efficiency can theoretically be defined from the intensity of the entropy source, which is zero for an open cell and attains a maximum value for a short-circuited cell. The short-circuit entropy source is obtainable by extrapolating the work performed by the cell on the external circuit and determining the current passing through a device of known resistance (external load). The necessary equations can be obtained from the equation for the enthalpy of the system (4.44) if it is somewhat simplified to allow for the working conditions of the cell. At constant pressure, the expression $\partial p/\partial t + \vec{v}\,\mathrm{grad}\,p$ can be disregarded; if there are no convection currents, the convection enthalpy $\mathrm{div}\left(\sum_i c_i H_i \vec{v}\right) = 0$ can also be ignored, and the partial derivative $\partial H/\partial t$ may be put equal to $\mathrm{d}H/\mathrm{d}t$. Denoting the energy lost through internal friction $\mathrm{d}Q_\eta/\mathrm{d}t$ by

$$\frac{\mathrm{d}Q_\eta}{\mathrm{d}t} = 2\eta(\mathrm{grad}\,\vec{v})^2 + \eta_v(\mathrm{div}\,\vec{v})^2 \tag{4.78}$$

one obtains the equation

$$\frac{\mathrm{d}H}{\mathrm{d}t} = -\mathrm{div}\vec{J}_q + \sum_i \vec{J}_i\vec{F}_i + \frac{\mathrm{d}Q_\eta}{\mathrm{d}t} \tag{4.79}$$

This can be written in the form of the first law of thermodynamics, remembering that $-\mathrm{div}\,\vec{J}_q$ describes the energy acquired by a volume element as heat $\mathrm{d_e}Q$ in unit time

$$\frac{\mathrm{d_e}Q}{\mathrm{d}t} = -\mathrm{div}\,\vec{J}_q \tag{4.80}$$

Next, components transferring charge (free solvated ions and ion triplets) as a result of potential differences at points close to both electrodes $\Delta\Phi$ and neutral components moving down chemical potential gradients, are separated from the sum $\sum_i \vec{J}_i \vec{F}_i$

$$\sum_i \vec{J}_i \vec{F}_i = \sum_{i=1}^{k} z_i F \frac{\mathrm{d}c_i}{\mathrm{d}t} \Delta\Phi - \sum_{i=k+1} \vec{J}_i \,\mathrm{grad}\,\mu_\iota \tag{4.81}$$

Moreover, the fluxes of neutral entities, i.e. solvent molecules and ion pairs, must balance in the electrolyte system placed between the electrodes, whereas the fluxes of charge-transferring components are equal to the products generated in the electrode reactions. Then, this sum can be written down with the aid of Faraday's constant and the extent of the electrode reaction (degree of conversion)

$$\sum_{i=1}^{k=q} z_i D \frac{\mathrm{d}c_i}{\mathrm{d}t} = nF \frac{\mathrm{d}\xi}{\mathrm{d}t} \tag{4.82}$$

where n is now the charge number of electrons participating in the unit (or elementary) electrode reaction.

The equation balancing the enthalpy of the system (the electrolyte solution and electrode phenomena) can be written as the sum of the energy acquired as heat in unit time, the work performed by the cell and the energy dissipated as heat d_iQ

$$\frac{\mathrm{d}H}{\mathrm{d}t} = \frac{\mathrm{d_e}Q}{\mathrm{d}t} + \frac{\mathrm{d_i}Q}{\mathrm{d}t} - zF\Delta\Phi \frac{\mathrm{d}\xi}{\mathrm{d}t} \tag{4.83}$$

or

$$\frac{\mathrm{d}H}{\mathrm{d}t} = \frac{\mathrm{d_e}Q}{\mathrm{d}t} + T\sigma - \Delta\Phi I \tag{4.84}$$

where the current $I = zF\,\mathrm{d}\xi/\mathrm{d}t$, $T\sigma = \mathrm{d_i}Q/\mathrm{d}t$, and σ is the entropy source.

The cell entropy source will now be assessed on the basis of van Rysselberghe's discussion [12]. For an open cell whose entropy is zero

$$\frac{\mathrm{d}H}{\mathrm{d}t} = \frac{\mathrm{d_e}Q}{\mathrm{d}t} - \Delta\Phi^0 I \tag{4.85}$$

For a cell working under an external load, the above value will be lower by the amount of energy dissipated in the cell, i.e. in the electrolyte solution and at the electrodes (equation (4.84))

$$\frac{\mathrm{d}H}{\mathrm{d}t} = \frac{\mathrm{d_e}Q}{\mathrm{d}t} + T\sigma - \Delta\Phi I$$

When a cell is short-circuited, its voltage falls to zero, $\Delta\Phi = 0$, so the value of the entropy source is at a maximum

$$\frac{\mathrm{d}H}{\mathrm{d}t} = \frac{\mathrm{d_e}Q}{\mathrm{d}t} + T\sigma_{\mathrm{sh}} \tag{4.86}$$

Comparing the enthalpy change in an open (4.85) and shorted cell (4.86), one obtains

$$T\sigma_{\mathrm{sh}} = |\Delta\Phi^0|\, I \tag{4.87}$$

Since the product of the internal resistance r_{int} and the current I for a quasi-stationary cell is $r_{\mathrm{int}}I = \Delta\Phi^0$, $T\sigma_{\mathrm{sh}} = r_{\mathrm{int}}I^2$, i.e. the energy dissipated in a shorted cell is the product of the internal resistance and the square of the current. At the same time, the sequence of equalities

$$T\sigma_{\mathrm{sh}} = zF\Delta\Phi^0 \frac{\mathrm{d}\xi}{\mathrm{d}t} = A^0 \frac{\mathrm{d}\xi}{\mathrm{d}t} = A^0 J_r \tag{4.88}$$

holds.

Since $T\sigma_{\mathrm{sh}} = \mathrm{d_i}Q/\mathrm{d}t = T\mathrm{d_i}S/\mathrm{d}t$, the sequence of equalities

$$\frac{1}{zF} T \frac{\mathrm{d_i}S}{\mathrm{d}\xi} = (\Phi_{\mathrm{a}}^0 - \Phi_{\mathrm{b}}^0) = \frac{A^0}{zF} \tag{4.89}$$

also holds, where $T(\mathrm{d_i}Q/\mathrm{d}\xi)$ is the uncompensated Clausius heat for an electrode reaction $\Delta\xi = 1$.

On the basis of equation (4.89) van Rysselberghe [12] calls the term A^0/zF the chemical tension of the cell because in an open cell it is equivalent to the standard e.m.f. of that cell. All the working regimes of a cell lie between these two extreme cases (open cell–shorted cell). When a cell is supplying a current of low intensity, less entropy is produced than by a shorted cell, so the potential difference (or cell terminal voltage) is greater. If this difference is denoted by $(\Phi_{\mathrm{a}} - \Phi_{\mathrm{b}})$, one can write

$$\frac{T}{zF} \frac{\mathrm{d_i}S}{\mathrm{d}\xi} = \frac{A^0}{zF} - (\Phi_{\mathrm{a}} - \Phi_{\mathrm{b}}) \tag{4.90}$$

and if $(\Phi_{\mathrm{a}} - \Phi_{\mathrm{b}}) \equiv E$, and $(\Phi_{\mathrm{a}}^0 - \Phi_{\mathrm{b}}^0) = A^0/zF \equiv E^0$,

$$\frac{T}{zF}\frac{\mathrm{d_i}S}{\mathrm{d}\xi} = E^0 - E \tag{4.91}$$

This corresponds to an entropy of

$$T\sigma = (E^0 - E)\,zF\,\frac{\mathrm{d}\xi}{\mathrm{d}t} \tag{4.92}$$

The difference $(E^0 - E)$ can be represented as the sum of the anodic and cathodic polarization potentials $(E^0 - E) = |\eta_a| + |\eta_c|$. By substituting this value in equation (4.92), the entropy source of a working cell can be described as the sum of the anodic and cathodic entropy souce ($\sigma = \sigma_a + \sigma_c$):

$$\sigma_a = \frac{|\eta_a|}{T} zF \frac{\mathrm{d}\xi}{\mathrm{d}t} \qquad \sigma_c = \frac{|\eta_c|}{T} zF \frac{\mathrm{d}\xi}{\mathrm{d}t} \tag{4.93}$$

This equation clearly links the polarization potentials with the intensity of irreversible processes at the anode and cathode. Let us return to the equation

$$\sigma = \frac{r_{int} I^2}{T} \tag{4.94}$$

which enables the entropy source of a cell to be calculated if the current, internal resistance and the currents defining the cell's work from a current of $I = 0$ (open cell) to that of a short-circuited cell where $\sigma_{sh} = r_{int} I_{sh}^2 / T$ are known. By defining the entropy source with the aid of equation (4.94), the efficiency of a cell working under a small external load can be assessed.

First, the theoretical efficiency coefficient of a cell working reversibly and supplying a limiting very low current must be defined. The efficiency coefficient η^0 is equal to the ratio of electrical work done in the external circuit to the enthalpy change during the electrode reaction when $\Delta\xi = 1$. In the simplest form

$$\eta^0 = \frac{\Delta G}{\Delta H} \tag{4.94a}$$

but after expansion

$$\eta^0 = \frac{\left(\frac{\partial A}{\partial \xi}\right)_p}{\left(\frac{\partial H}{\partial \xi}\right)_p} = \frac{zFE^0}{zF\left[E^0 - T\left(\frac{\partial E^0}{\partial T}\right)_p\right]} \tag{4.95}$$

When simplified, this expression becomes

$$\eta^0 = \frac{E^0}{E^0 - T\left(\dfrac{\partial E^0}{\partial T}\right)_p} \tag{4.96}$$

Assume further that the cell is connected to the external circuit and performs work there. This work is equal to the charge that has flowed through the circuit multiplied by the cell-terminal voltage. This voltage is lower than the e.m.f. by the potential drop across the internal resistance

$$\frac{\partial A}{\partial \xi} = zF(E^0 - Ir_{\text{int}})$$

Here, r_{int} consists of the polarization potentials $Ir_{\text{int}} = |\eta_a| + Ir_{\text{int}}^0 + |\eta_c|$. These depend on the current, hence Ir_{int} is not linear with respect to the current. The enthalpy change in the denominator of the right-hand side of equation (4.95) equals the heat of the electrode reaction

$$Q = zF\left[E^0 - T\left(\frac{\partial E^0}{\partial T}\right)\right]$$

When $\Delta\xi = 1$, this heat $Q = zFr_{\text{int}}I$. From the point of view of the thermodynamics of irreversible processes, $zFr_{\text{int}}I$ is equal to the energy dissipated in electrode reactions during the transport of ions and solvent molecules and in the external circuit. The cell's efficiency coefficient is therefore

$$\eta = \frac{E^0 - r_{\text{int}}I}{E^0 - T\left(\dfrac{\partial E^0}{\partial T}\right)_p} \tag{4.97}$$

4.5 CHARGE TRANSPORT IN ELECTROLYTE SOLUTIONS

At equilibrium, the ions of an electrolyte solution move about in a random manner, and the average velocity of an ion is zero. Each single ion moves through the solution with a velocity which varies depending on whether there is another ion in its immediate neighbourhood, and if so, whether the sign of this second ion is the same as or opposite to that of the first. The relative velocity of two such ions at any instant may exceed the average value $\vec{v}_{ij} > |\vec{v}_k|$. If the solution is in an external electrical field of gradient

X [V cm^{-1}], the ions start moving along force field lines: random motion of the ions has non-vanishing components in the direction of the field. The spherically symmetrical charge distribution in the ionic atmosphere becomes asymmetrical, displaying axial symmetry with respect to the lines of force of the external field. Since the relaxation time of the ionic atmosphere is finite, the concentration of is a little lower in front of the ion in question and a little higher behind it than would be the case if the movements were equalized. This asymmetry is responsible for the existence of an electrical retarding force called the relaxation. The asymmetrical distribution of ions in the ionic atmosphere gives rise to an additional retarding force.

The molar conductivity Λ of an electrolyte solution depends in a simple manner on the velocity of the ions $\vec{v}_i$ and the charge transferred by the number of ions n_i that appear in a given volume of 1 M electrolyte solution

$$\Lambda = \frac{\sum_i n_i z_i e_i \vec{v}_i}{cX} = \sum_i \frac{n_i}{c} z_i e_i \frac{\vec{v}_i}{X} \tag{4.98}$$

where n_i/c is the ratio of the number of i-th ions to the analytical concentration c of electrolyte in 1 cm^3, and $\vec{v}_i/X$, the ionic mobility, is the ratio of the velocity of an i-th ion to the external field gradient X. A quantity, which must be calculated, is the average velocity $\vec{v}_k$ which an ion attains in the external field. During the 1920s, Debye, Hückel and Onsager [20, 21] calculated the molar conductivity of simple binary electrolytes in a solvent treated as a continuum, expressing it in term of electrolyte concentration, and the temperature, viscosity and electrical permittivity of the medium. They treated ions as electrostatically interacting charged spheres moving in the liquid medium in accordance with hydrodynamic laws. By introducing the concept of viscous resistance ρ or its reciprocal $\omega = 1/\rho$, Onsager showed that the average velocity of the j-th ion, when the i-th ion was nearby, could be expressed as

$$\vec{v}_k = \vec{v}_{ji} | r_{ji} | = \vec{v}(r_j) + \omega_j \vec{K}_{ji} - kT \ \mathrm{grad}_j \ln f_{ji} \tag{4.99}$$

where $\vec{v}(r_j)$ is the velocity at the position r_j of the j-th ion, ω_j is the reciprocal of the coefficient of the resistance offered by the medium towards the motion of the j-th ion, $\vec{K}_{ji}$ is the resultant electrical force causing the j-th ion to move, $kT\,\mathrm{grad}_j \ln f_{ji}$ is the "diffusion force" due to the appearance of a chemical potential gradient. According to Onsager, $\vec{K}_{ij}$ consists of three quantities

$$\vec{K}_{ji} = e_j X(r_j) - e_j \,\mathrm{grad}_j\, \psi_j(a_1) - e_j \,\mathrm{grad}_j\, \psi_i(r_i) \tag{4.100}$$

where $e_j X(r_j)$ denotes the external field force, and $e_j \operatorname{grad}_j \psi_j(a_1)$ the force of the internal electrical field due to the asymmetrical ionic atmosphere of the j-th ion. The potential of an asymmetrical ionic atmosphere $\psi_j(a_1)$ at the position of the j-th ion, can be represented as the potential describing the asymmetry of the atmosphere superimposed on the potential of a spherical ionic atmosphere $\psi_j(a_0)$. The potential $\psi_j(a_1)$ can be treated as the perturbing potential. The gradient $\operatorname{grad}_j \psi_j(a_0) + \operatorname{grad}_j \psi_j(a_1) = \operatorname{grad}_j \psi_j(a_1)$ since $\operatorname{grad}_j \psi_j(a_0) = 0$ at equilibrium. Hence $e_j \operatorname{grad}_j \psi_j(a_1)$ is equal to the electrical force due to the relaxation effect of the j-th ion's own ionic atmosphere retarding the movement of an ion; $e_j \operatorname{grad}_j \psi_j(r_i)$ is the force exerted by the i-th ion and its atmosphere on the j-th ion. Assuming electroneutrality can write

$$\operatorname{div}_j (f_{ji} \vec{v}_{ji}) + \operatorname{div}_i (f_{ij} \vec{v}_{ij}) = 0 \tag{4.101}$$

When solving the differential equations describing the motion of ions, Onsager [21] showed that three retarding forces appeared during stationary processes like charge transport. These three forces are electrical (relaxation), diffusional (concentration gradient) and hydrodynamic (electrophoretic effect) in nature. The first-order retarding forces, i.e. the relaxation and electrophoretic effects, give rise to second-order forces which partly cancel out the first-order ones. For example, concentration inhomogeneities lead to non-uniformities in osmotic pressure which, by the feed-back principle, bring about flows that balance the first-order fores to a certain extent. This mechanism therefore demonstrates that the stationary state of ionic motion is attained as a result of the complex interactions of an external force, the first-order decreasing forces which it has induced, and second-order forces partly cancelling out the first-order ones. Electrophoretic forces brought about by the countercurrent of solvent make the greatest contribution to the forces decreasing electrical conductivity.

The influence of the electrophoretic forces will now be examined in greater detail. According to hydrodynamic principles, every ion moves in a countercurrent of oppositely-charged ions which carry a volume element of solvent along with them; the further this volume element is from the ion, the weaker the counter-current. The following discussion is based on the reasoning of Fuoss and Accascina [8].

Let the forces acting on the ions [8] be denoted by $\vec{k}_j$ with respect to a single ion. The overall force acting on unit volume of solution is then $\sum n_j \vec{k}_j$, where n_j is the number of ions contained in unit volume. This force is transferred to the solvent molecules. If $\vec{k}_0$ is the averaged force acting on a single solvent molecule, and n_0 is the number of solvent molecules in a given volume of solution, then there are no resultant forces at equilibrium

$$n_0 \vec{k}_0 + \sum_j n_j \vec{k}_j = 0 \tag{4.102}$$

However, a resultant force does appear at distances of the order the depth of the Debye ionic atmosphere, κ^{-1}; for a small volume element dV near the j-th ion this force is equal to $\sum_i n_{ij} \vec{k}_i \mathrm{d}V$ (different from the average for larger volumes of solutions). The force acting on the solvent in a small volume dV is $n_0 \vec{k}_0 \mathrm{d}V$. The resulting force acting on a given volume element near the j-th ion is therefore

$$\left(\sum_i n_{ij} \vec{k}_i + n_0 \vec{k}_0 \right) \mathrm{d}V = \sum_i (n_{ji} - n_i) \vec{k}_j \mathrm{d}V \tag{4.103}$$

The force acting on a spherical sheath of depth dr around the ion is

$$\mathrm{d}\vec{F} = \sum_i \vec{k}_i (n_{ji} - n_i) 4\pi r^2 \mathrm{d}r \tag{4.104}$$

This force acts evenly on the whole envelope in the direction of the applied external field. From the hydrodynamic point of view this is equivalent to a force $\mathrm{d}\vec{F}$ acting on a rigid sphere of radius r in a liquid medium of viscosity η. As a result, the solvated ion moves with a velocity $\vec{v}_j$

$$\mathrm{d}\vec{v}_j = \mathrm{d}\vec{F} / 6\pi r \eta \tag{4.105}$$

in accordance with Stokes' equation. Using distribution functions

$$\mathrm{d}\vec{F} = 4\pi r^2 \sum_i [-n_i e_i e_j \exp(-\kappa r)/\mu D k T r +$$

$$+ n_i e_i^2 e_j^2 \exp(-2\kappa r)/4\mu^2 D^2 k^2 T^2 r^2] X e_i \mathrm{d}r \tag{4.106}$$

the force $\mathrm{d}\vec{F}$ should now be coupled with the electrostatic potential of the ionic atmosphere ψ calculated from Debye–Hückel theory. The total velocity $\Delta\vec{v}_j$ due to the presence of all the ions in the j-th ion's atmosphere is given by the integral of $\mathrm{d}\vec{v}_i$. The velocity of an ion together with its solvation sheath in symmetrical electrolytes is

$$\Delta\vec{v}_j = -X e_j \kappa / 6\pi \eta (1 + \kappa a) \tag{4.107}$$

where, according to the Debye–Hückel theory, a denotes the distance of closest approach between the centres of two oppositely-charged ions. Geometrically a denotes the diameter of a spherical ion. Fuoss and Accascina [8] have suggested that a should be identified with the hydrodynamic radius r of the solvated ion in Stokes' equation.

Because the ionic atmosphere is distorted by the electrical field gradient, the disturbed ionic atmosphere potential ψ, $\psi_j = (\psi_j^0 + \psi)$, must be allowed for when solving the equations for $d\vec{F}$, where ψ_j^0 is the undisturbed ionic atmosphere potential. They yield a corrected value of $\Delta\vec{v}_j$. This is the velocity of the ion with its solvent envelope, which together are a kinetic unit moving in a viscous medium. Beyond the envelope moving at the velocity of the ion, there are velocity fields of further solvent layers, which gradually disappear with increasing distance from the centre of the sphere, i.e. the point where ion j is located.

Fuoss and Onsager [22] extended the molar conductivity equation to take account of not-very-dilute solutions:

$$\Lambda = \Lambda^0 - S\sqrt{c} + Ec\log c + Jc - F\Lambda^0 c + 0(c^{3/2}) \tag{4.108}$$

The expressions $Ec\log c + Jc$ have arisen because higher approximations of the relaxation and electrophoretic effects, expressions coupling these effects, and the finite dimensions of ions have been taken into consideration. The expression $F\Lambda^0 c$ describes the increase in the solution viscosity when c moles of symmetrical electrolyte are dissolved in 1 dm^3 of solvent. It includes the viscosity coefficient η_s of an electrolyte solution that Fuoss and Onsager approximated by means of Einstein's equation [8]

$$\eta_s = \eta_0(1 + \tfrac{5}{2}\varphi) \tag{4.109}$$

where η_0 is the viscosity coefficient of the solvent and φ the volume fraction of dissolved electrolyte. The expression $0(c^{3/2})$ collects all terms proportional to the cube of the square root of the concentration; they are small in comparison with the other terms.

The viscosity of the solution may be lower than that of the solvent in solutions of inorganic salts of Group I metals at medium concentration in conditions dependent on the temperature, electrolyte concentration and composition of mixed solvents (mainly hydrogen-bonded solvents) [35, 36].

The concept of ionic atmosphere allows for the possibility that two ions may collide for a brief instant without alterating the thermodynamic properties of the solution or the character of irreversible processes being visibly affected. By their very nature, statistical theories must take such events into account, and the equations to which these theories lead do this. However, if an ion-ion collision produces an ion pair whose half-life exceeds the relaxation time of the ion cloud (ca. 10^{-8} s), a dynamic dipole like this acts as a neutral molecule. Such a pair does not participate in charge transport. The question then arises, in which systems can the formation of fairly stable ion pairs be expected? The smaller the dimension parameter of ions d and the lower the relative electrical permittivity of the solvent, the higher the electrostatic energy of interaction between two ions. In solvents of low electrical permittivity, the potential energy of the two ions forming a pair may be lower than the mean kinetic energy ($\approx kT$) and then the half-life of such a pair is sufficiently long.

In 1957–1958 Fuoss [23] derived an equation for the molar conductivity in electrolyte solutions in which ions associate into pairs. If c_i is the concentration of free ions, $c_i = c\gamma$, where γ is the degree of dissociation of the electrolyte equal to $\gamma = 1 - K_A c\gamma^2 f_\pm^2$ (K_A is the ion-pair association constant)

$$\Lambda = \Lambda^0 - S\sqrt{c\gamma} + Ec\gamma\log c\gamma + Jc\gamma - K_A c\gamma^2 f_\pm^2 \Lambda \qquad (4.110)$$

Association into ion triplets [9, 25, 54] has also been observed in solvents of low and medium relative electrical permittivity; spectroscopic investigations [26] and computer simulation [27] have confirmed these observations. Fuoss treated association into ion triplets as a chemical reaction between an ion and an ion pair and derived a formula for the ion-triplet association constant [28]. From purely electrostatic premises, he deduced that linear triplets would be the most stable and that they could be either positively (+ – +) or negatively (– + –) charged.

The Fuoss electrical conductivity equation [8] for electrolyte solutions in which solvated free ions, ion pairs and triplets coexist takes the form

$$\Lambda\sqrt{c_i}\, g(c_i) = \Lambda^0\sqrt{K_D} + \frac{\sqrt{K_D}}{K_T}\Lambda_T^0\left(1 - \frac{\Lambda}{\Lambda^0}\right)c_i \qquad (4.111)$$

where

$$g(c_i) = \frac{\exp\left[-\ln 10\beta''\sqrt{c_i\Lambda/\Lambda^0}\right]}{1 - S(\Lambda^0)^{-3/2}(c_i\Lambda)^{1/2}(1 - \Lambda/\Lambda^0)^{1/2}}$$

and K_D and K_T are the respective dissociation constants of ion pairs and triplets: $K_D = 1/K_A$ and $K_T = 1/K_A^T$.

Furthermore, there are papers where the Fuoss interpretation has not been used, namely in the cases where the $\Lambda(c^{1/2})$ function is non-linear and has a minimum; the increase of molar conductance at higher concentration is then explained as a result of redissociation of ion pairs to separate ions [29].

In 1978–1979, Fuoss [41, 42], and Lee and Wheaton [43] published new papers in which they laid greater emphasis on ion-dipole, dipole-dipole and ion-dipole-ion interactions in the solvation envelopes of anions and cations. Lee and Wheaton drew particular attention to the dynamics of interaction in the Gurney sphere [41]. The hierarchy of ion-dipole-ion and dipole-dipole interactions evidently collapses here, and the energies of these interactions becomes comparable. Short-range potentials were introduced into the calculations of ion-surroundings interactions in the Gurney sphere, and soon after the parameter R – the size of the solvated ion – appeared in equations. The Lee–Wheaton and Fuoss equations are:
– Lee–Wheaton [43] for symmetrical electrolytes

$$\Lambda = \gamma \left\{ \Lambda_0 [1 + C_1(\kappa R)\varepsilon\kappa + C_2(\kappa R)(\varepsilon\kappa)^2 + C_3(\kappa R)(\varepsilon\kappa)^3] - \frac{\rho\kappa}{(1+\kappa R)} [1 + C_4(\kappa R)\varepsilon\kappa + C_5(\kappa R)(\varepsilon\kappa)^2 + \kappa R/12] \right\} \quad (4.112)$$

where

$$\varepsilon = \frac{|z|^2 e^2}{DkT}, \qquad \rho = F\xi e|z|/3\pi\eta_0$$

$$\xi = 1/299.7925, \qquad \kappa = \left[\frac{8\pi N_A z^2 e^2 c\gamma}{1000\,DkT}\right]^{1/2}$$

and $C_1 - C_5$ are complex functions of (κR) which are given in the original paper [43b]; all other symbols have their usual meanings;
– the Fuoss equation from 1978 [41, 42]

$$\Lambda = \gamma[\Lambda_0(1 + RX) + EL] \quad (4.113)$$

where RX and EL could be expressed as functions of $t = \kappa R$

$$RX = -\varepsilon\kappa/6(1+q)(1+t)(1+qt) + \varepsilon^2\kappa^2[\ln(t)/12 + FH2 + \varepsilon\kappa(FH3)] + DXV$$

$$DXV = \beta_0\varepsilon\kappa\sqrt{c\gamma}[(H2)/16 - 0.125\ln(t)]/\Lambda_0$$

$$EL = -\beta_0\sqrt{c\gamma}\{1/(1+t) + \varepsilon\kappa[\ln(t)/8 + (H1)/2]\}$$

interpolating polynomials for the functions $H1, H2, FH2$ and $FH3$ valid over the range $0 < t < 0.8$, are given in the appendix of reference [41b].

Note that, like Fuoss, Lee and Wheaton treated association as a quasi-chemical reaction. The association of ions into ion triplets is not taken into consideration in equations (4.112) and (4.113). Lee and Wheaton's papers present the means of calculating Λ_0, K_A and R simultaneously from the experimentally determined relationships of $\Lambda(c)$.

Among others, one of the authors of the present book (L.W.) has used the Lee–Wheaton equation to calculate Λ_0, K_A and R, and has compared these results with the solutions of other electrical conductivity equations for non-aqueous electrolytes applicable in lithium cells [46].

Byung Chang Fu [47] used kinetic theory as a basis for his theory of non-equilibrium states of electrolyte solutions acted upon by an external electrical field. He generalized Kirkwood's integral equation to allow for non-equilibrium states and local chemical potentials, and obtained equations containing the coefficients of diffusion and ionic mobility, and the entropy flux. The fact that the Onsager–Fuoss conductivity equations also emerge from his theory can be regarded as evidence of its correctness.

In the literature, three classes of transport equations can be found: empirical extensions of equations for dilute solutions, molten salt approaches, and empirical equations for fitting measured data [48, 49]. Mention should also be made of the increasing number of papers making use of quantum mechanics, and of MSA (mean spherical approximation) and HNC (hypernetted chains) to describe non-equilibrium states in electrolyte solutions [50–53].

4.5.1 *The model of non-aqueous Li/CuS cell*

Consider a cell with a lithium anode and a porous semiconductor cathode of copper sulphide. The electrolyte is $LiClO_4$ in tetrahydrofuran (THF). This solvent is assumed to be inert towards both electrodes – it is neither oxidized nor reduced – but it dissolves $LiClO_4$ quite well. The overall cell reaction is $2Li^0 + CuS \rightarrow Cu^0 + Li_2S$. As a result of this reaction, the values of ΔG^0, ΔH^0 and ΔS^0 are as follows [32]: $\Delta G^0_{298} = -431\ \text{kJ mol}^{-1}$; $\Delta H^0_{298} = -439.7\ \text{kJ mol}^{-1}$; $\Delta S^0_{298} = -30\ \text{J K}^{-1}\ \text{mol}^{-1}$. The standard e.m.f. of this cell is [32]:

$$E^0 = -\Delta G^0_{298}/zF = 431\,000/2 \times 96\,485 = 2.234\ \text{V}$$

The temperature coefficient is [32]:

$$\left(\frac{dE^0}{d\tau}\right)_p = \Delta S^0/zF = -30/2 \times 96485 = -1.55 \times 10^{-4}\ \text{V K}^{-1} = -155\ \mu\text{V K}^{-1},$$

and the coefficient of the theoretical efficiency η^0 is

$$\eta^0 = \frac{\Delta G}{\Delta H}; \qquad \eta^0 = \frac{2.345}{2.234 + 298.15 \times 1.55 \times 10^{-4}} = 0.98$$

This value is large due to the fact that the entropy of the electrode reaction is low, i.e. ca. $-30\ \mathrm{J\,K^{-1}\,mol^{-1}}$.

To a first approximation the cell e.m.f. should not depend on the kind of solvent used, provided that it does not react with the lithium anode or the CuS semiconductor cathode. However, the e.m.f. measured experimentally during the first, fairly short discharge step is higher than the calculated value [44, 45]. So the lithium anode may well be passivated by a film which may contain solvent decomposition products and small quantities of lithium oxide, hydroxide and nitride. Assuming that the surface structure is slightly disturbed so that the work function of the Li^+ decreases by 1 eV and that the protection offered by the passivating film is kinetic in nature, an e.m.f. of 3.234 V is obtained [44, 45]. When the protecting layers are removed, the e.m.f. should fall to 2.343 V. In fact, the cell reaction proceeds in two steps

$$1.\ \mathrm{Li + CuS_{por.} \rightarrow LiCuS;} \qquad \mathrm{e.m.f.}^0 = 2.42\ \mathrm{V}$$
$$2.\ \mathrm{Li + LiCuS \rightarrow Li_2S_{solid} + Cu^0;} \qquad \mathrm{e.m.f.}^0 = 2.04\ \mathrm{V}$$

and the calculated value of 2.234 V is the mean $1/2\,(2.42 + 2.04) = 2.23$ V [44, 45]. Indeed on the discharge curves there is a potential drop of ±0.38 V half-way through the discharge period, which is evidence for a two-step electrode reaction (cf. Chapter 8, Figs. 8.4, 8.5). In theory, the e.m.f. ought not to depend on solvent properties; however, dynamic characteristics like the mechanisms of electrode reactions, the kinetics of ion associations, conductivity, diffusion and viscosity are to a large extent determined by solvent properties.

4.5.2 *Ion association processes in tetrahydrofuran*

It has been assumed that solutions of $LiClO_4$ in tetrahydrofuran (THF) are used in the model non-aqueous cell. Let us consider the processes of ion association in this solvent and their effect on cell phenomena.

The solubility of lithium perchlorate in THF is about 1 mol dm^{-3}. Conductivity studies have shown that in THF, lithium and perchlorate ions are strongly associated into ion pairs and triplets. The experimental association constants [33] are

$$K_{A2} = \frac{\{\mathrm{Li^+ClO_4^-}\}}{\{\mathrm{Li^+}\}\{\mathrm{ClO_4^-}\}} = 2.47 \times 10^7\ \mathrm{dm^3\,mol^{-1}}$$

$$K_{A3} = \frac{\{\mathrm{Li^+ClO_4^-Li^+}\}}{\{\mathrm{Li^+ClO_4^-}\}\{\mathrm{Li^+}\}} = 41.49\ \mathrm{dm^3\,mol^{-1}} \qquad (4.114)$$

However, it is hard to distinguish $Li^+ClO_4^-Li^+$ triplets from $ClO_4^-Li^+ClO_4^-$

triplets in conductivity experiments. In fact, the value of 41.49 $dm^3 mol^{-1}$ refers to the equilibrium

$$K_{A3} = \frac{[\{Li^+ClO_4^-Li^+\}\{ClO_4^-Li^+ClO_4^-\}]^{1/2}}{\{Li^+ClO_4^-\}\{Li^+\}\{ClO_4^-\}]^{1/2}}$$

which is the "geometric mean" of the constants K_{A3} for both kinds of ion triplet. The plots of the proportions of free ions γ_1, ion pairs γ_2 and ion triplets γ_3 as a function of electrolyte concentration in Bu_4NClO_4–THF solutions (Fig. 4.2) show that at moderate concentrations, large numbers of perchlorate ions are associated into pairs. At $LiClO_4$ concentrations of ca. 6×10^{-3} mol dm^{-3} the proportions of free ions and ion pairs we suppose are the same, but as the concentration increases from that point, the proportion of ion pairs far exceeds that of unassociated ions. Current is therefore carried by free ions and ion triplets. Triplet formation is insignificant in $LiClO_4$–γ-butyrolactone solutions and free solvated ions predominate, even in concentrated solutions [34, 55].

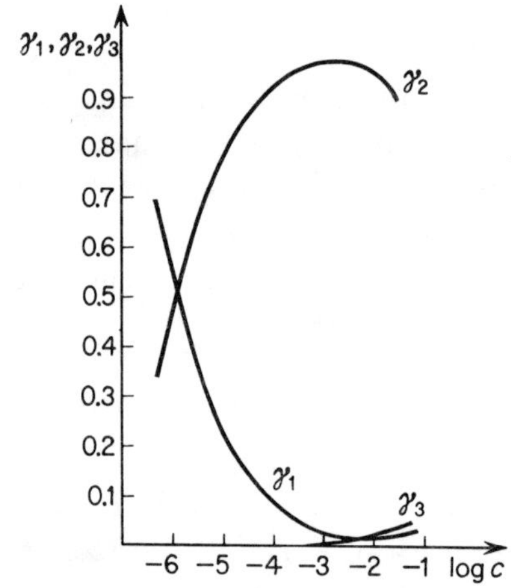

Fig. 4.2 Proportions of free ions γ_1, ion pairs γ_2 and ion triplets γ_3 with respect to the concentration of Bu_4NClO_4–THF solutions at 30°C.

In order to complete the description of a cell working under load, one would have to know the rate constants of the formation and dissociation of ion pairs and triplets

$$\frac{d[Li^+ClO_4^-]}{dt} = k_{A2}[Li^+][ClO_4^-] - k_{D2}[Li^+ClO_4^-]$$

$$\frac{d[Li^+ClO_4^-Li^+]}{dt} = k_{A3}[Li^+][ClO_4^-Li^+] - k_{D3}[Li^+ClO_4^-Li^+] \quad (4.115)$$

$$\frac{d[ClO_4^-Li^+ClO_4^-]}{dt} = k_{A3}[Li^+ClO_4^-][ClO_4^-] - k_{D3}[ClO_4^-Li^+ClO_4^-]$$

However, an inspection of the literature shows that these rate constants have not been so far determined experimentally although recent studies on the molecular level of solvation and chemical dynamics by ultra-short (picosecond) laser pulses indicate that such processes are very rapid [68, 69] so that overall rates for ionic equilibria are diffusion-controlled [70].

4.5.3 *Kinetics of processes at the lithium anode*

At the lithium anode, lithium ions are exchanged between the metallic phase and the solution. In the latter, lithium ions exist as free solvated ions, solvated pairs and solvated triplets. The lithium ion is solvated, probably by one or two THF molecules, when it interacts with the free pair of electrons on the oxygen in the THF molecule. Ion triplets are also solvated, and are assumed to be linear. A $Li^+ClO_4^-Li^+$ triplet is probably solvated by two THF molecules, but a $ClO_4^-Li^+ClO_4^-$ triplet merely polarizes the solvent. The lithium anode surface also interacts with the solvent through the lone electron pairs on the THF oxygen atoms.

If one wished to describe the elementary transfer of a lithium ion from the metal to the solution and vice versa, one would need to examine separately lithium ions at the smooth parts of the surface, as well as those at external and internal linear edges and at the edge borders. In each of these cases the surface ion interacts differently with the bulk of the metallic phase and solvent molecules: it is less strongly held at edges than at a smooth surface. A lithium ion in the anode boundary layer adjoining the solution can interact with a solvent molecule or a perchlorate anion or an ion pair. The approach of an ion towards the surface from the solution side can be shown in the model as a sequence of activated jumps. In the bulk of the solution, the average length of an activated jump is equivalent to the dimensions of a solvated ion, but near the electrode, this length shortens and the ion is desolvated.

When the solvated lithium ion enters the double layer of the strongly inhomogeneous electric field, partial desolvation occurs. Desolvation requires energy which is compensated by a lowering of the potential energy of the lithium ion. The motion of a lithium ion in an ion pair or triplet is similar. The high value of the ion-pair association constant is indicative of the strong interaction between a lithium and a perchlorate ion. The free energy of formation of an ion pair is

$$\Delta G^0_{A2} = -RT \ln K_{A2} = -8.3144 \times 298 \ln 2.47 \times 10^7 = -42.2 \text{ kJ mol}^{-1}$$

and corresponds to one quarter of the strength of an average chemical bond. It may therefore be assumed that the deposition at the anode surface of lithium ions torn away from ClO_4^- ions will not compete with the arrival there of free ions.

The kinetics of lithium ion transfer across the interface are clearly affected by the potential gradient in the double layer, i.e. the electrical potential difference between the anode surface Φ_a^M and the bulk of the solution Φ^R, $\Delta\Phi_a = \Phi_a^M - \Phi^R$. If $\Delta\Phi_a < 0$, the rate at which Li^+ ions pass from the metal into solution decreases exponentially with the value of $\Delta\Phi_a$, but the reverse processes are accelerated. The rate of Li^+ ion transfer across the interface can be written in the general form

$$\frac{dc_l}{dt} = k_a' c_l^M - k_c' c_l \tag{4.116}$$

and for this particular case

$$\frac{d[Li^+]}{dt} = k_a' [Li^+]^M - k_c' [Li^+]$$

where $c_l = [Li^+]$, $c_l^M = [Li^+]^M$ denotes the lithium ion concentration at the metal surface and k_a' and k_c' are the temperature- and potential-difference-dependent rate constants of the anodic and cathodic reactions. These relationships are exponential

$$k_a' = k_a^0 \exp[-\mathcal{E}_a^{\neq}/RT]\exp[\alpha F\Delta\Phi/RT]$$

$$k_c' = k_c^0 \exp[-\mathcal{E}_c^{\neq}/RT]\exp[(1-\alpha)F\Delta\Phi/RT] \tag{4.117}$$

where $\mathcal{E}_a^{\neq}$ and $\mathcal{E}_c^{\neq}$ are the activation energies of the anodic and cathodic reactions. The appearance of a potential difference between the minimum potential energy of a lithium ion at the anode surface and one in the solution gives rise to a relative change in the minimum height of $\delta\Phi$ where, at a distance equal to the position of the barrier peak, the increments are $\alpha F\delta\Phi$ on one side and $(1-\alpha)F\delta\Phi$ on the other. If the change in electrical potential between the two minima were linear and the barrier peak were situated exactly between the two minima, α would then be equal to 0.5 and $(1-\alpha) = 0.5$. But this is not usually the case, and α is greater or smaller than 0.5. The position of the minima and the barrier shape of elementary lithium-ion transfer from ion pairs or triplets to the metal may differ from the values characteristic of free ions.

4.5.4 *Mechanism of elementary processes at the lithium anode*

The exchange of a cation between the metallic phase and the solution is a basic elementary process; another is the solvation of such a cation, and a third is its association with an anion. Resolving these complex processes into sequences of simpler processes makes it easier to compile a set of differential equations describing the process as a whole. Suggesting a mechanism is by no means straightforward: on the one hand it depends on intuition, and on the other, whether the proposed mechanism can be tested.

Let us assume that the transfer of a lithium cation to the $LiClO_4$–THF solution consists of three parallel processes. The first is the activated jump of Li^+ cation from the metal surface to the equilibrium position corresponding to a Li^+ cation solvated by one molecule

(1)

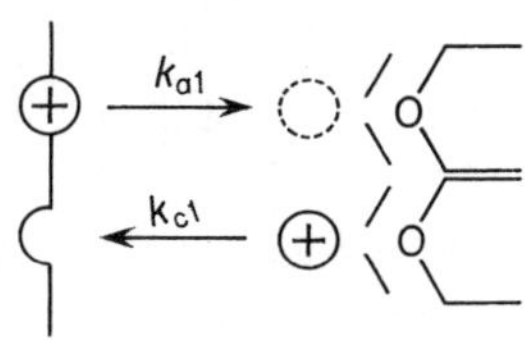

$$\frac{d[Li^+]}{dt} = k_{a1} \exp[\alpha_a F \Delta\Phi/RT][Li^+]^M -$$

$$- k_{c1} \exp[-\alpha_c F \Delta\Phi/RT][Li^+] \qquad (4.118)$$

The subscripts a and c appended to the rate constants indicate the anode and cathode reactions; $[Li^+]^M$ is the concentration of lithium ions in the metal phase – it is constant and has been introduced so that all rates have the same dimension (s^{-1}). Obviously, these rate constants take different values in different processes.

The second elementary process is that of a Li^+ ion passing from the metal into the neighbourhood of a ClO_4^- anion and vice versa

(2)

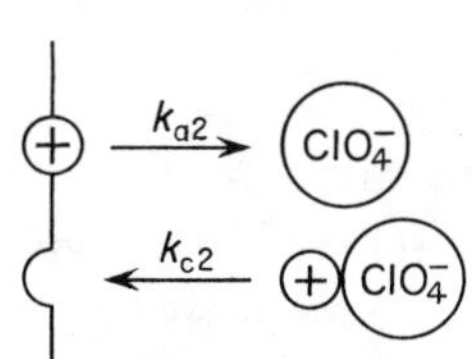

$$\frac{d[Li^+]}{dt} = k_{a2} \exp[\alpha_a F \Delta\Phi/RT][Li^+]^M -$$

$$- k_{c2} \exp[-\alpha_c F \Delta\Phi/RT][Li^+ClO_4^-] \qquad (4.119)$$

The third such process is that of a Li^+ ion moving into the environment of a $ClO_4^-Li^+$ ion pair

(3)

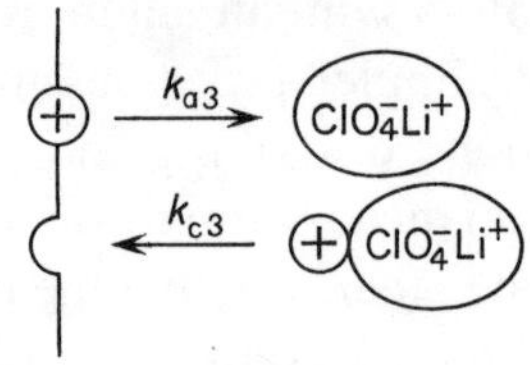

$$\frac{d[Li^+]}{dt} = k_{a3}\exp[\alpha_a F\Delta\Phi/RT][Li^+]^M - $$

$$- k_{c3}\exp[-\alpha_c F\Delta\Phi/RT][Li^+ClO_4^-Li^+] \quad (4.120)$$

The processes by which lithium ions pass into solution are accompanied by the association and dissociation of ion pairs and triplets, regardless of the above-mentioned electrode processes. Elementary processes should be understood as a reaction between two individual species, as a result of which energy is exchanged and a change of state ensues.

(4)
$$Li^+ClO_4^- \underset{k_{D3}}{\overset{k_{A3}}{\rightleftharpoons}} Li^+ + ClO_4^-$$

$$\frac{d[Li^+ClO_4^-]}{dt} = k_{A3}[Li^+][ClO_4^-] - k_{D3}[Li^+ClO_4^-]$$

(5)
$$Li^+ + ClO_4^-Li^+ \underset{k_{D4}}{\overset{k_{A4}}{\rightleftharpoons}} Li^+ClO_4^-Li^+ \quad (4.121)$$

$$\frac{d[Li^+ClO_4^-Li^+]}{dt} = k_{A4}[Li^+][ClO_4^-Li^+] - k_{D4}[Li^+ClO_4^-Li^+]$$

(6)
$$ClO_4^- + Li^+ClO_4^- \underset{k_{D5}}{\overset{k_{A5}}{\rightleftharpoons}} ClO_4^-Li^+ClO_4^-$$

$$\frac{d[ClO_4^-Li^+ClO_4^-]}{dt} = k_{A5}[ClO_4^-][Li^+ClO_4^-] - k_{D5}[ClO_4^-Li^+ClO_4^-]$$

Rewriting all these equations in a more compact form provides a set of equations defining the formation of ions, ion pairs and triplets in the anode layer:

$$\frac{dc_1}{dt} = \sum_{i=1}^{3} (k_{ai}\exp[\alpha_a F\Delta\Phi/RT]\, c_i^M - k_{ci}\exp[-\alpha_c F\Delta\Phi/RT]\, c_i)$$

$$\frac{dc_3}{dt} = k_{A3}c_1c_2 - k_{D3}c_3$$

$$\frac{dc_4}{dt} = k_{A4}c_1c_3 - k_{D4}c_4)$$

$$\frac{dc_5}{dt} = k_{A5}c_2c_3 - k_{D5}c_5 \quad (4.122)$$

This set can be further simplified if some of the kinetic constants introduced here are several orders of magnitude smaller than others. The former constants can then be ignored. If the rates of association are much lower than the rates at which ions move between the metallic phase and the solution, then, according to steady-state approximation, the rates dc_2/dt, dc_4/dt and dc_5/dt can be taken to be zero. Moreover, assuming that $ClO_4^-Li^+ClO_4^-$ triplets (c_5) form much less often than $Li^+ClO_4^-Li^+$ (c_4) triplets, one can write $k_{A5} << k_{D5}$ and $k_{A5} \approx 0$. Having made these assumptions, one ends up with a set of two equations

$$k_{A3}c_1c_2 - k_{D3}c_3 = 0$$
$$k_{A4}c_1c_3 - k_{D4}c_4 = 0$$

Since $k_{A3}/k_{D3} = K_{A2}$ and $k_{A4}/k_{D4} = K_{A3}$

$$k_{A3}c_1c_2 = k_{D3}c_3 \qquad \text{and} \qquad c_3 = \frac{k_{A3}}{k_{D3}}c_1c_2 = K_{A2}c_1c_2$$

and

$$k_{A4}c_1c_3 = k_{D4}c_4 \qquad \text{and} \qquad c_4 = \frac{k_{A4}}{k_{D4}}c_1c_3 = K_{A3}K_{A2}c_1^2c_2$$

which lead to the following linear equations:

$$\frac{dc_1}{dt} = [k_{a1}\exp(\alpha_{a1}F\Delta\Phi/RT) + k_{a2}\exp(\alpha_{a2}F\Delta\Phi/RT) + k_{a3}\exp(\alpha_{a3}F\Delta\Phi/RT]\,c_1^M - k_{c1}\exp(-\alpha_{c1}F\Delta\Phi/RT)\,c_1 + k_{c2}\exp(-\alpha_{c2}F\Delta\Phi/RT)\,c_2 + k_{c3}\exp(-\alpha_{c3}F\Delta\Phi/RT)\,c_3 + k_{c4}\exp(-\alpha_{c4}F\Delta\Phi/RT)\,c_4$$

If $\alpha_{a1} \approx \alpha_{a2} \approx \alpha_{a3}$, $k_a = k_{a1} + k_{a2} + k_{a3}$ and $\alpha_{c1} \approx \alpha_{c2} \approx \alpha_{c3} \approx \alpha_{c4}$, then the final linear equation is

$$\frac{dc_1}{dt} = k_a\exp(\alpha_aF\Delta\Phi/RT)c_1^M - \exp(-\alpha_cF\Delta\Phi/RT)(k_{c1}c_1 + k_{c2}c_2 + k_{c3}K_{A2}c_1c_2 + k_{c4}K_{A2}K_{A3}c_1^2c_2) \tag{4.123}$$

In non-stationary processes occurring under heavy external loadings, the cell relationship become non-linear (exponential) and a concentration gradient of the electrolyte builds up in the solution.

These considerations are valid if no compact passivation films are formed on lithium. When the cell is under heavy drain the passivating layers may collapse and the Li^+ ions have unimpeded access to the bulk of the solution and vice-versa – the anions from the solution can then enter the double layer at the lithium electrode surface.

4.5.5 *Electrode reactions at the CuS cathode*

When the external circuit of a charged $Li/LiClO_4/CuS$ cell is closed, the electrons [18, 37, 38, 56, 57] from the conduction band of the lithium anode move to the copper sulphide cathode, which is an electron sink (see Fig. 4.1). There they fill the conduction band of the CuS and react with the holes, which are virtual positive charges in the valence band. Then the positive charges of the copper(II) ions present in the nodes of the crystal lattice are partly neutralized. This happens in particular in the electrical double layer on the side of the semiconductor CuS electrode.

As a result, the copper–sulphur bonds are weakened and new bonds between copper atoms are formed. Eventually the electron pair from the largely covalent Cu – S bond is shifted to the sulphur atom and a free S^{2-} anion appears. Its charge is balanced by the Li^+ cations arriving from the solution side, and at practical current densities Li_2S is precipitated, unless complexing agents such as crown others are added to increase the solubility of Li_2S.

Simultaneously, Li^+ cations leave the lithium anode surface in the exact numbers needed to neutralize the electrolyte anions (here ClO_4^-) left in solution when some of the lithium cations are removed by the sulphide anions just formed as insoluble Li_2S.

From the mathematical point of view, expressions for the anodic and cathodic current densities for stationary processes (with light loads) will constitute flat sources of ions and flat ion flows (sinks) whose efficiencies are defined by the kinetic constants of the redox processes taking place at the lithium anode and semiconductor cathode.

4.5.6 *Description of a model non-aqueous cell – summary*

This description aims to link electrode processes with transport processes in the solution and the electrode spaces.

The model cell contains an electrolyte solution in a non-protonating solvent of medium or low electrical permittivity, chemically inert towards the anode, but strongly promoting ion association; the electrode system comprises a lithium metal anode and a porous semiconductor cathode.

The basic condition for selecting electrode materials in accordance with the above requirements is that the electrochemical affinity of the substrates participating in the electrode reaction should be high, and that the entropy of this reaction should be low. Then, the free enthalpy of reaction can be converted to electrical energy in a reversible process almost without loss of heat. Such a cell supplying a moderate current works stably when the entropy production of the cell processes is at a minimum. This quantity is estimated by using Onsager's approach to the principles of the thermodynamics of irreversible processes. The model cell has several interfaces: the lithium metal anode, the electrolyte solution and the semiconductor cathode made of porous CuS, for example, on an electrolytic copper base. The description of these phenomena is complex. Especially difficult to model and describe mathematically is the reaction at the porous CuS semiconductor cathode together with ion transport in the cathode space and the pores, and the transfer of electrons and holes within the semiconductor. Hence, the electrode processes and their kinetics have been described to a first approximation by assuming that equations like the Butler–Volmer–Erdey–Gruz equation hold. These equations have been formulated for a one dimensional half-space for a flux of Li^+ ions passing from anode to solution. It reacts with sulphide ions produced during reduction, and is deposited as lithium sulphides on the cathode (together with metallic copper), mainly in its pores. It has been assumed that the ions passing from the metal anode to the solution and vice versa are unassociated, and that ion association into pairs and triplets and their dissociation proceed consecutively and concurrently with the current-generating reactions at the electrodes. Similar assumptions have been made for the reduction processes at the CuS cathode only to simplify the kinetic equations. A set of equations has been derived to define the sources of ions, ion pairs and triplets in the layers adjacent to the electrodes and their influence on the principal process – that of lithium ions crossing the interface. The description of the dynamics of these processes is limited by the fact that kinetic constants for the formation and dissociation of ion pairs and triplets have not been determined experimentally. The dissociation constants of ion pairs and triplets determined experimentally by one of the authors [33] show that the proportions of free ions, ion pairs and triplets in solutions depend on the electrolyte concentration. It is these constants which are the deciding factor as regards charge transfer in the bulk of the solution. The transport of ions, pairs and triplets in the bulk of the solution is described by the electrostatic theories of Debye, Hückel, Onsager, Fuoss, Lee and Wheaton, Ebeling and Falkenhagen.

The transport processes formulated for the three spatial coordinates are described by a closed set of differential equations balancing cell components, charge and enthalpy, and defining the entropy source and the overall quantity of energy lost while the cell is working. Scalar and vector fluxes, expressed as transport equations, describe the reaction kinetics, migration and diffusion of n-component solution (composed of anions, cations, ion pairs and triplets), and solvent molecules (usually mixed). These fluxes depend on thermodynamic forces, the most important of which for a galvanic cell is the electrical potential gradient. The equation describing the potential gradient as a function of distance from the electrode surface closes the set of equations and enables it to be solved once appropriate boundary conditions have been laid down.

A suitably ordered set of equations defining the concentration fields of components and the distribution of electrical field intensity under given boundary conditions illustrates the changes occurring during stationary processes in a working cell. The closed system of differential equations with boundary conditions constitute a mathematical model of the cell. In principle, this set of equation lends itself to computer programming and can be solved numerically. Cell efficiency has also been assessed and a formula for determining the efficiency coefficient derived.

4.6 TRANSPORT IN CONCENTRATED ELECTROLYTE SOLUTIONS

Transport phenomena in concentrated solutions of electrolytes deserve special mention in this book, as dilute solutions are hardly ever used in practical cells or rechargeable batteries.

Classical thermodynamic theories differ from contemporary theories based on statistical mechanics basically in that the former treated solutions as a structureless continuum in which point charges are inbedded, while the latter consider the solutions as a collection, or set, of a vast number of individual particles such as molecules and ions, each of which has its appropriate individual movement and trajectory, thus being a separate kinetic entity.

Such a closed system can be described, or represented by a Hamiltonian $H_{r,p}$ which contains the equation of motion for N particles, each of which (index i, j) is characterized by a mass m_i, momentum p_i and position r_i in a multidimensional phase space [58, 59]

$$H_{(r,p)} = \sum_{i=1}^{N} \frac{p^2}{2m_i} + \frac{1}{2} \sum_{i=1}^{N} \sum_{j=1}^{N} U_{ij}(|r_i - r_j|) \tag{4.124}$$

The first term on the right-hand side of equation (4.124) describes the kinetic energy, while the second refers to the potential energy expressed as the overlapping of interactions between two neighbouring particles. This approximation is validated by the fact that almost all intermolecular interactions can be separated into bimolecular interactions that prevail at any given moment.

Of course, it is not possible to define and evaluate the instantaneous values of $r_i p_i$ for all individually moving particles in a macroscopic system, but one can try to see whether the observable experimental quantities correspond to a point in the phase space which describes the whole system. Using the partition function ρ, the mean values of a particular variable $\overline{Y}_t$ at time t may be evaluated on the basis of Liouville's theorem. Then

$$\bar{Y}_t = \int\int Y(r,p)\,\rho(r,p;t)\,\mathrm{d}r\,\mathrm{d}p \tag{4.125}$$

where $\mathrm{d}r\,\mathrm{d}p = \mathrm{d}r_1\,\mathrm{d}r_2\ldots\mathrm{d}r_N\,\mathrm{d}p_1\,\mathrm{d}p_2\ldots\mathrm{d}p_N$ is an element of the phase space.

The partition function ρ approaches a limit, which is determined by a time-independent state, or the state of equilibrium, ρ_{eq}:

$$\rho_{\mathrm{eq}} = \frac{\exp\left[-H(r,p)/kT\right]}{\int\int \exp\left[-H(r,p)/kT\right]\mathrm{d}r\,\mathrm{d}p} \tag{4.126}$$

The methods of statistical mechanics enable the derivation of the properties of electrolyte solutions by means of bimolecular partition functions obtained from the general equations (4.125) and (4.126), and the correlation functions g_{ij} which describe the force acting between ions i and j as function of the interionic distance r_{ij} to yield finally a potential $W_{ij} = -kT \ln g_{ij}$.

When the activity coefficients a_i and molar concentrations c_i of the electrolyte are introduced, the chemical potentials μ_Z for the electrolyte Z are obtained [60–62]

$$\mu_Z = \mu_+ + \mu_- = \mu_Z^\infty 2RT \ln(a_\pm c_\pm) \tag{4.127}$$

where μ_Z^∞ is the chemical potential of electrolyte Z at infinite dilution.

In dilute solutions of electrolytes it is sufficient to take into account the long-range electrostatic forces acting between solvated ions, i.e. those, which are separated by at least two solvation layers of two ions, each being independently solvated. In this simple case the classical theories of conductivity are adequate.

Transport phenomena in solution, such as diffusion and electrolytic conductivity, depend not only on the number of charge carriers expressed as concentrations of ions, c_{ij} but also on their velocities v_{ij}. The latter are determined by the external forces acting on the ions, causing the ions to migrate, which means that to the chaotic, brownian movement of ions a field vector is applied which tends to shift them in a preferred direction with a mean velocity called the electrophoretic velocity.

In concentrated solutions of electrolytes, in particular those in solvents with low electric permittivity, ionic pairs, and triplets and even more complex aggregates may exist in significant amounts.

Such clusters behave as separate kinetic entities, and to account for their presence and concentration, short-range forces must be considered. As a consequence of their operation an additional energy of short-range interaction appears between ions bound in a cluster.

A quantitative approach requires that the finite dimensions and electrical properties of molecules and charged particles are accounted for. The first theory which considered the association of ions was that of Bjerrum in 1926, but it was still limited to fairly dilute solutions. The introduction of additional variables, such as the reduced temperature together with the Debye and Hückel approximations permitted more precise calculations [63, 64].

In a natural and logical way these theories were extended to cover more concentrated solutions [65]. The mean spherical approximation (MSA) was applied to calculate the partition functions for solutions of 1:1 and 2:2 electrolytes. The MSA approximation reduces to the Debye and Hückel approximation for small interionic distances.

Still better results are given by the hypernetted chain (HNC) approximation, in particular for concentrated solution, however the calculations are cumbersome. Therefore both the MSA and HNC approaches are applied [51, 52, 66].

The current MSA model treats the electrolyte solution as a mixture of incompressible spheres with point dipoles (solvent molecules) and incompressible spheres with point charges (the ions). It can be applied to solutions of any, even the highest attainable concentrations of electrolytes [53].

On the other hand HNC theory is being continuously refined. It now includes interactions of ions not only with dipolar, but also multipolar molecules [50]. Moreover, it has been used to describe solution structure near a charged phase boundary, which can be the double layer at a charged electrode [67]. So far this approach has been applied only to aqueous solutions, but the theory is quite general and one can expect its application to other solvents including those used in high-energy cells and batteries.

All these papers discuss the structure of electrolyte solutions at equilibrium or quasi-equilibrium, if a low-intensity external field is added. A theory based on non-equilibrium states of electrolytes placed in an external electric field has been developed only recently [47]. The Kirkwood integrals are generalized and partition functions are introduced for an N-dimensional space containing N particles. Hence equations are derived which contain the ionic mobilities, diffusion coefficients and entropy fluxes. In addition, one can derive from these relations the Fuoss–Onsager equations, which may be considered as a strong verification of the new theory. There is now great activity in statistical mechanical theories for associating liquids: The HNC approximation is very successful and accurate in describing 1:1 electrolytes (and other Coulombic systems) as shown also by computer simulations [71–74], but for highly charged ions (and colloids) a new integral equation theory has been developed. It contains a modified soft-sphere MSA direct correlation function [71, 75]. Explicit predictions made for the fraction of monomers, dimers and trimers in a 2:2 electrolyte are in agreement with computer simulations [75, 76] also for highly asymmetric electrolytes [77]. More complete equations of irreversible thermodynamics have been used as transport equations [78].

Summarizing, one can see a steady advance and development in the theories of electrolyte solutions. New concepts are being introduced, but also the time-honoured, classical theories are being continually refined and extended. This provides an improved and welcome theoretical background for the practical efforts to derive better electrolytes for galvanic and fuel cells.

REFERENCES

[1] L. Onsager, *Phys. Rev.*, **37**, 405 (1931); *ibid.*, **38**, 2265 (1931).

[2] S.R. De Groot and P. Mazur, *Non-Equilibrium Thermodynamics*, North-Holland Publ. Co., Amsterdam (1962).

[3] R. Haase, *Thermodynamik der irreversiblen Prozesse*, Dietrich Steinkopff Verlag, Darmstadt (1963).

[4] B. Baranowski, *The Thermodynamics of Non-Equilibrium Processes* (in Polish), PWN, Warszawa (1974).

[5] L.A. Kvasnikov and R.G. Tazietdinov, *Regenerative Fuel Elements* (in Russian), Atomizdat, Moskva (1978).

[6] J. Prigogine, *Introduction Thermodynamics of Irreversible Processes*, J. Wiley, New York (1962).

[7] K.J. Vetter, *Elektrochemische Kinetik*, Springer-Verlag, Berlin–Göttingen–Heidelberg (1961).

[8] R.M. Fuoss and F. Accascina, *Electrolytic Conductance*, Interscience Publishers Inc., New York–London (1959).

[9] L. Werblan, A. Suzdorf and J. Lesiński, *Roczniki Chem.*, **51**, 2233 (1977).

[10] P. Marcelin, *Ann. Phys.*, **3**, 158 (1915).

[11] Th. De Donder, *L'affinité* , Gauthier-Villars, Paris (1927).

[12] P. Van Rysselberghe, in: *Modern Aspects of Electrochemistry* (J. O'M. Bockris, ed.), Vol. 4, Ch. 1, Plenum Press, New York (1966).

[13] J. Wiśniewski, B. Staniszewski and R. Szymanik, *The Thermodynamics of Non-Equilibrium Processes* (in Polish), PWN, Warszawa (1973).

[14] A.V. Lykov, *Heat-Mass Transfer – a Handbook* (in Russian), 2nd ed., Energiya, Moskva (1978).

[15] D.D. Fitts, *Non-Equilibrium Thermodynamics*, McGraw-Hill, New York (1962).

[16] K. Gumiński, *The Thermodynamics of Irreversible Processes* (in Polish), 2nd ed., PWN, Warszawa (1983).

[17] H. Gerischer, in: *Physical Chemistry. An Advanced Treatise* (H. Eyring, D. Henderson and W. Jost, eds.), Vol. 9A, Ch. 5, Academic Press (1970).

[18] M. Green, in: *Modern Aspects of Electrochemistry no 2* (J.O'M. Bockris, ed.), Ch. 5, Butterworths, London (1959).

[19] J.S. Newman, *Electrochemical Systems*, Prentice-Hall Inc., Englewood Cliffs, NJ (1973).

[20] P. Debye and E. Hückel, *Physik. Z.*, **24**, 185 (1923).

[21] L. Onsager, *Physik. Z.*, **28**, 277 (1927).

[22] R.M. Fuoss and L. Onsager, *J. Phys. Chem.*, **61**, 668 (1957); *ibid.*, **62**, 1339 (1958).

[23] R.M. Fuoss and C.A. Kraus, *J. Am. Chem. Soc.*, **79**, 3304 (1957).
[24] R.M. Fuoss and C.A. Kraus, *J. Am. Chem. Soc.*, **55**, 476 (1933).
[25] P. Beronius and T. Lindback, *Acta Chem. Scand.*, **A32**, 423 (1978).
[26] P. Bacelon, J. Corset and C. de Loze, *J. Solution Chem.*, **9**, 129 (1980).
[27] P.J. Ressky, J.B. Dubowicz, B.L. Tembe and H.L. Friedman, *J. Chem. Phys.*, **73**, 3372 (1980).
[28] R.M. Fuoss and C.A. Kraus, *J. Am. Chem. Soc.*, **55**, 2387 (1933).
[29] A.M. Sukhotin and Z.N. Timofieeva, *Zh. Fiz. Chim.*, **33**, 1602 (1959).
[30] R.M. Fuoss, L. Onsager and J.F. Skinner, *J. Phys. Chem.*, **69**, 2581 (1965); C.A. Kraus, *J. Phys. Chem.*, **58**, 673 (1954); *ibid.*, **60**, 1929 (1956).
[31] H. Falkenhagen, W. Ebeling and W.D. Kraeft, in: *Ionic Interaction* (S. Petrucci, ed.), Vol. 1, Academic Press, New York–London (1971); W. Ebeling and M. Grigo, *J. Solution Chem.*, **11**, 151 (1982); J.-C. Justice and W. Ebeling, *J. Solution Chem.*, **8**, 809 (1979); M. Grigo, *J. Solution Chem.*, **11**, 527 (1982).
[32] J.G. Gibson and J.L. Sudworth, *Specific Energies of Galvanic Reactions*, Chapman and Hall Ltd., London (1973).
[33] L. Werblan, A. Suzdorf and J. Lesiński, *Bull. Acad. Polon. Sci., ser. sci. chim.*, **28**, 627 (1980).
[34] L. Werblan, A. Suzdorf and E. Lin, *Pol. J. Chem.*, **56**, 1071 (1982).
[35] L. Werblan, A. Rotowska and S. Minc, *Electrochim. Acta*, **16**, 41 (1971).
[36] L. Werblan, *Bull. Acad. Polon. Sci., ser. sci. chim.*, **27**, 903 (1979).
[37] A. Pomianowski and P. Nowak, Material from the 5th Scientific Session on Problem MR I-11, *The Electrochemistry of Interfaces (Phase Boundaries) and Ionogenic Media* (in Polish), University of Łódź (1984), pp. 93–112.
[38] E. Ghali and A. Lewenstam, *Proceedings of the 3rd Symposium on ISE*, Budapest (1981), p. 235; A. Hulanicki, T. Krawczyński and A. Lewenstam, *Anal. Chim. Acta*, **158**, 343 (1984).
[39] J. O'M. Bockris and A.K.N. Reddy, *Modern Electrochemistry, An Introduction to an Interdisciplinary Area*, Plenum Press, New York (1970).
[40] V.G. Levich, *Physicochemical Hydrodynamics* (in Russian), Moskva (1955), pp. 248–255.
[41] R.M. Fuoss, *Proc. Nat. Acad. Sci.*, **71**, 4491 (1974); **75**, 16 (1978); **77**, 34 (1980).
[42] R.M. Fuoss, *J. Phys. Chem.*, **79**, 525 (1975); **82**, 2427 (1978).
[43] W.H. Lee and R.J. Wheaton, *J. Chem. Soc., Faraday Trans. 2*, **74**, 743 (1978); **74**, 1456 (1978); **75**, 1128 (1979).
[44] L. Werblan and J. Lesiński, *Pol. J. Chem.*, **53**, 2571 (1979).
[45] L. Werblan, J. Lesiński, A. Suzdorf, J. Izydorek and R. Drachal, *J. Power Sources*, **9**, 409 (1983).
[46] A. Bałkowska, G. Szymański and L. Werblan, *J. Electroanal. Chem.*, **287** (1990), 229.
[47] Byung Chang Eu, *J. Chem. Phys.*, **87**, 1220 (1987).
[48] J. Barthel, R. Wachter and H.J. Gores, in: *Modern Aspects of Electrochemistry*, Vol. 13 (B.E. Conway and J. O'M. Bockris, eds.), Plenum Press, New York (1979).
[49] J.F. Casteel and E.S. Amis, *J. Chem. Eng. Data*, **17**, 55 (1972).
[50] P.G. Kusalik and G.N. Patey, *J. Chem. Phys.*, **88**, 7715 (1988).
[51] D. Wei and L. Blum, *J. Chem. Phys.*, **87**, 2999 (1987).
[52] H.R. Corti, F. Prini and L. Blum, *J. Phys. Chem.*, **87**, 3052 (1987).
[53] J.S. Hye and E. Lomba, *J. Chem. Phys.*, **88**, 5790 (1988).
[54] J. Barthel, R. Gerber and H.-J. Gores, *Ber. Bunsenges. Phys. Chem.*, **88**, 616 (1984).
[55] M. Salomon and E.J. Plichta, *Electrochim. Acta*, **30**, 113 (1985).
[56] S.R. Morrison, *Electrochemistry at Semiconductor and Oxidized Metal Electrodes*, Plenum Press, New York (1980).

[57] W. Shockley, *Electrons and Holes in Semiconductors*, Van Nostrand, New York (1950).

[58] J. Barthel, *Ber. Bunsenges. Phys. Chem.*, **83**, 252 (1979).

[59] J. Barthel, *Chem.-Ing.-Techn.*, **50**, 259 (1978).

[60] J. Barthel, *Ionen in nichtwässringen Lösungen*, Steinkopf-Verlag, Darmstadt (1976).

[61] M.-C. Justice, *J. Solution Chem.*, **5**, 543 (1976).

[62] W. Ebeling and M. Grigo, *J. Solution Chem*, **11**, 527 (1982).

[63] W. Ebeling, *Z. Phys. Chem. (Leipzig)*, **247**, 340 (1971).

[64] H.L. Friedman and B. Larsen, *J. Chem. Phys.*, **70**, 92 (1979).

[65] A. Tani and D. Henderson, *J. Chem. Phys.*, **79**, 2390 (1983).

[66] J.C. Rasaiah, *J. Chem. Phys.*, **56**, 3071 (1972).

[67] G.M. Torri, P.G. Kusalik and G.N. Patey, *J. Chem. Phys.*, **88**, 7826 (1988).

[68] G.R. Fleming and P.G. Wolynes, *Physics Today*, **40** (5), 36 (1990).

[69] M. Maroncelli, *J. Chem. Phys.*, **94**, 2084 (1991).

[70] S. Sridharan, J. A. McCammon and J.B. Hubbard, *J. Chem. Phys.*, **90**, 237 (1989).

[71] V. Vlady, T. Ichiye and A.D.J. Haymet, *J. Am. Chem. Soc.*, **113**, 1077 (1991).

[72] J.M. Caillol and D. Levesque, *J. Chem. Phys.*, **94**, 594 (1991).

[73] Jian Zhu and J.C. Rassiah, *J. Chem. Phys.*, **94**, 3141 (1991).

[74] K. Olaussen and G. Stell, *J. Stat. Phys.*, **62**, 221 (1991).

[75] Yu.V. Kalyuzhnyi and M.F. Holovko, *J. Chem. Phys.*, **95**, 9151 (1991).

[76] D.E. Smith, Yu.V. Kalyuzhnyi and A.D.J. Haymet, *J. Chem. Phys.*, **95**, 9165 (1991).

[77] Per Linse, *J. Chem. Phys.*, **94**, 8227 (1991).

[78] F.O. Rainieri and E.O. Timmerman, *J. Chem. Soc. Faraday Trans.*, **86**, 1057 (1990).

CHAPTER 5

Anode Materials

Generally speaking, any substance that can be electrochemically oxidized is a suitable anode material. However, every high-energy galvanic cell must contain an electrode system having a high negative free energy ($-\Delta G$) and the lowest possible equivalent mass. High specific energies are theoretically available with most metals in the first, second and third principal groups of the Periodic Table, but only metallic lithium [1–3] and lithium alloys [4–7] have found wide application, although calcium has been utilized [8, 9].

The energy properties of metallic lithium are the most advantageous because its standard potential of -3.01 V [1] with respect to the standard hydrogen electrode is the highest and its atomic mass is small. Of all the alkali metals, it is also the least reactive towards organic solvents.

Even so, electronegative electrode materials with a high negative standard potential are generally very reactive, especially towards water and water-like solvents (protonating solvents, like alcohols, containing an exchangeable, i.e. mobile hydrogen – see Section 7.1). In practice, this precludes the use of water as solvent and imposes the requirement that the non-aqueous solvents used should be very carefully dried. In practical applications, the fact that lithium forms compounds with oxygen and nitrogen must also be taken into account, especially as reaction with the latter already takes place at -50°C. The other alkali metals and the alkaline earths react with oxygen, but not with nitrogen at room temperature. Table 5.1 sets out some of the properties of a lithium electrode [10].

The stability of lithium in most solvents is due to surface effects. Passivating films of sparingly soluble lithium salts protect the metal from further corrosion and render it largely independent of the electrolyte solution, except when the most reactive solvents such as water and water-like solvents are involved [7]. Traces of water in non-protonating and aprotic solvents contribute to the formation of passivating films which are sufficient-

Table 5.1. Standard potential of the lithium electrode E^0 in selected pure solvents: limiting conductivity $\lambda^0_{Li^+}$ and solvation number of the lithium ion Li^+ [10] (SHE – standard hydrogen electrode)

Solvent	E^0, V	Reference electrode and temperature, °C	$\lambda^0_{Li^+}$, $\Omega^{-1}cm^2mol^{-1}$ (temp. 25°C)	Solvation number of Li^+ (temp. 25°C)
Water	–3.01	SHE 25	38.68	3.5–7
Methanol	–3.095	Ag/AgCl 25	39.0	4–7.5
Acetonitrile	–0.06	Rb/Rb^+ 25	59.3	5.5–6
NH_3 (liquid)	–2.99	SHE –50		
Hydrazine	–0.19	Rb/Rb^+ 25		
Ethanol			14.9	3.6–6.5
Acetone			65.0	ca. 4
Nitromethane			56.0	1–1.3
Pyridine				3.4–5

ly permeable for lithium ions to pass them through into the solution when the cell is operating. The passivating films forming on Group II and III metals are much more compact, thereby isolating the electrode from the electrolyte solution and preventing the electrode reaction.

Thus, the chemical stability of lithium in solutions of organic solvents is not based on thermodynamic properties, because its standard potential is high and negative [11–14] (like that of the other alkali metals), implying high reactivity.

The reversible potentials of metal electrodes in equilibrium with their ions (electrodes of the first kind) referred to the standard ferrocene electrode in propylene carbonate and aqueous solutions are given in Table 5.2, and the standard potentials in acetonitrile [15] in Table 5.3.

The stability of lithium in solutions, confirmed experimentally but inconsistent with its high negative standard potential, is due to surface effects ensuing from the formation of passivating films. The nature of the latter depends on the solvent involved, as they mostly consist of sparingly-soluble lithium salts resulting from reduction by lithium of the solvent or its impurities such as water, and in some cases of polymerized second-

Table 5.2 Standard potentials E^0 of metallic electrodes referred to standard ferrocene electrode in propylene carbonate and aqueous solutions [7] (at 25°C)

Solvent	Electrode, E^0, V					
	Li/Li^+	Na/Na^+	K/K^+	Rb/Rb^+	Cs/Cs^+	Ca/Ca^{2+}
Propylene carbonate	–3.26	–3.05	–3.35	–3.38	–3.40	–3.05
Water	–3.43	–3.10	–3.32	–3.32	–3.32	–3.26

Table 5.3 Standard potentials E^0 of alkali metals and calcium in acetonitrile [15] at 25°C, measured with respect to the silver electrode in the system Cu|M|MX 0.01 $mol dm^{-3}$ || || 0.01 $mol dm^{-3}$ $AgNO^3$|Ag|Cu

System M/MX	Li/LiCl	Na/NaI	K/KI	Rb/RbI	Cs/CsI	$Ca/Ca(NO_3)_2$
E^0 [V]	–2.476	–2.259	–2.344	–2.327	–2.272	–2.140

ary products. Sometimes, these materials are formed as a consequence of processes which are the same as or similar to those taking place when the cell discharges, for instance in cells containing oxyhalides ($SOCl_2$, SO_2Cl_2 etc.) which act both as solvents and as cathode materials.

The kinetics of processes at alkali metal electrodes have been studied fairly extensively in a number of solvents, such as propylene carbonate, dimethylformamide, ethylene carbonate and dimethylene sulphide [16–18]. Surface layer formation inhibits mass transfer, but the behaviour of alkali metal and alkaline earth electrodes can nevertheless be described by the Nernst equation, at least at low current densities. Essentially, they are reversible electrodes at which the reaction kinetics are determined by sufficiently high exchange currents [7].

The charge-transfer activation energy of a lithium electrode is quite high, i.e. ca. 42 kJ mol^{-1}. Typical experimental exchange current densities (i_0) in fairly concentrated, well-conducting electrolyte solutions are ca. 1 mA cm^{-2}, while the theoretical values (i.e. those of electrodes without passivating films, e.g. in liquid ammonia – see Chapter 6) may be higher by two or three orders of magnitude [19]. Therefore the above-mentioned activation energy is more a reflection of the overall transfer process. The experimental activation energy $\mathcal{E}_{exp}$ is the sum of the low activation energy required to transfer a charge across the electrical double layer $\mathcal{E}_H$ which can be determined, e.g. in liquid ammonia, and of the energy $\mathcal{E}_d$ needed to transfer ions across the passivating film, which is sometimes quite high as the film often has semi-permeable membrane properties and acts as a highly viscous diffusion layer. Then $\mathcal{E}_d = zF\Delta E_d$, where ΔE_d is the potential drop across this layer. Therefore, $\mathcal{E}_d$ is not actually the activation energy but the energy lost in overcoming the internal resistance, which is what this layer of sparingly soluble deposits in fact represents. If it is crystalline (as in the case of LiCl), closing the cell circuit decreases the potential of the Li anode and causes the injection of Li^+ ions into the interstices of the lattice. These ions then move as Frenkel defects through the layer increasing the discharge current, which leads to the disintegration of the passivating film, and enhancement of the exchange current; the overall anodic process is thereby improved.

The factor limiting the actual exchange currents as opposed to the theoretical ones is the presence of passivating films on the anode surface. Therefore, too much importance should not be attached to contamination of the electrolyte by traces of water [20], since passivating films are formed anyway, during even the slightest reaction between the anode material and the solvent. For example, instantaneous exchange currents as high as 20 mA cm^{-2} have been measured in solutions containing ca. 0.05% water (500 ppm). Such currents were obtained by the anodic pulse method [21] at partly-depassivated electrodes.

As passivating films limit the current densities when the cell is operating, attempts have been made to reduce cell anode passivation by adding transition metal chlorides and hexamethylphosphoramide (HMPA) to the electrolyte solutions. Even very small amounts of these compounds are enough to bring about marked improvement [7].

5.1 REVERSIBLE LITHIUM ELECTRODES

Lithium can be electrodeposited from solutions of its salts in organic solvents with a 100% yield. However, this does not mean that such an electrode can be used in a storage battery offering a large number of charge/discharge cycles, because the metal is often deposited as a loose, inhomogeneous dendrite layer contacting badly with the operating electrode (the remaining mass of metal). Moreover, some of the deposited metal may simply fall away from the electrode surface to the bottom of the cell. So the main problem when looking for a reversible, multiple-use lithium cell is to find a suitable lithium anode at which lithium can be deposited in a compact, electron-conducting form at least several hundred times, i.e. an anode with a high service life, which means almost 100% useful recovery. In order to get a reversible anode, a suitable electrolyte solution must be found [6, 21–24] which promotes the formation on the lithium of a conducting passivating film that simultaneously hinders the formation of loose microcrystalline dendrites on the lithium surface.

The initial results with heterocyclic compounds like tetrahydrofuran or propylene carbonate have not been encouraging, as lithium fairly readily opens the ring in these compounds by reducing the ether (or carbonate) bond to the corresponding anion which forms sparingly soluble salts with the lithium ion. These surround the lithium grains and contribute to the formation of loose dendrites when the metal is being electrodeposited. The properties of the deposits strongly depend on the organic solvent used, e.g. the yield of deposited lithium increases on progressing from propylene carbonate, through tetrahydrofuran, to 2-methyltetrahydrofuran, in which lithium is least reactive. The methyl substituent in position 2 of the tetrahydrofuran ring considerably reduces the latter's reactivity towards lithium by raising the energy level of the lowest unfilled molecular orbital of the oxygen atom [11]. 2-Methyltetrahydrofuran can be reduced by lithium in several ways, but since these reactions compete with the electrochemical reactions, it would be advantageous to reduce the reactivity of lithium (cf. Section 7.5).

The consequence of these reactions is that lithium forms organic salts which are deposited on the anode as a very thin ionically-conducting passivating film which prevents dendrite formation. In the extreme case, a monomolecular film is sufficient to prevent the lithium anode from corroding, and enables a compact layer of metal to be deposited evenly during charging, primarily by separating the lithium ions from the electrolyte at the anode. The decrease in electrolyte concentration is compensated by the simultaneous dissolution of the lithium salt deposited at the cathode.

Another way of avoiding dendrite formation is to add to the electrolyte solution substances, such as bromides, which are capable of regenerating the lithium lost when dendrites are formed. In a multiple-use cell, bromine appears at the anode during charging and reacts with loose dendrites: $Br_2 + 2Li \rightarrow 2Li^+ + 2Br^-$. The lithium cations pass into solution and during charging are redeposited at the forming anode (where cathode processes take place when a storage battery is being charged); the lithium in the electrically-isolated dendrites redissolves, etc. until a compact film of metal (an electrode ready for the discharging cycle) has been produced.

A more general way of improving the durability of a lithium anode is to reduce the chemical potential, and hence the reactivity, of the metal. This can be achieved by alloying lithium with a more noble metal. Of the many combinations that have been tested, the lithium-aluminium alloy has the optimum properties [5, 11, 25, 26]. Anodes of this alloy can be charged and discharged many times without side reactions such as the rapid decomposition of the organic electrolyte or dendrite formation taking place; but a fairly stable, thin passivating film permeable to ions still adheres to the anode. Even after a large number of charge/discharge cycles, when the electrolyte composition has stabilized, small exchange currents i_0 of ca. 0.2–1 mA cm^{-2}, corresponding to an activation energy of ca. 20 kJ mol^{-1}, indicate the existence of this film.

The production of a lithium-aluminium electrode depends principally on formation of a non-stoichiometric intermetallic compound called β-LiAl. This phase is crucial to the functioning of the lithium-aluminium anode. The apparently simple reaction $Li^+ + Al^0 + e \rightarrow LiAl$ is in fact very complex [6]. At first, a solid solution of lithium in metallic aluminium is formed, although initially the crystalline structure of the aluminium does not change. This solution of lithium in aluminium is called the α-Al(Li) phase. As more lithium is deposited at the aliminium anode, a new phase – β-LiAl – appears, the quantity of which increases as lithium is added (phases α and β coexist). When the lithium concentration reaches ca. 50 mol %, the intermetallic compound is almost pure phase β, which is the most highly charged form of the lithium-aluminium anode. Phase β is a non-stoichiometric intermetallic compound and exists when the lithium content ranges from 48 to 56 mol %; the symbol for lithium is then placed first: β-LiAl.

Discharging is a reversible process in which β-LiAl grains first change into α-Al(Li) containing up to 8% lithium. Further discharge continues to dilute the solid solution until the latter contains much less than 8% lithium. During this whole process the structure of the metallic grains is such that the lithium atoms can diffuse reversibly, into and out of the bulk of the grains [6].

As in the case of lead storage batteries, for which charged anodes are manufactured, a satisfactory anode of pure, fine-grained β-LiAl was obtained by the usual pyrometallurgical process. When press-moulded, this material acquires a microporous structure with an extensive surface area between the electrolyte and the electrode material, so quite high charge and discharge current densities can be obtained (ca. 20 mA cm^{-2}). The optimum composition of a charged lithium-aluminium anode is 48–50 mol % lithium in the β-LiAl form. At higher concentrations, the activity of the deposited lithium rises rapidly and approaches that of the pure metal; the advantages of the Li–Al system are thereby lost. Although the β-LiAl phase is present up to a content of 56 mol % lithium, overcharging the anode with more than 50 mol % lithium brings no benefits, despite an increase in capacity (Fig. 5.1).

One valuable property of the β-LiAl phase is the high diffusion coeffi-

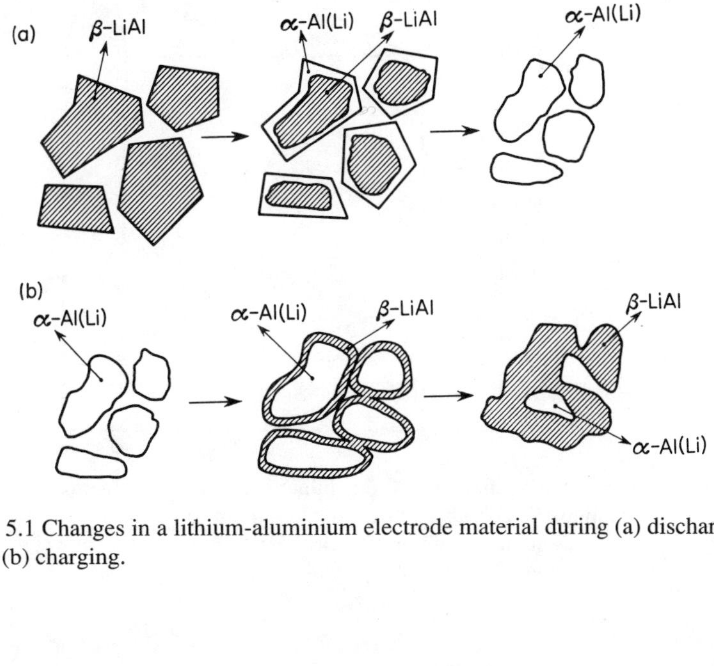

Fig. 5.1 Changes in a lithium-aluminium electrode material during (a) discharging and (b) charging.

cient of lithium, 10^{-4} $cm^2 s^{-1}$, whereas in the α-Al(Li) phase it is 10^{-10} $cm^2 s^{-1}$ at 415 °C. The lithium discharge reaction proceeds far more readily in the β-LiA than in the α-Al(Li) phase. The volume of the LiAl anode decreases by ca. 30–35% on passing from β-LiAl to the α-Al(Li) phase, and this fact must be taken into consideration in the design of a cell and its electrode system.

When a storage battery is being charged, lithium ions precipitated at the positive electrode are redissolved and transferred across the electrolyte to be deposited on the surface of α-Al(Li) particles, with which they react to yield a thin film of β-LiAl. The formation of β-LiAl increases the volume, which improves contact with neighbouring grains of anode material. This reaction resembles the process that occurs when metals are welded together. The thickness of the β-LiAl film must not exceed a critical value, otherwise internal stresses will cause it to crack. This happens because β-LiAl is a brittle, intermetallic compound, whereas the stresses are induced by the large differences in volume between the two phases. It has been found that anodes of this material can undergo 1000 charge/discharge cycles without their mosaic structure changing (Fig. 5.2).

Replacing the lithium anode by a lithium-aluminium alloy reduces the negative potential of the electrode by ca. 350 mV [11]. The resulting lithium activity in the LiAl alloy is sufficient to reduce the latter's reactivity towards organic solvents. Low-temperature storage batteries can therefore be constructed with this material.

The kinetics and mechanism of charge-discharge processes at a lithium-

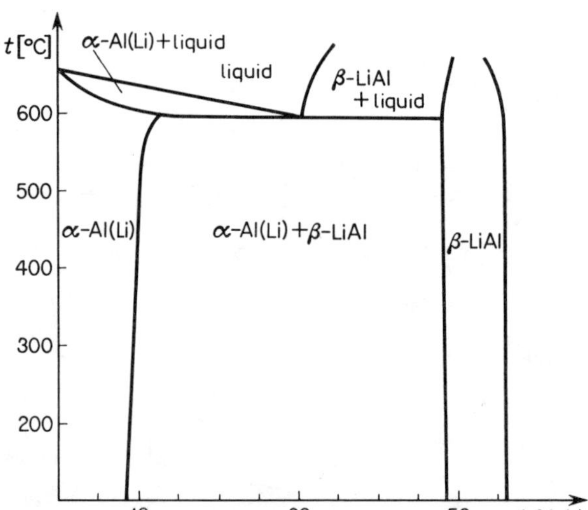

Fig. 5.2 Phase diagram of the lithium–aluminium system; α-Al(Li) – phase containing from 0 to 8% lithium in aluminium, β-LiAl – phase containing from 48 to 56% lithium in aluminium, α-Al(L) + β-LiAl is the region where the two phases coexist: liquid – Li–Al alloy in the liquid phase.

alloy anode have been studied at the Argonne National Laboratory (USA) by Battles *et al.* [6], and by Fawcett [11] at Guelph University in Canada. During charging, the anode reaction is $Li^+ + Al^0 + e \rightarrow LiAl$. When lithium is reduced on a fresh aluminium electrode at a controlled potential, three stages can be discerned on the chronoamperometric curves of this process. At first, the electrical double layer is charged, then the current gradually increases as lithium metal is deposited in the alloy phase. In the third stage the current is approximately constant, since its density is determined by mass transfer processes and the electrode potential (these studies were carried out in propylene carbonate).

Non-stoichiometric β-LiAl forms during charging. As a result, the potential of the quasi-equilibrium state (i.e. of the open circuit) varies slowly in time over a range of > 20 mV, hence the voltage of the whole cell also varies within these limits as long as it is operating.

As a rule, the activity of a metal decreases when it forms an alloy or intermetallic compound because it is diluted, and the same thing happens in lithium-magnesium and calcium-aluminium/silicon alloys. This is why the standard electrode potentials of electrodes made of alloys of alkali metals and their intermetallic compounds with elements from subsequent groups in the Periodic Table are more positive than the pure lithium electrode potential, from 0.1 V for lithium-magnesium alloys [7, 27–30] to 1 V for lithium-antimony intercalation compounds [28].

A new kind of anode material has been introduced recently [31–33]. It consists of lithium, the principal electroactive material, and Wood's alloy (Sn, Pb, Bi, Cd); the latter reduces the activity of lithium in the same way as aluminium does in the lithium-aluminium alloys described earlier. Also other metals are used for lithium alloys, such as Cu, Sn, In, Pb, Ag. Corrosion of lithium can be reduced by coating the anode with a thin Li/Al foil. The open-circuit voltage (OCV) is thereby reduced by 11% but the shelf life is increased [50]. For special applications in microelectronics ultra-thin ($\leq 10\ \mu m$) Li anodes are prepared by high-vacuum vapour deposition [51, 52].

5.2 ELECTRODES MADE OF OTHER MATERIALS

One of the first non-lithium alloys to be studied was the calcium–aluminium system, which soon proved to be useless for making electrodes, particularly for storage batteries, because it utilized only 50–60% of the available electrical capacity. This meant that systems using this alloy had a low specific energy. So the effect of adding a variety of metals and non-metals in the hope of improving the efficiency of these electrodes was studied [8, 34]. Silicon had the best properties of all these substances when its content in the aluminium matrix was 20 %, the remaining 80% being aluminium. The electrolytic deposition of calcium gave rise to the ternary Ca–Al–Si system which contained a variety of intermetallic compounds, including $CaAl_4$, $CaAl_2Si_2$, CaAlSi, $CaAl_2$ and CaSi, depending on the amount of calcium deposited. The optimum silicon-to-aluminium atomic ratio in this electrode material was 1:3. The specific energy obtained was close to that of a lithium-aluminium electrode.

A calcium-magnesium-silicon electrode was also tested. Here, the principal anode material was the intermetallic compound $CaMg_2Si$, but no applications were found for it [8] because its energy capacity was low even at low current densities (ca. a few mA cm^{-2}).

The Ca–Si alloy is an inexpensive substitute for the lithium-aluminium alloy, but it is not as energy efficient. Calcium is quite reactive, so it can be used in electrodes in alloy form only. Ca–Si is a multiphase system; i.e. solid solutions of variable stoichiometry as well as the compounds Ca_2Si, CaSi and $CaSi_2$ can form, depending on the percentage composition. In the initial stages of their operation (i.e. when fully charged), negative electrodes do not reach the maximum calcium content indicated by the formula Ca_2Si but always contain a mixture of Ca_2Si + CaSi; the composition is thus given by the non-stoichiometric formula $Ca_{1.34-1.28}Si$. Such a cell with, say, a FeS_2 cathode supplies a voltage of nearly 2 V. The calcium electrode potential remains almost constant as long as Ca_2Si is present, but when only CaSi is left, the cell voltage drops abruptly to ca. 1.6 V. A new compound $CaSi_2$ is formed if discharging continues, but the voltage plateau falls to around 1.3 V, at which point the practical range of application of the calcium electrode in such a cell comes to an end [8]. This kind of anode material has been successfully used in storage batteries, mainly in the high-temperature ones that use molten salts.

Pure calcium and magnesium were used quite a long time ago as anode materials in non-aqueous cells for special applications, both at room temperature and below zero [15]. As the cell discharges, both metals are oxidized to their bipositive ions in a one-step redox process: $M \rightarrow M^{2+} + 2e$.

The calcium anode has recently been used successfully in high-energy non-aqueous cells containing thionyl or sulphuryl chloride and $Ca(AlCl_4)_2$ as the electrolyte. Calcium chloride is precipitated from this solution as a passivating film when the cell is being discharged, especially at high current densities. Passivation is prevented by the use of a mixed solvent and SO_2-saturated electrolyte solution [35]. The use of calcium in low-temperature cells has been described [36].

As metallic sodium is extremely reactive, especially in organic solvents, it is not used alone in cheap non-aqueous cells, but attempts are being made to reduce its reactivity by using sodium amalgam anodes [37]. Another way of overcoming this difficulty is to use gelatinous electrolytes with poly(ethylene oxide) (PEO) and sodium perchlorate [38]. As the polymeric matrix is chemically inert and the sodium ions are highly mobile, it has been possible to produce a cell capable of undergoing a large number of charge/discharge cycles.

The prospects of aluminium anodes are promising as the equivalent mass of this metal is small – only 9, i.e. 27.98/3 – in the reaction $Al \rightarrow Al^{3+} + 3e$. There are problems, however, in that the standard potential of this reaction is rather low, $E^0 = -1.67$ V, and that compact, impervious passivating films are formed rather easily. The latter drawback has been largely overcome by the use of molten salts of a special composition whose melting points are well below zero [39]. These are mixtures of aluminium chloride and "onium" salts (ammonium and others; see Chapter 7). Also aluminium alloys with a variety of metals, such as Mg, In, Ga, Ag have been tried. Higher discharge voltage and a better anode utilization was obtained [53].

The problem of corrosion in alkali metal anodes, which causes self-discharge of the cell, has been tackled recently by coating the anode with a semipermeable membrane of a neutral polymer resistant to and insoluble in the electrolyte solution. The results of trials with polyethylene coatings have been encouraging [40].

The films that form on the surface of reactive anode materials like lithium or magnesium when they react with electrolyte solutions may be at once beneficial, in that they reduce the self-discharging of a cell and hence prolong its lifetime, and detrimental, because they increase its internal resistance, which in turn reduces the current density. The properties of these surface layers are being studied [41, 42].

Magnesium alloy electrodes have also been used quite frequently in reserve cells, because their potentials are low enough for hydrated electrolytes to be used. However, they are aqueous systems, so no more will be said about them here [15, 43].

Besides strongly negative anodes, more noble anode materials such as copper have been tried [44, 45]. Copper was oxidized to copper(I) ions in the discharge half-reaction. When a storage battery was being charged, the reverse reaction proceeded with almost 100% current efficiency because the system was stabilized by the formation of a complex ion by Cu(I) with a non-aqueous solvent such as acetonitrile.

The mechanism by which metals are deposited in non-aqueous solvents has been greatly studied. The matrix has been found to influence strongly the formation of crystallization nuclei and the first layers of deposited metal; moreover, the charge transfer kinetics are affected by metal–solvent complexes [46].

Another group of anode materials includes organic polymers with extensive conjugated double bond systems in which the π-electrons are delocalized to such an extent that the material acquires good electron conductivity and some metallic properties, and graphite, which has a layer structure of condensed aromatic rings with good electrical conductivity.

Three kinds of polymer have already been tested as electrode materials: polyacetylene, in which the $(-CH=)_x$ group is the repeat unit (R) of the polymer, polyphenylene $(-C_6H_4-)_x$ and polyphenylquinoline $(C_{15}H_9N)_x$. Since the numerous π-electrons are delocalized, a conduction band arises which can be filled by additional electrons derived from the electroreduction [47]. Negatively-charged anode materials are obtained in this way; their additional charge is balanced out by cations entering between the layers or chains of polymer when the anode is being charged and diffusing back to the electrolyte solution during discharge. When charged, the anode material is a polymeric salt of the type $(R)_x^{n-} \cdot nM^+$, where M^+ is a monovalent cation (usually lithium or the tetra-alkylammonium cation [48]), and n is the number of the charges introduced per repeat unit (monomer) R of the polymer.

Anodes of a rather high degree of reduction have been prepared: the mean number of added negative charges per monomer R is 0.7 for polyphenylquinoline, 0.5 for polyphenylene and 0.167 for polyacetylene. However, this is not much in comparison with, say, lithium; the difference is all the more evident when the number of faradays of electrical charge available from 1 kg of anode material on dividing 1000 (g) by the molecular mass of the monomer R is compared with the value for lithium. This number is $0.7 \times 1000/203 = 3.45\ \mathrm{F\,kg^{-1}}$ for polyphenylquinoline, $0.5 \times 1000/76 = 6.58\ \mathrm{F\,kg^{-1}}$ for polyphenylene, $0.167 \times 1000/13 = 12.85\ \mathrm{F\,kg^{-1}}$ for polyacetylene, but $1 \times 1000/6.94 = 144\ \mathrm{F\,kg^{-1}}$ for lithium. The current efficiency of these polymers is therefore at least one order of magnitude less than that of lithium. These figures would be smaller still if the mass of cations bound up in salt form were taken into consideration. The voltages obtained are 1–2 V less in comparison with the lithium anode. The theoretical specific energies in $\mathrm{W\,h\,kg^{-1}}$ are therefore of the same order as those of modern lead storage batteries. Furthermore, only a small number of charge/discharge cycles are available [49]. A carbon fibre negative electrode with a graphite structure was prepared as anode matrix for rechargeable Li batteries. Lithium is deposited in the matrix during charging and is released during discharge. $8.2\ \mathrm{F\,kg^{-1}}$ and 100% coulombic efficiency were obtained [54].

In view of these facts, the prospects of polymeric materials at the present stage of development appear to be poor compared with those of metallic lithium and its alloys.

Nonetheless, the fact that they can be used in storage batteries is worthy of mention, because positively-charged cathode materials can be obtained from the same polymers when the direction of current is reversed; they can be reduced during discharge. These materials will be discussed further in the next chapter.

Further research into anode materials appears to be concentrating on improving lithium alloys in order to increase the number of charge/discharge cycles, because it is on this that the practical, large scale introduction of non-aqueous storage batteries with high current and energy densities will depend. Complete cells of this kind are described in Chapter 8.

REFERENCES

[1] R. Bauer, *Chemie-Ing.-Techn.*, **44**, 147 (1972).

[2] A. Brenner, *Advances in Electrochemistry and Electrical Engineering* (P. Delahay and Ch. W. Tobias, eds.), Vol. 5, Interscience, New York (1968).

[3] W.R. Fawcett, in: *Proceedings of the Third Symposium on Electrode Processes* (S. Bruckenstein, J.D.E. McIntyre, B. Miller and E. Yeager, eds.), The Electrochemical Society, Princeton, USA (1980).

[4] J.O. Besenhard, *J. Electroanal. Chem.*, **94**, 77 (1978).

[5] E.M.L. Rao, R.W. Francis and H.A. Christopher, *J. Electrochem., Soc.*, **124**, 1490 (1977).

[6] J.E. Battles, F.C. Mrazek and N.C. Otto, Report ANL-80-130, Argonne National Laboratory (1980).

[7] J.O. Besenhard and G. Eichinger, *J. Electroanal. Chem.*, **68**, 1 (1976).

[8] L.E. Ross, S.K. Preto, N.C. Otto, C.C. Sy and M.F. Roche, *Proceedings of the 15th Intersociety Energy Conversion Engineering Conference*, Seattle, USA (1980).

[9] *1980 Annual Report for Review Committee,* Argonne National Laboratory, Chemical Engineering Division, January 1981, p. 37.

[10] L.E. Ross, S.K. Preto, N.C. Otto, C.C. Sy and M.F. Roche, *Gmelins Handbuch der anorganischen Chemie*, 8 Aufl. Syst. No 20, "Lithium, Hauptwerk, u. Ergänzunsband", Gmelin-Institut, Frankfurt a /Main (1960); Th. Moeller, *Inorganic Chemistry*, PWT, Warszawa (1959).

[11] W.R. Fawcett, *Proceedings of the Electrochemical Symposium,* Polish Chemical Society and Polish Academy of Sciences, Jachranka near Warsaw (1981).

[12] M.L. 'Her and J. Courtot-Coupez, *Bull. Soc. Chim., France*, **1972**, 3645.

[13] M. Niro, U. Kisaburo and T. Zenichi, *Bull. Chem. Soc. Japan,* **47**, 813 (1974).

[14] D.P. Boden and L.M. Mukherjee, *Electrochim. Acta*, **18**, 781 (1973).

[15] G.W. Heise and N.C. Cahoon, *The Primary Battery*, Vol. 1, J. Wiley and Sons, New York (1971), p. 357.

[16] E. Constantinescu, *Rev. Roum. Chim.*, **17**, 1819 (1972).

[17] R. Jasiński, in: *Advances in Electrochemistry and Electrochemical Engineering* (P. Delahay and Ch.W. Tobias, eds.), Vol. 8, Interscience, New York (1971).

[18] J. Jorné and Ch.W. Tobias, *J. Electrochem. Soc.*, **121**, 994 (1974).

[19] R. Jasiński, *Collect. Czechoslov. Chem. Commun.*, **36**, 1079 (1971); *ibid.*, **36**, 77 (1971).

[20] R. Jasiński, *High Energy Batteries*, Plenum Press, New York–London (1967).

[21] W.H. Tiedemann and D.N. Bennian, *J. Electrochem. Soc.*, **120**, 1624 (1973).

[22] A.N. Dey, *J. Electrochem. Soc.*, **118**, 1574 (1971).

[23] R.D. Rank and S.B. Brummer, *Electrochim. Acta*, **22**, 75 (1977).

[24] J. Broadhead and F.A. Trumbore, *Power Sources 5, Proceedings of the 9th*

International Symposium, Brighton 1974 (D.H. Collins, ed.), Academic Press, London (1975), Abstract No. 41.

[25] J.O. Besenhard, *J. Electroanal. Chem.*, **94**, 78 (1978).

[26] J.O. Besenhard, *1980 Annual Report for Review Committee*, Argonne National Laboratory, Chemical Engineering Division, January 1981, p. 21.

[27] M.M. Nicholson, *J. Electrochem. Soc.*, **121**, 734 (1974).

[28] A.N. Dey, *J. Electrochem. Soc.*, **114**, 823 (1967).

[29] D. Sam and J.H. Amburs, 146th Meeting of the Electrochemical Society, New York (1974), Extended Abstract No. 22.

[30] K. Fujita, M. Mizumoto, H. Sugimoto, A. Toyama, N. Ebata and S. Matsuda, Jap. Pat. 63,136,476 (08.06.1988).

[31] Y. Toyoguchi, S. Nankai, J. Yamaura, T. Matsui and T. Iijima, *24th Battery Symp. Japan 1983*, pp. 205, 209.

[32] N. Koshiba, H. Hayakawa and K. Nomose, *26th Battery Symp. Japan 1985*, p. 145.

[33] Y. Toyogushi, T. Matsui, J. Yamaura and T. Iijima, *Progr. in Batt. and Solar Cells*, **6**, 58 (1987).

[34] D.L. Barney, M.F. Roche, S.K. Preto, L.E. Ross, N.C. Otto and F.J. Martino, Report ANL-81-14, Argonne Natl. Lab. (1981).

[35] Ch. Walker, W.L. Wade and M. Binder, *J. Electrochem. Soc.*, **135**, 2471 (1988).

[36] C.W. Walker Jr., *J. Power Sources*, **25**, 13 (1989).

[37] S. Sarkar and I.N. Basumallick, *J. Electrochem. Soc., India*, **37**, 63 (1988).

[38] K. West, B. Zachau-Christiansen, T. Jacobson, E. Hiorst-Lorenzen and S. Skaarup, *Br. Polym. J.*, **20**, 243 (1988).

[39] S.D. Jones and G.E. Hlongren, US Pat. 4, 764, 440 (Eveready Battery, appl. 05 May 1987).

[40] T. Kitano and I. Hirasawa, Jap. Pat. 63,207,051 (26.03.1988).

[41] I.A. Kedrinsky, I.V. Murygin, V.E. Dmitrenko, O.E. Abolin, G.I. Sukhova and I.I. Grudyanova, *J. Power Sources*, **22**, 99 (1988).

[42] A.K. Vijh, *J. Power Sources*, **24**, 243 (1988).

[43] K. Wiesener, W. Glaesner and R. Peltz, *Power Sources 5, Proceedings of the 5th International Symposium, Brighton 1974*, Academic Press, London (1975), Abstract No. 26.

[44] B. Kratochvil and K.R. Betty, *J. Electrochem. Soc.*, **121**, 851 (1974).

[45] T. Sotomura, T. Kambaru and S. Sekido, *Progr. Batteries Solar Cells*, **6**, 25 (1987).

[46] G. Hills and L. Peter, *Electrochemical Power Sources,* 29th Meeting of the International Society of Electrochemistry, Varna (1977), Extended Abstract No. 136.

[47] L.W. Shacklette, M. Manfield, S. Gould, J.F. Wolf, T.R. Jew and R.H. Baughman, *Proc. Int. Conf. on Sci. and Techn.* (ICMS 86), Kyoto, Japan, 1–6 June 1986, Part II, p. 611 (Synth. Metals, Vol. 18).

[48] Y. Kobayashi, T. Shishikura, H. Nakamura and H. Kenuba, *Synth. Metals*, **18**, 619 (1987).

[49] P. Passiniemi and J.-E. Österholm, *Synth. Metals*, **18**, 637 (1987).

[50] Ch. Ishibashi, K. Moriwaki and S. Furukawa, Jap. Pat. 91-37, 964 (from *Chem. Abstr.*, 115, 163079).

[51] G.M. Raynaud, G. Regazzoni, G. Nussbaum and M. Reboul, Eur. Pat. Appl. EP 426, 579 (08 May 1991).

[52] B.J. Seddon, M.J. Eddowes, A. Firth, A.E. Owen and H.H.J. Girault, *Electrochim. Acta*, **36**, 763 (1991).

[53] C. Ishibashi, K. Moriwaki and S. Furukawa, Jap. Pat. 91-17, 961.

[54] M. Morita, N. Nishimura, H. Isutsumi and Y. Matsuda, *Chem. Express*, **6**, 619 (1991).

CHAPTER 6

Cathode Materials

Several kinds of substances can be used as cathode materials, from pure metals and semiconductors to ionic and covalent compounds, all of which can exist in the solid, liquid or gaseous state when pure. Materials with semi-conducting properties are used most often as they conduct current electronically. Ionic and covalent compounds with insulating properties, e.g. CuF_2, nitrobenzene, quinone, Ag_2CrO_4, ought to be even slightly soluble, so that their ions or molecules can react at the interface at which electrons are supplied from the anode via an electrical conductor and there transferred to these particles during reduction. The transfer of electrons across this interface is the essence of the electrode reaction. The cathode material must lie within reach of the double layer of the interface to make electron transfer possible. If both the cathode material and the reaction products are insulators, then the electrical conductor very soon becomes isolated from the rest of the cathode material and the cathode reaction is impeded. Such a situation is prevented by the material dissolving in the electrolyte solution; as a result, the electroactive material's molecules diffuse to the surface where the cathode reaction occurs. The higher the cell's current load is to be, the more soluble the cathode material must be in order to ensure an adequate supply of material to the interface. However, the solubility of the cathode material may cause the cell to discharge if the electroactive molecules diffuse to the anode. Passivation at the anode prevents self-discharge, as does the use of suitable membranes separating the cathode and anode spaces. On the other hand, semiconductors, which conduct electronically, are not and should not be soluble in the solution since electron conductivity facilitates electron migration into the bulk of the material. The ions, atoms or molecules present in the crystal lattice may be reduced at the surface. Intercalation compounds, which will be discussed later, in which the components being reduced are between the layers of the electronically conducting material also belong to this class of cathode materials. Thus, electrical conductivity is assured throughout the mass of cathode material and its solubility in the solution is superfluous and may even be harmful.

6.1 SULPHIDE CATHODE MATERIALS

These semiconducting compounds, often non-stoichiometric, were among the earliest to be used in non-aqueous cells [1–7], and they are still being improved. This class of compound is continually being extended by development of new types of material such as intercalation compounds.

The specific conductivity of sulphides which are semiconductors is high enough for them to be applied as cathodes without additives, e.g. graphite. In the literature, this type of compound is usually treated as stoichiometric, but one should be aware of the fact that the overall stoichiometric formula may often cover a whole range of non-stoichiometric compounds of approximately similar composition, especially in natural minerals. For instance, the sulphur–copper system includes not only Cu_2S and CuS, but also Cu_9S_5 and Cu_7S_5. Synthetic products used as cathode materials are generally stoichiometric, in accordance with Dalton's law (daltonides).

If one starts from the simple Lewis–Kossel rules, the atoms in stable compounds ought to have the noble gas electronic configuration. Atoms can usually attain such a configuration by forming an ionic or a covalent bond. When an ionic bond is formed, the metallic element, whose valence electrons are weakly bound, transfers these to the non-metallic element, i.e. to the element which requires one or more electrons to complete its noble gas shell.

Consider iron sulphide – FeS. Its formula written as $Fe^{2+}S^{2-}$ means that two of the outermost *s* electrons pass from the iron atom (configuration $3d^6 4s^2$) to the sulphur atom ($3s^2 3p^4$) to yield the argon configuration (S^{2-} – $3s^2 3p^6$). But a problem arises here: a cation is always stable, whereas an anion formed in this way is not quantum-mechanically stable. So the anion must be stabilized by the ions of the crystal lattice, and this is exactly what happens in this particular case. On the other hand, the electrons of the iron are affected by the negative potential of the anions so formed. Ligand-field theory can therefore be applied: the degeneracy of the *d* electrons is partly or totally removed and the 10 *Dq* parameter plays a decisive role in the formation of the octahedral structure around the central iron ion. In this case, the interatomic distances are determined by the crystallographic ionic radii.

In covalent bond formation, an electron is also transferred from atom 1 to atom 2, but an electron is simultaneously transferred from atom 2 to atom 1, and the two electrons yield an electron pair of opposite or compensated spins. Negative charge accumulate between the atoms which in principle remain electrically neutral. Ligand-field theory is inapplicable to the description of interatomic distances in such a case. They are much shorter than the sum of the ionic radii: Fe^{2+} – 76 pm; S^{2-} – 184 pm; S^0 – 104 pm; Fe^0 – 116 pm. Hence the Fe^0–S^0 distance is 220 pm, whereas the $Fe^{2+}S^{2-}$ distance, equal to the sum of the ionic radii, is 260 pm. One could write $Fe^{II}S^{II}$ as Fe = S, where the dashes indicate the double bond formed by 4 electrons, two from the iron atom and two from the sulphur atom. As a result, the iron atom has a full configuration of electrons with compensated spins. Figure 6.1 shows the two extreme electronic configurations, one for the purely ionic bond (Fig. 6.1a), the other for the covalent bond (Fig. 6.1b).

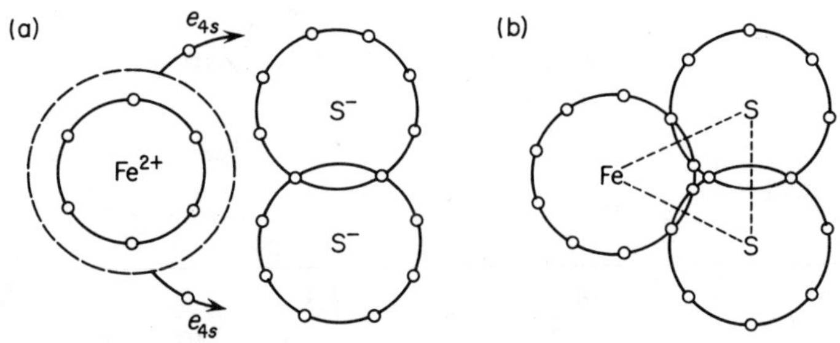

Fig. 6.1 Valence electron configuration of (a) the ionic $Fe^{2+}(S_2)^{2-}$ structure, (b) the traditional covalent structure of FeS_2.

Notice that in both structures the electrons participating in the bond are localized. Actually, the electronic structure is intermediate between the two extremes. Additionally, interactions between atoms of the same element, i.e. Fe–Fe and S–S, must be included. In nature, FeS occurs in three crystalline forms – hexagonal, cubic and tetragonal [1]. Structural studies have shown that the Fe–S distance is the shortest (223 pm) in the tetragonal form, which corresponds to a covalent FeS structure. In the hexagonal form the distance is 249 pm, while that calculated for the purely ionic structure is 260 pm. The slight difference of 11 pm indicates that ionic bond character prevails in hexagonal FeS. In the cubic, blende-type FeS structure, the distance is 234 pm, intermediate between the distance calculated for purely ionic and purely covalent bonds. This implies an intermediate kind of interatomic interaction in which the *d* orbitals of the metal overlap the *p* orbitals of the anion yielding antibonding orbitals.

The tetragonal and cubic (blende-type) forms are metastable and are converted into the hexagonal form on heating. The thermodynamic stability of the latter is due to the increase in the coordination number to 6; in the cubic and tetragonal forms the coordination number is 4. Thus, the most stable form of FeS is the most polar one, corresponding to the classical description of this compound as an ionic salt.

The formulae of formally stoichiometric sulphides are set out in Table 6.1. However, the specific resistances given alongside cover a wide range of numerical values which can be attributed at least in part to the occurrence of non-stoichiometric structures in these substances. The wide range

Table 6.1 Resistances of metal sulphides [1]

Metal sulphide	Resistance, Ω m
CuS	$(0.3–83) \times 10^{-6}$
$CuFeS_2$	$(150–9000) \times 10^{-6}$
Cu_5FeS_4	$(1.6–6000) \times 10^{-6}$
$Fe^{II}S_2^{-I}$	$(1.2-600) \times 10^{-3}$
Fe_7S_8	$(2–160) \times 10^{-6}$
$(Fe, Ni)_9S_8$	$(1–11) \times 10^{-6}$
NiS	$(2–4) \times 10^{-7}$
MnS_2	10–20
Ag_2S	$(1.5–2) \times 10^{-3}$
Bi_2S_3	3–570
PbS	6.8×10^{-6}
ZnS	from 2.7×10^{-3} to 1.2×10^4

of resistances of each sulphide is due to small differences in composition and to inhomogeneities in the semiconducting solid phases. Most copper, iron, nickel, silver and lead sulphides have low resistances, comparable with those of metals (ca. 10^{-6}–10^{-7} Ω m), and can be used as solid electrodes through which currents of the order of 100 mA cm^{-2} can flow. There is an enormous difference between the conductivities of the sulphides just mentioned and those of a second group of sulphides, of manganese, bismuth and cadmium, whose conductivities are close to those of germanium (ca. 10^{-3} $\Omega^{-1} cm^{-1}$). Graphite should be added to such materials to improve their conductivity. These examples of conductivity in sulphides and the differences between them refer to sulphide minerals.

The most general method of synthesizing sulphides is heating the two elements together. Special care must be taken to remove all the oxygen from the reaction vessel, otherwise oxidation products are formed and alter the properties of the sulphide material. By using copper sulphides and mixed iron and copper sulphides (chalcopyrite), it has been demonstrated that sulphides can undergo both reduction and oxidation; these processes are reversible or quasi-reversible [1–3]. The cathodic reduction of CuS proceeds in several steps via intermediate non-stoichiometric sulphides, i.e. $Cu_{1.1}S$, $Cu_{1.75}S$, $Cu_{1.8}S$ (digemite) and $Cu_{1.97}S$, until Cu_2S is formed. When oxidized, Cu_2S is converted back into CuS in the reverse order via the same intermediate sulphides. In principle, the electrolytic reduction of $CuFeS_2$ proceeds in the same way, although there are other reaction products besides Cu_2S, depending on the reaction conditions. The overall redox reactions for copper sulphides are:

(1) $10CuS + 2e \rightarrow Cu_{10}S_9 + S^{-2}$ or $9Cu_{1.1}S + S^{2-}$

$Cu_{10}S_9 = Cu_8^{II}Cu_2^{I}S_9^{-II}$

(2) $7CuS + 6e \rightarrow Cu_7S_4 + 3S^{2-}$ or $4Cu_{1.75}S + 3S^{2-}$

$Cu_7S_4 = Cu^{II}Cu_6^{I}S_4^{-II}$

(3) $9CuS + 8e \rightarrow Cu_9S_5 + 4S^{2-}$ or $5Cu_{1.8}S + 4S^{2-}$

(4) $2CuS + 2e \rightarrow Cu_2S + S^{2-}$

Further reduction of Cu_2S may yield metallic copper and sulphide ions directly:

$$Cu_2S + 2e \rightarrow 2Cu^0 + S^{2-}$$

If CuS were used as a storage cell cathode material, the process of charging, i.e. oxidation, would proceed with the formation of intermediate products in the reverse sequence. If the metal sulphide (e.g. Li_2S) is poorly soluble in the solution, as is the case with lithium perchlorate solutions in propylene carbonate, 1,2-dimethoxyethane (DME) [10] or their mixtures, the equilibrium concentration of sulphide ion in the solution is slight, despite all the sulphur having passed into solution as free sulphide ions during the reduction of CuS and Cu_2S to copper metal. That is why charging is strongly inhibited: the reaction sequence from Cu_2S to CuS is impossible in the absence of available S^{2-} ions. Under such conditions, a sulphide cathode operates irreversibly, like a primary cell, because of the very low solubility of Li_2S.

Recharging a sulphide electrode requires the solubility of the sulphide in the electrolyte to be increased, a fact to which Eichinger and Fritz [4] drew attention. There must either be a sufficiently large supply of sulphide ions in the solution, or the S^{2-} ions lost due to their reprecipitation as copper sulphides must be replaced fairly quickly by the dissolution of a sulphide present in excess (Na_2S or Li_2S). This increase in solubility can be achieved by adding a Lewis base to bind the alkali metal cations which act as Lewis acids here, i.e. as acceptors of electron pairs consisting of *n*-type electrons. To this end, Eichinger and Fritz [4] used crown polyethers [12] which turned out to be ligands strongly solvating Li^+ ions, and thereby increasing the solubility of Li_2S; in fact, a new lithium complex, soluble in non-aqueous solvents like DME, was formed. Thus the desired solubility increase and a higher sulphide ion concentration in the solution is attained. In other words, adding a crown ligand shifts the equilibrium in the direction of the ionized components

$$Li_2S \rightleftharpoons 2Li^+_{solv} + S^{2-}, \qquad K = \frac{[Li^+_{solv}]^2\,[S^{2-}]}{[Li_2S]}$$

Besides copper sulphides, which have been among the most thoroughly studied, many other simple and complex sulphides have been used as cathode materials, e.g. iron, manganese, nickel, busmuth, silver and lead sulphides (see Table 6.1). The oxidation state of sulphur in sulphides can vary. It has been shown using Mössbauer spectroscopy that iron is present in pyrite as Fe(II), hence sulphur occurs as the disulphide ion S_2^{2-}, whereas it occurs as the sulphide ion S^{2-} in chalcopyrite [1] because iron is again present as Fe(II).

Cadmium sulphide is probably the best-known compound in this group of sulphides (Fig. 6.2). It is an *n*-type semiconductor with an energy gap of 2.4 V between the principal valence band and the conduction band. It passes a negligibly low current in the dark, but up to 3 mA cm^{-2} on being illuminated, although the saturation current is attained already at 1 V. The filled principal valence band contains the anions, the unfilled conduction band the cations. At the instant a cadmium ion passes into solution, it leaves behind electrons bound to sulphur ions in the valence band. When illuminated, these electrons gain sufficient energy to pass into the excited state, i.e. into the conduction band.

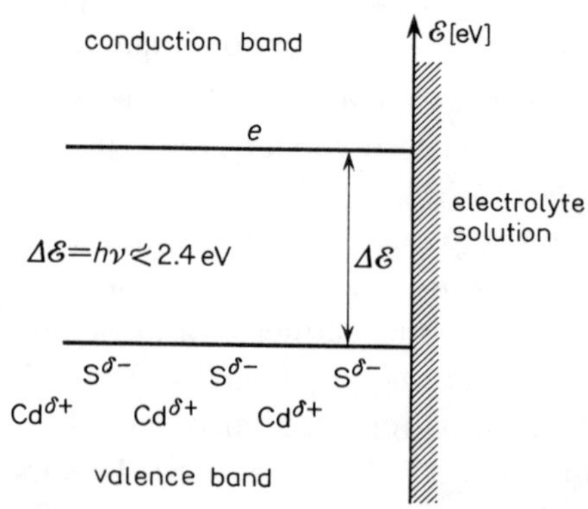

Fig. 6.2 Energy band diagram in an *n*-type intrinsic semiconductor (CdS).

In other sulphides, where the energy gap is narrower, the thermal energy at room temperature is enough to excite electrons from the principal valence band to the conduction band; as a result, the pure material acquires intrinsic-semiconductor properties and neither the effect of light, nor limiting current values with increasing polarization (applied voltage) are observed. This is the case with FeS_2: $\Delta\mathcal{E} = 1\,eV$, $CuFeS_2$: $\Delta\mathcal{E} = 0.53\,eV$ and PbS: $\Delta\mathcal{E} = 0.37\,eV$. If the energy gap is much wider, the material becomes an insulator; if it becomes extremely narrow or overlaps the valence band, it displays the properties of a metal – it becomes a metallic substance. Hence the important condition that the energy gap in cathode materials should not be too wide, less than ca. 1.5 eV, which corresponds to the long-wave end of the visible light spectrum. The energy gap $\Delta\mathcal{E}$ can be reduced by adding impurities or by using non-stoichiometric materials.

6.2 NON-STOICHIOMETRIC CHALCOGENIDE CATHODE MATERIALS

Non-stoichiometric compounds can be formed in many ways. Non-integral stoichiometric coefficients in, for example, sulphur compounds, describe the overall composition of these cathode materials and indicate that either the element in it is present in different oxidation states, or atoms of one element have been isomorphically substituted by atoms of another of similar size and charge. Last but not least, atoms of some other element can enter the interlattice space of a compound, which may lead to a change in the oxidation state of a cathode material component or the formation of compounds or intermetallic phases of the kind discussed in the chapter on anode materials (alloys and intermetallic compounds of lithium).

Copper sulphides belong to the first group of non-stoichiometric compounds because the composition of the sulphur–copper system is variable over a broad range; many of these compounds are non-stoichiometric – a few examples are presented in Section 6.1. For instance, the compound of formula $Cu_{1.1}S$ or $Cu_{10}S_9$ can be regarded as copper sulphide CuS in which 1/5 of the copper atoms have been reduced to monovalent copper: $Cu_8^{II}Cu_2^{I}S_9^{-II}$. As a result, the overall number of positive charges at the lattice nodes has decreased by 10%, and a corresponding number of sulphide ions have been released. The latter diffuse away into the solution and can produce sulphides, e.g. with alkali metal ions, if any are present in the solution. The fact that Cu_2S is the reduction product proves that the reduction of Cu(II) to the metal is a two-step rather than a one-step process. Likewise, the compound $Cu_{1.75}S = Cu_7S_4$ can be considered as an isomorphous system in which only one copper atom out of seven is in oxidation state (II) and the other six in oxidation state (I), i.e. $Cu_7S_4 = Cu_1^{II}Cu_6^{I}S_4^{-II}$.

A different case is the reduction of pyrites FeS_2, which contains Fe(II) and disulphide ions. The latter can be reduced: $S_2^{2-} + 2e \rightarrow 2S^{2-}$. Of the two sulphide ions thus formed, one can remain combined with iron as FeS (also a non-stoichiometric compound), while the other can dissolve or react at once with the lithium ions to form a more or less sparingly soluble deposit, depending on the solvent used (sulphur atoms in disulphide ions are covalently bound): $Fe^{II}S_2^{-I} + 2e \rightarrow Fe^{II}S^{-II} + S^{2-}$ (cf. Fig. 6.1).

Isomorphousally substituted sulphides, sometimes called mixed sulphides, are structurally similar. Pentlandite $(Fe,Ni)_9S_8$ is an example: here, iron atoms are isomorphousally replaced by nickel atoms, of similar radius and charge, and the crystal lattice is not distorted:

$$Fe_9S_8 \xrightarrow{\text{nickel substituted}} (Fe,Ni)_9S_8$$

This kind of compound is stable provided the lattice energy is high enough; the stoichiometric composition is of secondary importance as long as the compound is electrically neutral. Thus it is hard to establish how many Fe(II) and Fe(III) ions there are in pentlandite $(Fe,Ni)_9S_8$ without resorting to Mössbauer spectroscopy. As the Ni^{3+} ions are rather unstable, it is assumed that it is the oxidation state of iron that mainly varies (atomic radius Fe – 116 pm, Ni – 115 pm), although Ni^{+} ions may also be present.

The oxidation states of the elements in non-stoichiometric compounds are being investigated and opinions on this particular subject are changing. The widely studied Fe–S system is typical in this respect.

It is known that there is no such thing as a stable iron(III) sulphide, Fe_2S_3, which implies that not all Fe atoms in a crystal structure having approximately this composition appear as Fe(III). Instead, some of the sulphur atoms attain the S^- configuration, which can also be explained as an electron gap in the conduction band. If the concentration of S^- is high enough, S_2^{2-} ions are formed, as in pyrites. Thus, the compound of composition Fe_2S_3 may in part correspond to the double sulphide $Fe^{II}S \cdot Fe^{II}S_2$ in which iron is bivalent. The electronic structure of selenides such as Fe_3Se_4 may formally correspond to the formula $Fe^{III}Fe_2^{II}Se_3^{-II}Se_1^{-I}$ so that the stoichiometric valency of -8 for the total of 4 Se^{2-} anions and $+8$ for the cations in $Fe_2^{III}Fe^{II}Se_4^{-II}$ is not attained [13]; this is most simply explained by the existence of covalent bonds in the compound's structure. For chalcopyrite, partial charges of $q = 1.3$ residing on equivalent sulphur atoms, have been found, as well as $q = 1.6$ for the iron atom, which leaves $q = 1$ for the copper atom [8].

The marked difference in the electronegativities of the elements in the cations and anions, as well as the large size of cations in lower oxidation states (which increases the electron-donor character of cations) stabilizes the higher valencies of transition metal ions like Fe(III). These factors partly stabilize Fe(III) in chalcopyrite, but the interaction with sulphide ions gives rise to formal partial charges of $q < +2$ for Fe and $q > -2$ for S. This reduction in the formal oxidation state is due to the covalent bonds in these materials, which are not purely ionic salts. In cubanite, $CuFe_2S_3$, a semiconductor, both iron atoms are equivalent, but they occur as Fe^{V+} pairs with one electron shared, and the formal oxidation state is thus fractional at + 2.5. Copper is monovalent in this compound.

6.3 INTERCALATION COMPOUNDS

Cathode materials belonging to a third group of non-stoichiometric compounds, intercalation compounds, have aroused the greatest interest at present, as these are systems of the highest reversibility, particularly suited for non-aqueous storage cells.

They are usually MY_2-type Group IVA, VA and VIA transition-metal dichalcogenides, where M is a metal and Y is a chalcogen [14]. The following disulphides and diselenides have been studied so far:

TiS_2	$TiSe_2$		VSe_2		
ZrS_2	$ZrSe_2$	NbS_2	$NbSe_2$	MoS_2	$MoSe_2$
HfS_2	$HfSe_2$	TaS_2	$TaSe_2$	WS_2	WSe_2

They all have a characteristically layered "sandwich" structure, in which the two pieces of bread are layers of chalcogen atoms and the meat a layer of metal atoms. This elementary unit – the sandwich – is held together by chemical bonds. In a crystal, the sandwich units are separate from one another, the only bonds between them being weak van der Waals interactions; hence, the gaps between the units are called van der Waals gaps [14] (Fig. 6.3).

Within the sandwich units there are a number of substructures, characteristic of particular compounds whose electronic structures are already quite well known [15–22]. The chalcogenides of group IVA and VIA metals are semiconductors, whereas those of the group VA metals (V, Nb, Ta) are metallic [23]; the differences are satisfactorily explained by valence bond theory [15].

The layered structure of such sulphides makes possible the formation of compounds, or rather of specific structures, called intercalation compounds. The process consists of a reversible penetration of ions (in some cases of molecules or atoms) into the host crystalline lattice, causing a structural rearrangement as the number of guest molecules or atoms in-

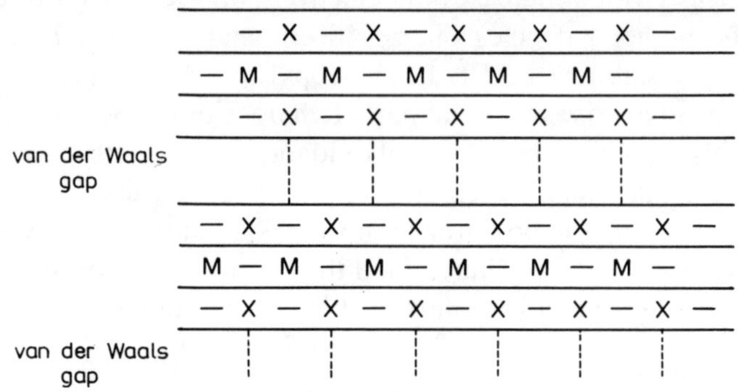

Fig. 6.3 Arrangement of structural units in transition metal dichalcogenides with a sandwich structure; M – metal atoms, X – chalcogen atoms (S,Se,Te).

creases. This kind of structure is quite widespread in nature: zeolites are well-known minerals which can have a layered or channelled structure enabling water molecules to be taken up. Synthetic zeolites, known as molecular sieves, are widely applied to absorb water and to dry gases or organic solvents. Others selectively absorb organic molecules larger than those of water. The chemistry of the Group IVA metals – Ti, Zr, Hf – is well known. The highest oxidation state is the common one, and metals in this state have weakly oxidizing properties. In lower oxidation states, e.g. titanium (III), they are quite strong reducing agents.

The best-known intercalation compound is Li_xTiS_2, in which $0 \le x \le 1$, although Li_2TiS_2 probably also exists. $Li_{1.0}TiS_2$ ($x = 1$) has an accurately determined crystal structure. If the lithium-titanium ratio is different, the compound has a highly heterogeneous structure. The sandwich units move apart along an axis perpendicular to them, i.e. along the crystallographic c axis, as lithium ions are gradually introduced, while the other unit cell parameters change insignificantly. This means that only the van der Waals gap widens (see Fig. 6.4) whereas the other interatomic distances remain practically unaltered (Table 6.2).

The ionic nature of lithium intercalated into the van der Waals gap of titanium disulphide is evidenced by a small chemical shift in NMR measurements (3–12 ppm). The shift displayed by lithium atoms is much greater – ca. 240 ppm [24]. The intercalation process is illustrated diagrammatically in Fig. 6.4 for titanium disulphide.

The process of filling of vacant spaces in the van der Waals gap between the disulphide layers can be described in terms of a two-dimensional model like that used to describe chemisorption at interfaces. In this particular case, "chemisorption" refers to the monolayer pressed in between the two sulphide surfaces. As in chemisorption, the activity of the incorporated atoms may be altered as a result of changes in the configurational

Table 6.2 Increase in *c*-lattice axis in transition metal disulphides and diselenides (in Å) [14]

Compound	*c* axis	*c* axis for $Li_{1.0}MY_2$	Δc^*
TiS_2	1×5.696	1×6.195	0.50
ZrS_2	1×5.835	3×6.25	0.42
HfS_2	1×5.856	3×6.375	0.52
NbS_2	3×6.00	3×6.421	0.42
TaS_2	2×6.04	2×6.475	0.44
MoS_2	2×6.15	2×6.40	0.25
WS_2	2×6.18	–	–
$TiSe_2$	1×6.004	1×6.480	0.48
$ZrSe_2$	1×6.160	1×6.648	0.49
$HfSe_2$	1×6.160	1×6.642	0.48
VSe_2	1×6.10	1×6.356	0.26
$NbSe_2$	2×6.27	2×6.772	0.50
$TaSe_2$	2×6.348	2×6.817	0.47
$MoSe_2$	2×6.50	–	–
WSe_2	2×6.50	–	–

* Δc – the difference between the value of the *c*-lattice axis in the pure compound and after intercalation with lithium.

entropy of the atoms filling the van der Waals gap and of changes in the partial molar energy due to bond formation between the sulphide layers when positively-charged lithium ions appear between them. The partial molar energy includes a term, depending on the stoichiometric composition, resulting from a change in the chemical potential of a Fermi-level electron which causes the electron density to increase with the intercalated ion concentration. This electron density increase means that the titanium ions at the lattice nodes are partly reduced, usually from oxidation state IV to III, which is electrically compensated by the positive charge of the lithium ions entering the gap between the TiS_2 layers.

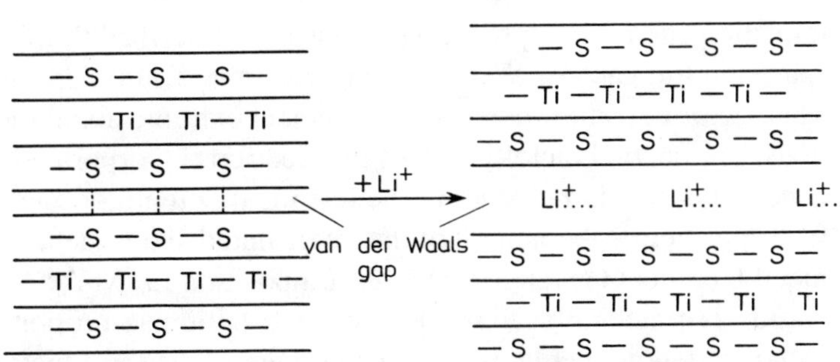

Fig. 6.4 Intercalation of lithium ions into the van der Waals gap between sulphur layers in TiS_2.

Another energy term in the chemical potential of the intercalated ions includes the energy required to push apart or bring together the chalcogenide layers in accordance with the number of atoms entering or leaving the van der Waals gap. The composition-dependent changes in the chemical potential of lithium and other alkali metals introduced into the disulphides mentioned in Table 6.2 have been measured electrochemically. These changes are large and negative, $\Delta G < -96$ kJ mol^{-1}, and are usually a linear function of the chemical composition. The magnitude and changes in chemical potentials of these alkali metals are important parameters in the assessment of the possibilities of using intercalated dichalcogenides as cathode materials in non-aqueous cells.

Nagelberg and Worrel [25] have demonstrated that there are two thermodynamic quantities determining the changes in the chemical potential of alkali metals as a function of their content in sulphide intercalation compounds, namely, the energy of interaction between the alkali metal atoms (cation repulsion) and the changes in the chemical potential of electrons as the alkali metal concentration changes. The chemical potential of electrons in semiconducting or metallic compounds is directly related to the Fermi level (see Chapter 3) and should therefore be treated as the electrochemical potential of electrons in such materials.

Two-phase systems exist within the low alkali-metal concentration range, i.e. from MY_2 to ca. $L_{0.1}MY_2$, where L is the alkali metal, M the transition metal and Y the chalcogen. At a site where an alkali metal ion is incorporated, the sandwiches are pushed apart along the crystallographic c axis; elsewhere, the layers are still in contact and constitute a separate phase. When the L content is greater than about 20%, i.e. in L_xMY_2, where $x > 0.2$, the original MY_2 disulphide phase disappears and only the intercalated phase L_xMY_2 is left, in which the dimensions of the new van der Waals gap now remain almost constant, even though the alkali metal concentration continues to increase, filling the gap.

In the final analysis, one must take into account four factors affecting the chemical potential of an alkali metal ion intercalated into a layered transition metal disulphide. The first one describes the filling of vacant spaces in the van der Waals gap between the disulphide sandwich layers. This factor becomes the most important at extreme alkali metal concentrations. The second factor involves the electrons incorporated in the sulphide conduction band: a change in the chemical potential of an electron μ_e affects the chemical potential of the intercalated alkali metal μ_L over a broad metal L content (x) range up to the composition $L_{0.8}MY_2$. As the derivative $d\mu_e/dx$ is practically constant, the contribution is proportional to the L metal content in the intercalation compound. The third factor is the mutual repulsion between the intercalated atoms, which are considered to be totally ionized when present in the van der Waals gap [24]; this is the interaction energy between the alkali metal atoms. The last factor is the energy needed to push apart the MY_2 sandwich layers, i.e. the energy required to expand the dichalcogenide lattice as the *c*-lattice parameter increases. The magnitude of this factor can be estimated only indirectly from changes in the disulphide's compressibility.

Interaction energies and the derivative $d\mu_e/dx$ cannot be determined independently using only the chemical potential equation, but it is possible to estimate them indirectly [25]. The chemical potential itself, however, can be measured electrochemically. The plot in Fig. 6.5 shows that, for metal L contents from $x = 0.1$ to $x = $ ca. 0.8, intercalating an alkali metal makes the chemical potential μ_e change quasi-linearly. This indicates that the formation of such intercalation compounds is reversible, so they can be used as cathode materials in storage batteries.

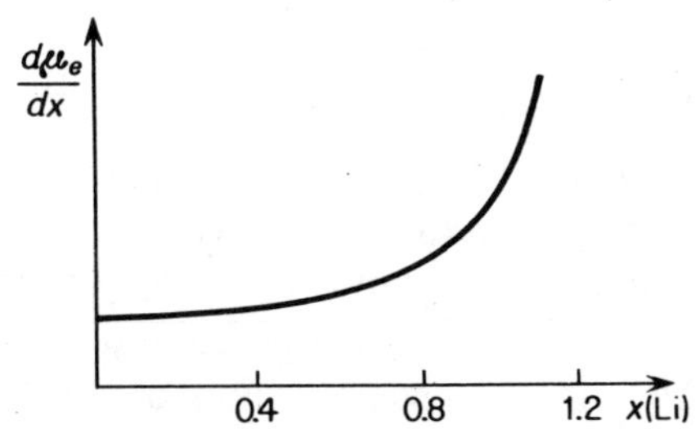

Fig. 6.5 The calculated derivative of the chemical potential of an electron vs. alkali metal content in transition metal disulphides.

Li_xTiS_2 [26] is an extremely interesting compound: it has been demonstrated experimentally that the chemical potential of lithium increases almost linearly with its content in this intercalation compound. This means that as the lithium content in Li_xTiS increases, the e.m.f. of a battery using this cathode material varies linearly (Fig. 6.6) from ca. 2.5 V for the unconsumed cathode material (TiS_2) to ca. 1.9 V for $Li_{1.0}TiS_2$ [26]. The discharge reactions in this cell are

$$Ti^{IV}S_2 + e + Li^+_{solv} \rightarrow LiTi^{III}S_2 \quad \text{(cathode reaction)}$$

$$Li^0 - e + \text{solvent} \rightarrow Li^+_{solv} \quad \text{(anode reaction)}$$

Over such a concentration range, this intercalation compound is suitable as a cathode material in multiple-use reversible cells, i.e. for storage cells. The e.m.f. temperature coefficient is largely independent of the cathode material composition, which is beneficial in practical applications.

The mobile cation intercalated between the sandwich layers of intercalation compounds may also be Cu^+, as for example, in the cathode

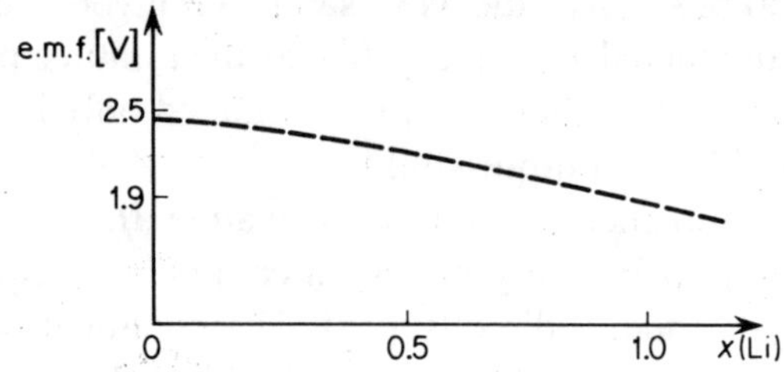

Fig. 6.6 E.m.f. of a Li/TiS_2 cell vs. lithium content (x) in the intercalation cathode material Li_xTiS_2.

material of general formula Cu_xTiS_2. The electrode reaction is

$$TiS_2 + xCu^+ + xe^- \text{ (discharge)} \rightarrow Cu_xTiS_2$$

It is reversible when $0 \leq x \leq 0.17$ [27].

From this point of view, the properties of most other sulphides are somewhat inferior, e.g. the reduction in e.m.f. is much greater in tantalum sulphide. With TaS_2, the e.m.f. is 2.9 V, but only ca. 1.5 V in a reversibly discharged cell with $Li_{1.0}TaS_2$ as cathode material. Even so, the e.m.f. varies linearly with increasing lithium content in both cases. Li_xMoS_2 [28] is another promising material, together with Li_xNbS_2 [75]. Electronic structure in relation to the reactivity of non-stoichiometric layered compounds was studied by Molenda [76], and interaction of Li^+ by Chevrel phase sulphides by Wakinara and Uchida [77].

6.4 NON-STOICHIOMETRIC CARBON COMPOUNDS

Graphite-based electrode materials have been known for a good number of years now; graphite is a good conductor of electricity and provides a structural framework for the electroactive substances bound at its surface or within its lattice [29]. The first such compound to be used in cells was an insoluble, polymeric fluorographite of empirical formula $(CF_x)_n$ where n is a large number and x is less than unity. There are compounds in which x is greater than 1, e.g. $(CF_{1.5})_n$, but they are insulators and do not display the electronic conductivity characteristic of graphite. For instance, the polymeric compound of the empirical formula $(CF_{1.0})_n$ is a white, non-conducting substance in which every carbon atom is covalently bound to a fluorine atom and to three adjacent carbon atoms by three sp^3-hybridized bonds to form a network as shown in Fig. 6.7. As the carbon atoms are arranged tetrahedrally, the structure assumes a corrugated layer form.

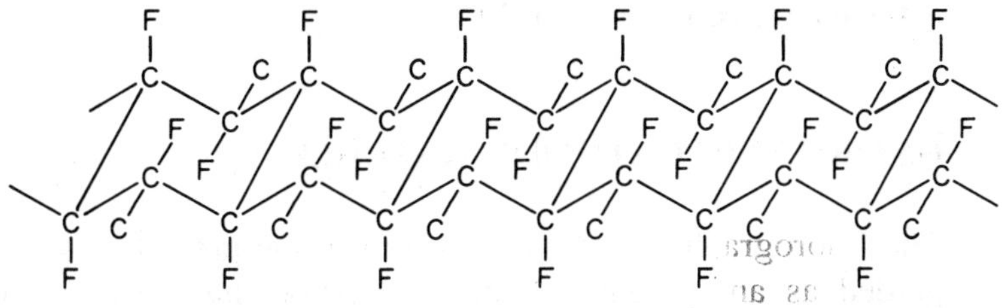

Fig. 6.7 The structure of polymeric fluorographite with the empirical formula $(CF_{1.0})_n$. Spatial projection.

Fluorographites containing less than one fluorine atom per carbon atom are conductors as they contain non-bound electrons, which are responsible for the electrical conductivity of graphite. Higher fluorine contents ($x \geq 1$) yield insulating cathode materials which must be mixed with additives like pure graphite if they are to be conductors. During discharge, electrical contact in these materials deteriorates, so that the materials best suited to cells are those with a stoichiometric coefficient x from 0.8 to 0.99 and a molecular formula from $(CF_{0.80})_n$ to $(CF_{0.99})_n$ [30] or C_2F [31].

Cathode materials like these are suitable for cells but not for storage batteries because the electrochemical reactions in which they participate are irreversible:

$$(CF_x)_n + ye \rightarrow c_n + yF^- \qquad \text{where } y = nx$$

The fluoride ions pass into the solution where they meet cations from the supporting electrolyte solution, usually lithium ions, with which they form lithium fluoride. As this is insoluble in non-aqueous solvents, it precipitates in the cathode material pores and breaks up the latter's structure, yielding a mixture containing very fine particles of graphitic carbon. Reconstituting the polymeric structure of $(CF_x)_n$ by reversing the current direction, i.e. by charging the cell, is no longer possible; such cathode materials can be used only in primary systems.

Thus, it appears that substances in which $x \geq 1$ are not intercalation or interstitial compounds, although they are often referred to as such. In genuine intercalation compounds, the host sandwich layer structure is not destroyed but expands or shrinks reversibly. Consequently, they can be used in storage batteries.

The laminar structure of graphite also makes it possible to expand considerably and change into an activated charcoal surface, a good electrical conductor. Applying an electrical potential to graphite saturated with electrolyte solution causes large numbers of ions of the opposite sign to be absorbed, which is equivalent to obtaining an electroactive material acting as a condenser. The two sides of the double layer at which charges have accumulated are the condenser plates. This is how a positive electrode of activated charcoal works [32].

6.5 ORGANIC CATHODE MATERIALS

The fluorographite ($x < 1$) discussed in the preceding section can be regarded as an inorganic material, despite the covalently-bound carbon atoms it contains. There are many organic compounds suitable as cathode materials, including aliphatic and aromatic compounds with nitro $-NO_2$, nitroso –N=O, azo –N=N– and azoxy –N=N–O groups, quinones, thioquinones and halogen derivatives substituted at nitrogen (N–Cl, Br) or oxygen [33–37]. In non-aqueous cells, oligomers such as nonabenzhexaquinone [38] can be used instead of monomeric quinones.

Many such organic substances have been investigated with regard to their usefulness in cells. Although it is theoretically possible to transfer large numbers of electrons onto these compounds during their electrochemical reduction – a potential source of considerable specific energies – such reactions are difficult to apply in practice. One of the most serious problems is the necessity of including a proton donor in order to supply the hydrogen ions required to balance the electrochemical reduction in all the cases where double bonds are not formed. Although hydrogen ions participate in the electrolytic reduction of all the functional groups mentioned earlier, water is not always among the reaction products. A few of these electrode reactions will now be discussed to illustrate the situation. The nitro group is reduced in different ways, depending on the medium. In an acidic medium, a six-electron reduction takes place with the participation of hydrogen ions

$$(1) \qquad R\text{–}NO_2 + 6e + 6H^+_{solv} \rightarrow R\text{–}NH_2 + 2H_2O$$

In a neutral medium, all protonating solvents, like water and alcohols (see Chapter 7), become proton donors. In such reactions, a single nitro group is reduced by the addition of four electrons:

$$(2) \qquad R\text{–}NO_2 + 4e + 3H_2O \rightarrow R\text{–}NHOH + 4OH^-$$

In a strongly alkaline, or in a totally aprotic medium, i.e. that pertaining to most non-aqueous solvents, e.g. cyclic ethers, reduction involves one electron only and yields an organic radical-anion:

$$(3) \qquad R\text{–}NO_2 + e \rightarrow R\text{–}NO_2^{-}\cdot$$

In the absence of protons, the current efficiency is usually much lower, since the reduction products are ordinary anions. The consequence of this is at least a sixfold decrease in the cell's energy capacity. Reactions with other substituents include:

(4) 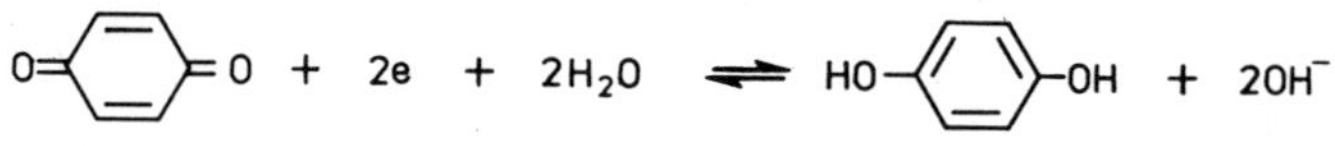

(5) $-N{=}O + 4e + 3H_2O \longrightarrow -NH_2 + 4OH^-$

(6) $-N{=}N- + 2e + 2H_2O \longrightarrow -\underset{H}{N}-\underset{H}{N}- + 2OH^-$

(7) $>N-Cl + 2e + H_2O \longrightarrow >N-H + OH^- + Cl^-$

As the solutions become strongly alkaline during these reactions, the former must be at least partially buffered. Both these factors bring about an increase in the mass of the electrolyte solution or cathode material while simultaneously limiting the use of strongly electronegative anode materials like lithium and other alkali metals.

There are only a few cases where hydrogen ions do not take part in electrode reactions – an example might be the reduction of 1,2-disubstituted halogen derivatives

$$(\text{–CCl–NCl–}) + 2e \rightarrow (\text{–C} = \text{N–}) + 2\text{Cl}^-$$

Another difficulty with these organic substances is that they must be at least partially soluble if they are to react electrochemically in the electrolyte solution (they themselves do not conduct electricity). Membranes separating the cathode and anode spaces must then be used to prevent the cell from self-discharging. In some cases, when the reaction products are sparingly soluble in the electrolyte solution, they are deposited on the anode as a passivating film. This rarely happens with soluble organic cathode materials, because the products are fairly soluble in organic solvents and do not prevent anode corrosion, as the above examples show. The self-discharge of a cell as a result of strongly negative anodes reacting with the cathode material is a drawback causing yet more problems in the application of organic materials.

Attempts have been made to overcome the adverse effects of the excessive solubility of organic compounds by polymerizing the cathode material, but then new problems have arisen. Now the material is indeed insoluble, but securing electrical contact within it to enable electrons to reach functional groups has turned out to be difficult because, by their very nature, these compounds are non-conductors. Thus, graphite must be added to improve conductivity, which increases the cathode mass and reduces the energy efficiency of the system. High-energy organic compounds have not been widely used because of all these disadvantages. Nevertheless, it has been found possible to use organic polymers as electrode materials.

Acetylene, one of the simplest hydrocarbons, has recently been polymerized [39–42]. In the pure state, this macromolecular material is a semiconductor. By electrochemically incorporating appropriate ions, this semiconductor can be converted into a conducting electrode material. The conductivity thereby increases to such an extent as to become equal to that of metals (ca. $10^3\ \Omega^{-1}\,cm^{-1}$). Moreover, the material takes on the gold-silver colouration and high reflection coefficient characteristic of metals.

This polymer occurs as thin fibres over twenty nanometres in length which are densely tangled up, forming a compact mass with a felt-like structure. Its specific volume fills only one third of the total volume, so that it can be soaked with electrolyte solution. The polymerization of acetylene gives rise to unsaturated polymers with an extensive conjugated double bond system, and the delocalized π-electrons of this polyene system ensure metallic conductivity

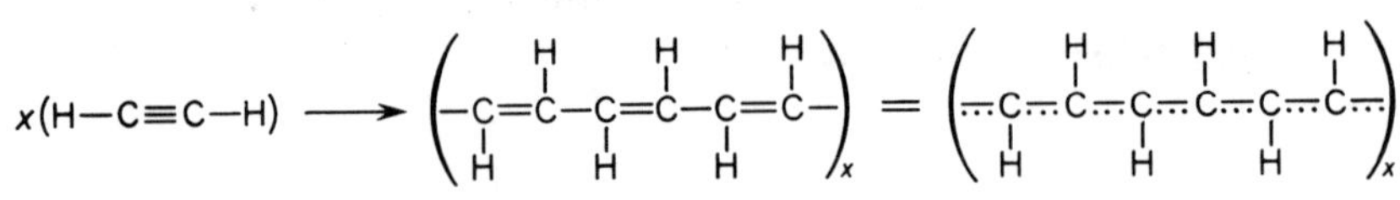

Polyacetylene foils $(CH)_x$ ca. 0.1 mm thick can be used as cathode or anode materials depending on whether they are cathodically or anodically polarized on being immersed in the electrolyte solution. As a result of polarization, ions of the charge opposite to that of the polyacetylene electrode are incorporated into the polymeric structure where they electrostatically balance out the external charge accumulated on the polymer fibres. A non-stoichiometric compound is formed with an ion in this way. If the electrode is positive, negative ions are incorporated and a cathode material results. Negative polarization yields an anode material with incorporated cations. Thus, charging produces conducting polyacetylene electrodes whose charging (oxidation) states are equal but opposite in sign. If, for example, there are perchlorate ions in the solution, the cathode material has a composition of the type $CH^{y+}(ClO_4)^-_y$, $y = 0.06$, after charging, which is equivalent to a 6% addition to the structure without the latter becoming distorted. When ready to use, the polymeric structure of a material with metallic conductivity probably contains polycarbonium ions of the type $(CH^{y+})_x$ stabilized by anions in such a way that the total composition corresponds to the general formula $[(CH^{y+})Y^-_y]_x$.

Discharging is allowed to continue for as long as the polyacetylene has metallic properties, so its anion content must not fall below 2%. The admissible overall discharging reaction in a cell with a lithium anode is then

$$(CH^{0.06+}A^{-}_{0.06})_x + 0.04Li \rightarrow (CH^{0.02+}A^{-}_{0.02})_x + 0.04Li^{+} + 0.04A^{-}$$

Within such a range of stoichiometric composition, the system is reversible to a large extent. It was subjected to more than 300 charge/discharge cycles without the metallic structure of the polyacetylene having deteriorated. Disadvantages are, however, the low current efficiency during a single discharge cycle – no more than 60% of the energy input during charging – and the rapid drop in the working voltage during cell operation.

The mechanism by which this electrode material is obtained when present in the form of compact layers is quite similar to the charging of an electrolytic condenser in which the metal plate is polyacetylene, while the other plate consists of the oppositely-charged ions from the electrolyte solution. When negatively charged, $(CH)_x + e \rightarrow (CH^{-})_x$ (reduction) (anode material), polyacetylene can be regarded as being in the reduced state. When positively charged, i.e. when it is a cathode material, the reaction is $(CH)_x \rightarrow (CH^{+})_x + e$ (oxidation). The discharge reaction is $(CH^{+})_x + e \rightarrow (CH)_x$.

Cells with polyacetylene electrodes are continually being improved [43].

In recent years, new cathode materials have been tested; they are based on aromatic polymers such as polypyrrole and polyaniline [44]. In some ways they are also like interstitial compounds: on being charged they are oxidized, thus π-electrons from the conduction band (see Chapter 5, p. 130) are removed, which causes positive charges to appear. These are balanced out by anions such as BF_4^- entering the lattice from the electrolyte. These processes are reversed during discharging. Considerable voltages – 3.0–3.5 V – have been obtained with respect to lithium, but capacities (in $A\,h\,g^{-1}$ or $A\,h\,cm^{-3}$) are still low.

The operational parameters of cathode materials, especially of those used in storage batteries, are improved by altering structures and by mixing different cathode materials, e.g. $CuO + CuInS_2$ + fluorographite [45]. By using a cathode made from polyaniline and polyacetylene, it has been possible to increase the number of charge/discharge cycles to ca. 650 [46].

6.6 CATHODE MATERIALS WITH SOLVENT PROPERTIES

There are a number of covalent inorganic compounds which are liquids or even gases at room temperature and which can be used as cathode materials while acting as solvents. They include sulphur and phosphorus oxyhalides and sulphur oxides (Table 6.3), as well as some interhalogen compounds such as BrF_3 and ICl,with m.p. +14°C, b.p. +97°C.

Table 6.3 Physical constants of cathode materials with solvent properties

Compound	Melting point, °C	Boiling point, °C	Relative electrical permittivity, ε	Dipole moment, $\mu_D \times 10^{30}$, C m (μ_D [D])
Thionyl halides				
$SOCl_2$	−104.5	+78.8	9.05 (22°C)	5.28 (1.58)
SOFCl	−139.5	+12.2		
$SOBr_2$	−49.5	+137		
Sulphuryl halides				
SO_2Cl_2	−54.1	+69.1	9.15 (22°C)	
SO_2FCl	−124.7	+7.1		
Pyrosulphuryl halides				
$S_2O_5Cl_2$	−37.5	+152.5		
Phosphoryl halides				
$POCl_3$	+1.2	+105	13.7 (25°C)	
$POFCl_2$	−80	+53		
$POFBr_2$	−117	+110		
POF_2Br	−85	+30.5		
Thiophosphoryl halides				
$PSCl_3$	−35	+125		
$PSFBr_2$	−75	+125.5		
PSF_2Br	−137	+36		
Sulphur dioxide				
SO_2	−75.5	−10	12 (−20°C)	5.38 (1.61)

Three types of sulphur oxyhalides are known: those of tetravalent sulphur – the thionyl halides, of general formula SOX_2, where X is the halogen atom; those of hexavalent sulphur, i.e. the sulphuryl halides, of general formula SO_2X_2; and the pyrosulphuryl halides of formula S_2O_5X. Among the phosphorus compounds of this kind, there are the phosphoryl halides POX_3, in which phosphorus is pentavalent such as $POFBr_2$. Their sulphur analogues, the thiohalides of pentavalent phosphorus (thiophosphoryl halides) of formula PSX_3 display similar properties. The latter compounds are similar in structure to their oxygen analogues.

Of the six known sulphur oxides, only sulphur dioxide is readily liquefield at room temperature and can act both as cathode material and solvent. Sulphur trioxide exists in the liquid state over too narrow a range of temperatures and, moreover, the solid α-SO_3 (m.pt. 62°C) readily crystallizes out of liquid SO_3. Phosphoryl and sulphuryl halides do not react with free halogens. SO_2 reacts with fluorine and chlorine to give SO_2F_2 and

SO_2Cl_2 respectively, but yields stable solutions with bromine and iodine. All the oxyhalides are reduced electrolytically, regardless of the type of electric conductor (e.g. graphite), which is an essential part of the cathode.

Free energy changes indicate that lithium reacts violently with these substances, but in fact passivating films protect the metal from further spontaneous reaction during which the cell might discharge; even so, the films are thin enough for a current to flow in the cell. Sulphur dioxide behaves similarly in the presence of lithium [47–49]. As a cathode material, SO_2 is reduced to the hyposulphite ion $S_2O_4^{2-}$: $2SO_2 + 2e \rightarrow S_2O_4^{2-}$. Raman spectroscopy has revealed that a single bond joins the two sulphur atoms in the $S_2O_4^{2-}$ ion [50]. The hyposulphite ion formed at the cathode reacts with the cations in the electrolyte solution yielding fairly stable salts like $Na_2S_2O_4$, $Li_2S_2O_4$ and CaS_2O_4, whose solubilities depend on the composition of the electrolyte solution. Sulphur dioxide is one of the most thoroughly investigated aprotic solvents, comparable – as far as the volume of research done on it is concerned – with water or liquid ammonia (see Chapter 7). Even though the overall reaction is simple, the kinetics and mechanism of electrolytic reduction is a complex process depending on the electrolyte properties [51].

Several reaction mechanisms have been considered for the electrolytic reduction of thionyl chloride [52]:

(1) $2SOCl_2 + 4e \rightarrow 4Cl^- + SO_2 + S$

(2) $3SOCl_2 + 8e \rightarrow 6Cl^- + SO_3^{2-} + 2S$

(3) $4SOCl_2 + 8e \rightarrow 6Cl^- + S_2Cl_2 + S_2O_4^{2-}$

Research has shown that the first of these mechanisms represents what probably happens, as the reduction products were found in spent cells; S_2Cl_2 was not present – this is highly reactive. However, the possibility that the second mechanism plays some part as a side reaction cannot be ruled out. SO_2 and S can be further reduced to $S_2O_4^{2-}$ and S^{2-}.

When electrolytically reduced, phosphorus oxychloride yields chloride anions and a polymeric phosphorus oxide $(PO)_x$, the latter being deposited at the cathode as a passivating film. Since the polymeric product is precipitated, the electrolytic reduction of $POCl_3$ is irreversible; moreover, the internal resistance of the cell rises, reducing its efficiency

$$xPOCl_3 + 3xe \rightarrow (\text{–PO–})_x + 3xCl^- \quad \text{(cathode reaction)}$$

Thionyl and sulphuryl chlorides appear to be the most useful as they do not form polymeric products. Sulphuryl chloride can be reduced in three ways:

(1) $2SO_2Cl_2 + 6e \rightarrow S_2O_4^{2-} + 4Cl^-$ (via $2SO_2 + 2e \rightarrow S_2O_4^{2-}$)

(2) $3SO_2Cl_2 + 10e \rightarrow 2SO_3^{2-} + 6Cl^- + S$

(3) $2SO_2Cl_2 + 6e \rightarrow SO_4^{2-} + 4Cl^- + S$

The third reaction is accompanied by the greatest free energy change [53].

One of the latest improvements is the use of electroactive BrCl in $SOCl_2$ [54].

6.7 HALOGEN CATHODE MATERIALS

Much research was done in the 1960s on transition metal halides including those of nickel, copper, silver, cadmium and cobalt, as well as the possible applications of free fluorine, chlorine, bromine and iodine. It turned out that these materials were difficult if not impossible to use in practice although they were highly suitable from the point of view of energy. In combination with lithium, fluorine, for example, would be the most efficient cathode material, but the fact that it reacts with every substance except its reaction products precludes its use for the time being. Only fluorine compounds, e.g. ClF_3 and fluorographite can be used as cathode materials and they have been extensively studied from this aspect. There are two categories of fluorine compounds: one comprises quite reactive compounds like oxygen fluorides (OF_2 and O_2F_2), chlorine fluorides (ClF and ClF_3) and fluorographite, the other almost unreactive compounds such as polytetrafluoroethylene (teflon) $(-CF_2-CF_2-)_x$ and calcium fluorides, which can be used in cans or other structural elements of cells. When fluorides like CuF_2 of NiF_2 are the cathode material, the actual cathode material which undergoes electrode reactions is the transition metal ion; fluorine as the fluoride ion remains in the same oxidation state and merely participates in the consecutive reaction of cation exchange e.g.

(1) $Cu^IF^{-I} + e \rightarrow Cu^0 + F^-$

(2) $F^- + Li^+$ (from the electrolyte) $\rightarrow LiF$

Gaseous chlorine has been used as a cathode material [55], and also in high-temperature cells [56]; it has been used in a cell with zinc as chloral hydrate [57].

Bromine is the first of the halogens which readily dissolves in organic solvents and can be used in this form as a cathode material in low-temperature cells. The presence of bromides in the solution increases the solubility of bromine owing to the formation of the complex Br_3^- ion in the reaction $Br_2 + Br^- \rightarrow Br_3^-$. During the cathode reaction, an electron becomes reversibly attached, as is the case with the other halogens:

$Br_2 + 2e \rightarrow 2Br^-$; the $Br_3^- + 2e \rightarrow 3Br^-$ redox reaction takes place in the presence of bromides when the potential is roughly similar. Chlorine and bromine are added to sulphuryl and thionyl chlorides in lithium cells [58, 59].

Liquid BrF_3 (m.p. +9°C, b.p. +126°C) could be an interesting cathode material as it can act simultaneously as a solvent, but it is a hazardous material [78]. Its conductivity of ca. $10^{-3}\ \Omega^{-1}\ cm^{-1}$ is relatively high due to autodissociation: $2\,BrF_3 \rightleftharpoons BrF_2^+ + BrF_4^-$; moreover, it readily dissolves double salts such as $KBrF_4$ which dissociate in it. Bromine is in oxidation state III in these compounds and can therefore be reduced:

$$BrF_3 + 4e \rightarrow Br^- + 3F^-$$

$$KBrF_4 + 4e \rightarrow K^+ + Br^- + 4F^-$$

Like bromine, iodine can also be a cathode material [60]. Molecular iodine I_2 and its complex anion I_3^- readily dissolve in organic solvents, in which the iodine concentration may rise to 50%. This cathode material does not require membranes to be used, e.g. in lithium systems, because solid lithium iodine is formed at the surface:

$$2Li + I_2 \rightarrow 2LiI$$

The lithium ions in LiI (a solid electrolyte) are mobile and ensure electrical conductivity; they also ensure that further layers of metal will react if the electrical circuit is closed. When this is open, a film of solid electrolyte stops the lithium from corroding (self-discharging) [37, 61].

As cathode materials, transition metal halides are characterized by the fact that their cations change their oxidation state and are the source of current-generating reactions.

The problem is more complicated when fluorides are the compounds in question, as none of the common organic solvents dissolves them to such an extent that useful discharge current densities are obtained. Good solubility is essential here because these fluorides do not have semiconducting properties. Voltammetric studies have furnished evidence that this difficulty can be overcome by adding an electrolyte which forms a more soluble compound [11], e.g. with copper: $Cu^+ + PF_6^- \rightarrow CuPF_6$. However, even $CuPF_6$ is unstable, because the PF_6^- anion slowly dissociates, leaving almost insoluble CuF, thereby inhibiting the cathode reaction

$$PF_6^- \rightarrow PF_5(\uparrow) + F^-, \qquad Cu^+ + F^- \rightarrow CuF$$

Besides fluorides, a number of chlorides have been tested e.g. those of silver, copper (I) and (II) and nickel [11, 62, 63]. The solubility of these chlorides is usually low enough to prevent self-discharging, but good enough to provide useful current densities. Such a compromise is necessary if halides are to be used as cathode materials.

Chlorides, particularly silver chloride, may be sufficiently soluble for the reaction to be reversed, even at high current densities. This means they can be used in storage batteries [64, 65] containing suitable separators.

Solvents that are highly basic according to the Lewis theory and capable of complex formation with the metal ion in the cathode material are unsuitable for systems with halide cathodes. Silver and copper halides readily forming soluble complexes would be useful as cathode materials (they can be charged and discharged at high current densities), were it not for the fact that this same good solubility weakens the passivating layers on lithium and would cause the cell to discharge itself unless a semipermeable membrane were used to protect the anode, so their applications are limited. As a consequence, only cells and systems containing fluorides are stable enough, but they supply only low current densities.

There have been attempts to improve the conductivity of these materials by converting them into semiconductors; to this end rare earth cations, e.g. ytterbium, have been added to non-conducting CdF_2. However, it is clear that in this way a material is obtained that can be used only in primary systems because the whole crystalline structure of the cathode material decomposes irreversibly during discharge to yield the metal and the fluoride of the cation in the electrolyte, e.g. $CdF_2 + 2e \rightarrow Cd + 2F^-$; $F^- + Li^+ \rightarrow LiF(\downarrow)$.

6.8 OXIDE CATHODE MATERIALS

Metal oxides such as AgO, HgO, Cu_2O, CuO, MnO_2, Cr_2O_3, CrO_3, V_2O_5, Tl_2O_3, Fe_3O_4 and MoO_3 have been used as cathode materials for quite a long time. The structures of these compounds are various and are hard to classify consistently. In principle, they can be divided into two main groups. One of them comprises noble and semi-noble metal oxides like AgO, Ag_2O, CuO, Cu_2O and HgO (Hg_2O – mercury (I) oxide is unknown), which are reduced to the metal. Cations of the principal electrolyte, e.g. Li^+, participate in the reduction at the cathode. These oxides are reduced in the following way [67, 79]:

(1) $HgO + 2Li + 2e \rightarrow Hg + Li_2O$

(2) $CuO + 2Li^+ + 2e \rightarrow Cu + Li_2O$

(3) $AgO + 2Li^+ + 2e \rightarrow Ag + Li_2O$

Unlike free sulphide ions S^{2-}, free oxide ions O^{2-} cannot exist in organic electrolyte solutions, which is why researchers once believed that oxide cathodes could not operate reversibly [4].

The other group of oxides includes those which are electrolytically reduced to bronzes [4], intercalation compounds [68] or lower oxides [69] but not to pure metals. Alkali metal oxides like Li_2O do not normally occur among electrolytic reduction products. In the case of vanadium pentoxide and molybdenum trioxide, the reactions proceed as follows:

(1) $V_2O_5 + xLi^+ + xe \rightarrow Li_xV_2O_5$

(2) $MoO_3 + xLi^+ + xe \rightarrow Li_xMoO_3$

The products are molybdenum or vanadium bronzes. Values of x may be constant, as in stoichiometric compounds, among them many intermetallic compounds, or variable over quite a wide range, as in non-stoichiometric compounds like molybdenum bronzes and intercalation compounds.

It was long thought that this second group of oxides of variable x included alkali metal oxides among the reaction products. The first indication that the products of electrolytic reduction were different, were the experimentally measured e.m.f.s, which were far higher than those calculated for the reactions presumed to yield Li_2O or other alkali metal oxides. For the reaction

$$MoO_3 + 2Li \rightarrow Li_2O + MoO_2$$

the calculated e.m.f. was 2.2 V, whereas the experimental value was 2.8 V. In the case of V_2O_5 this difference was even greater: the e.m.f. calculated for the reaction

$$V_2O_5 + 2Li \rightarrow V_2O_4 + Li_2O$$

was 2.4 V, whereas the experimental value was 3.5 V. These results evidently indicated that some product other than Li_2O was being formed. The products turned out to be the intercalation compounds discussed earlier (reactions 1 and 2).

The considerable amount of research done in recent years has shown that lithium is readily incorporated into the lattice structure of TiO_2, V_2O_5, Ta_2O_5, Cr_3O_8, MoO_3 and MnO_2 and the non-stoichiometric tungsten trioxide. X-ray studies of LiV_2O_5 have shown that the crystal structure is stretched and distorted. The unit cell of $Li_{1.5}MoO_3$ is distorted by no more than 100 pm, but if potassium is incorporated, the distortion is greater and may be as much as 200 pm.

Removing alkali metal ions from the lattice of the product is sometimes difficult and less readily reversible than from chalcogenides such as MoS_2; in these cases the capacity is markedly reduced. So for the time being, cathode materials based on the oxides of vanadium, manganese, chromium and molybdenum are suitable for primary cells, but only to a limited extent for use in storage batteries and need further improvement. Bronnerite-type copper vanadium molybdenum oxide cathodes are promising in this respect [80]. The number of charge/discharge cycles of V_2O_5 can reach 400 but only with a shallow, 25% discharge; a greater discharge depth (up to 40%) reduces it to 40, and only small current densities, ca. 1 mA cm^{-2}, are then available. If larger currents are drawn, the cathode polarizes rapidly with the result that the voltage and the specific energy of the current-generating system fall considerably.

The results of the studies performed so far indicate that the incorporation and removal of mobile ions must be more readily reversible. In order to achieve this, compounds with a more open and regular structure must be found, compounds which contain, for example, straight channels similar to those in zeolites and hexagonal molecular sieves. Sandwich structures would also be acceptable. A further condition is an energy gap between the valence and conduction bands that is sufficiently narrow for these materials to act as semiconductors or metals.

Because of its high specific energy, intercalated manganese dioxide is still one of the best oxide cathode materials [69].

The properties of some other oxide materials have been improved by using mixed oxides such as $(Pb,Bi)O_x$ [70], $AgBi(CrO_4)_2$ [71], MoV_2O_8 [72] and $Cu_4O(PO_4)_2$ [73], and also mixed sulphides such as $CuInS_2$ [45].

Sulphur can be used in both high- and low-temperature cells [74]. It can be reduced to the sulphide ion S^{2-} via polysulphide ions S_xS^{2-} or S_x^{2-} in which $x \leq 8$.

REFERENCES

[1] D.F.A. Koch, "Electrochemistry of sulfide minerals" in: *Modern Aspects of Electrochemistry* (J.O'M. Bockris and B.E. Conway, eds.), Ch. 4, Plenum Press, New York–London (1975).

[2] W. Noddack, K. Wrabetz and W. Herbst, *Z. Electrochem.*, **59**, 752 (1955).

[3] K. Wrabetz, *Z. Electrochem.*, **60**, 7, 722 (1956).

[4] G. Eichinger and H.P. Fritz, *Electroanal. Chem. Interfac., Chem.*, **58**, 357 (1975).

[5] P. Ruetschi and R.F. Amlie, *J. Electrochem. Soc.*, **112**, 665 (1965).

[6] R. Jasiński and B. Burrows, *J. Electrochem. Soc.*, **116**, 422 (1969).

[7] F.W. Dampier, *J. Electrochem. Soc.*, **121**, 656 (1974).

[8] E.F. Bertaut, *Pure Appl. Chem.*, **52**, 73 (1979).

[9] L.E. Ross, S.K. Preto, N.C. Otto, C.C. Sy and M.F. Roche, *Proceedings of the 15th Intersociety Energy Conversion Engineering Conference*, Seattle, USA (1980).

[10] J.P. Gabano, G. Gerbier and J.F. Laurent, *Chemie-Ing.-Techn.*, **42**, 206 (1970).

[11] G. Clinckspoor, H. Schmidt and W. Vielstich, *Chemie-Ing.-Techn.*, **42**, 206 (1970).

[12] C.J. Pedersen and H.K. Frensdorf, *Angew. Chem.*, **84**, 16 (1972).

[13] F. Jellinek, *Inorganic Sulfur Chemistry*, Elsevier, Amsterdam (1968).

[14] W.B. Johnson and W.L. Worrell, *Synthetic Metals,* **4**, 225 (1984).

[15] B.G. Silbernagel, *Mater. Sci. and Eng.*, **31**, 281 (1977).

[16] C.W. Chu, V. Diatchenko, C.Y. Huang and F.J. Di Salvo, *Phys. Rev. B*, **5**, 1811 (1972).

[17] R.Z. Bachrach, M. Skibkowski and F.C. Brown, *Phys. Rev. Lett.*, **37**, 40 (1976).

[18] J.C. McMenamin and W.E. Spicer, *Phys. Rev. Lett.*, **29**, 1501 (1972).

[19] P. Moline, D. Jerome and A.J. Grant, *Phil. Mag.*, **30**, 1091 (1974).

[20] D.W. Fischer, *Phys. Rev. B.*, **8**, 3576 (1973).

[21] F.R. Shepherd and P.M. Williams, *J. Phys. Chem.*, **7**, 4416, 4427 (1974).

[22] J.E. Graebuer, *Solid State Commun.*, **21**, 353 (1977).

[23] A.H. Thompson, F.R. Gamble and R.F. Koehler, *Phys. Rev. B.*, **5**, 2811 (1972).

[24] B.G. Silbernagel and M.S. Whittingham, *J. Chem. Phys.*, **64**, 3670 (1976).

[25] A.S. Nagelberg and W.L. Worrell, *J. Solid State Chem.*, July 1981.

[26] M.S. Whittingham, *J. Electrochem. Soc.*, **123**, 315 (1976).

[27] S. Sekido and Y. Ninomiya, *Solid State Ionic*, **1981**, 153.

[28] F.C. Laman and K. Brandt, *J. Power Sources*, **24**, 195 (1988).

[29] W. Rüdolff, in: *Advances in Inorganic Chemistry and Radiochemistry* (H.J. Emeléus and A.G. Sharpe, eds.), Vol. 1, Academic Press, New York (1959), p. 224.

[30] D.T. Meshri and W.E. White, 24th Meeting of the International Society of Electrochemistry, Brighton (1974), Extended Abstracts, p. 251.

[31] R. Hagiwara, M. Lerner, N. Bertlett and T. Nakajima, *J. Electrochem. Soc.*, **135**, 2393 (1988).

[32] N. Koshiba, H. Hayakawa and K. Momose, *26th Battery Symp. Japan, 1985*, p. 145.

[33] C.K. Morehause and R. Glickman, *J. Electrochem. Soc.*, **104**, 467 (1957); **105**, 299 (1958); **106**, 288, 741 (1959); **107**, 717 (1960).

[34] R. Glickman, *J. Electrochem. Soc.*, **108**, 1, 922 (1961); **109**, 352 (1962); **110**, 353 (1963).

[35] M.A. Gutjahr and K.D. Beccu, *Chemie-Ing.-Techn.*, **42**, 202 (1970).

[36] L. Werblan and A. Cisak, *Wiadomości Chem.*, **27**, 843 (1973).

[37] C.C. Liang and P. Bro, *J. Electrochem. Soc.*, **116**, 1322, 1452 (1969).

[38] T. Boschi, R. Pappa, G. Pistoia and M. Tocci, *J. Electroanal. Chem.*, **176**, 235 (1984).

[39] A.G. McDiarmid and A.J. Haeger, *Synthetic Metals*, **1**, 101 (1979/80).

[40] P.J. Nigrey, A.G. McDiarmid and A.J. Haeger, *J. Chem. Soc., Chem. Commun.*, **1979**, 594.

[41] S.L. Hsu, A.J. Signorelli, G.P. Pez and R.H. Baughman, *J. Chem. Phys.*, **69**, 106 (1978).

[42] H.A. Resing, J.P. Reardon, D.W. Weber, P. Brant, F.L. Vogel, T.C. Wu, D. Billard and A. Pron, in: *Magnetic Resonance in Colloid and Interface Science* (J.P. Fraissard and H.A. Resing, eds.), D. Reidel Publ. Comp., Philadelphia (1980).

[43] Y. Kobayashi, T. Shishikura, H. Nakamura and H. Konuba, *Synth. Metals*, **18**, 619 (1987).

[44] K. Fujita, M. Mizumoto, H. Sugimoto, A. Toyama, N. Ebato and S. Matsuda, Jap. Pat. 63,136,476 (Appl. 26.XI.1986).

[45] C. Ishibashi, K. Moriwaki and S. Furukawa, Jap. Pat. 63,174,270 (18.VII.1988) Appl. 87/5, 241 (13.I.1987).

[46] Y. Kobayashi, M. Takeuchi, M. Kaneyama, M. Ohira, Y. Kikuta and T. Takahashi, Jap. Pat. Appl.87/56, 809 (13.III.1987).

[47] J.J. Auborn, A. Heller and K.W. French, *Proceedings of the 25th Power Sources Symposium*, p. 6.

[48] J.J. Auborn, K.W. French, G.I. Libermann, V.K. Shah and A. Heller, *J. Electrochem. Soc.*, **120**, 1613 (1973).

[49] W.K. Behl, J.A. Christopulos, M. Ramirez and S. Gilman, *J. Electrochem. Soc.*,

120, 1619 (1973).

[50] A. Simon and H. Kueler, *Z. anorg. Chem.*, **260**, 151 (1949).

[51] C.L. Gardner, D.T. Fauchard and W.R. Fawcett, *J. Electroanal. Chem. Interfac. Electrochem.*, **82**, 303 (1977).

[52] A.J. Hills and N.A. Hampson, *J. Power Sources*, **24**, 253 (1988).

[53] M.M. Nicholson and M.L. Iverson, 140th Meeting of the Electrochemical Society, Cleveland (1971), Conference materials, p. 115.

[54] E.S. Takeuchi, K.J. Takeuchi, C. Bessel and E. Eads, *J. Power Sources*, **24**, 229 (1988).

[55] N. Furuya and S. Motoo, US Pat., 4,748,095 (Appl. 06 III 1987).

[56] F.M. Bowers and J.H. Amburs, *Power Sources 5, Proceedings of the 9th International Symposium, Brighton 1974* (D.H. Collins, ed.), Academic Press, London (1975); Abstract No. 36.

[57] P.C. Symons and P. Carr, *Proceedings of the 8th Intersociety Energy Conversion Engineering Conference, 1973*, p. 72.

[58] K.M. Abraham, M. Alamgir and R.K. Reynolds, *J. Electrochem. Soc.*, **135**, 2917 (1988).

[59] C. Herscovici, A. Leo and A. Charkey, Eur. Pat. Appl. 268,397 (20.XI.1986).

[60] M. Lin, Y.X. Jin, Y.Z. Young, Y.F. Hou and S.D. Wang, *J. Power Sources,* **20**, 141 (1987).

[61] B. Scrosati and M. Torroni, *Electrochim. Acta,* **18**, 225 (1973).

[62] M. Eisenberg, *Proceedings of the 7th Intersociety Energy Conversion Engineering Conference, 1972*, p. 75.

[63] G. Eichinger and H.P. Fritz, *Z. Naturforsch.*, **28b**, 763 (1973).

[64] J.N. Butler, *J. Phys. Chem.*, **72**, 3286 (1968).

[65] R. Alexander, E.C.F. Ko, Y.C. Mac and A.J. Parker, *J. Am. Chem. Soc.*, **89**, 3703 (1967).

[66] M.M. Nicholson, *J. Electrochem. Soc.*, **118**, 1047 (1971).

[67] N. Marincić, *J. Appl. Electrochem.*, **6**, 463 (1976).

[68] M.S. Whittingham, *Progr. Solid State Chem.*, **12**, 41 (1978).

[69] N. Kamat, T. Kishi, T. Harada, K. Takada and H. Aihara, *Progr. Batt. Solar Cells*, **6**, 52 (1987).

[70] J.C. Boivin, D. Thomas, J. Cannone, A. Verbaere, A. La Mehaute and M. Broussely, *Mater. Res. Bull.*, **22**, 439 (1987).

[71] R. Messina, A. Tranchant, J. Perichon, A. Lecerf and M. Broussely, *J. Power Sources*, **8**, 277 (1982).

[72] A. Tranchant and R. Messina, *J. Power Sources,* **24**, 85 (1988).

[73] C.D. Jaeger, M. Broussely, Y. Jumel, J.P. Gabano and A. Lecert, *J. Power Sources*, **20**, 111 (1987).

[74] E. Peled and H. Yamin, *Progr. Batteries Solar Cells*, **5**, 56 (1984).

[75] N. Kumagai and K. Tanno, *Electrochim. Acta*, **36**, 935 (1991).

[76] J. Molenda, *Phys. Stat. Solidi B*, **165**, 419 (1991).

[77] M. Wakinara and T. Uchida, *Chem. Abstr.*, **115**, 259837 (1991).

[78] J.C. Bailar, H.J. Emeléus, R. Nyholm and A.F. Trotman, *Comprehensive Inorganic Chemistry*, Vol 2, Pergamon Press, Oxford, 1973, p. 1517.

[79] P. Nowak and P. Podhajecky, *J. Power Sources*, **35**, 235 (1991).

[80] Y. Takeda, K. Itoh, R. Kanno, T. Ichikawa, N. Imanishi and O. Yamamoto, *J. Electrochem. Soc.*, **138**, 2566 (1991).

CHAPTER 7

Solvents and Electrolyte Solutions

Selecting the appropriate electrolyte solution for use in a non-aqueous cell is of prime importance. Such a solution should:

(a) be chemically inert towards the electrode materials unless the solvent also acts as the depolarizer (cathode material);
(b) be electrochemically stable in the potential range equal at least to the potential difference between the electrode pair used in a cell or storage battery;
(c) have the best possible transport properties, i.e. a high specific conductivity and a coefficient of dynamic viscosity low enough to ensure appropriate ion migration and the course of diffusion processes, as well as to minimize polarization phenomena at the electrodes.

One way of obtaining suitable solutions is to search by selection and physicochemical measurements for pure or mixed solvents which extensively ionize and dissociate the electrolyte into freely moving ions [1], with optimum viscosity, and electrochemical stability versus the electrode system. The stability of the electrode materials towards the electrolyte solution determines whether a given battery type is constructable. A stable electrode system can be achieved in two ways. One is to select electrode materials which do not react at all or only exceedingly slowly with the components of the electrolyte solution under open circuit condition. The other is to allow the solution and electrode materials to react with each other, but in such a way that the reaction products form thin passivation layers on the surface of one or both electrodes, which hampers the progress of reaction and secures kinetic stability of the system in spite of its lack of thermodynamic stability [2, 3]. When the circuit is closed, the appearance of a potential difference (e.m.f.) sets the electrochemical mechanisms of the current-generating reaction in motion, and the passivating films are loosened or even removed, so that the (ionic) current flow proceeds uninhibited.

7.1 CLASSIFICATION OF SOLVENTS

There are many ways of classifying non-aqueous solvents. One method is to divide them according to their relative electrical permittivity which may be high ($\varepsilon > 50$), medium ($12 < \varepsilon < 50$) or low ($\varepsilon < 12$) [4–6]. In high-ε solvents the weak interactions between solvated ions and the strong ion-dipole solvation yield a high concentration of freely moving ions which ensures that the solution has a good conductivity. In low-ε solvents, ion-dipole interactions are limited and Lewis acid-base interactions become more important, i.e. interactions between the free electron pairs of solvent molecules, and electron-deficient acceptors such as Li^+ or $AlCl_3$ [7, 8]. It has been shown [9], that adding anhydrous aluminium chloride to pure liquid phosgene increases the latter's intrinsic conductivity about one hundred thousand times (the conductivity of pure phosgene is $7 \times 10^{-9}\ \Omega^{-1}\text{cm}^{-1}$). The following dissociation takes place when $AlCl_3$ is solvated by phosgene:

$$\underset{\text{base}}{COCl_2} + \underset{\text{acid}}{AlCl_3} \rightleftharpoons COCl \cdot AlCl_4 \rightleftharpoons \underset{\substack{\text{conjugated}\\\text{acid}}}{COCl^+} + \underset{\substack{\text{conjugated}\\\text{base}}}{AlCl_4^-}$$

Phosgene autodissociates as follows:

$$COCl_2 \rightleftharpoons \underset{\text{acid}}{COCl^+} + \underset{\text{base}}{Cl^-}$$

In tetrahydrofuran, a solvent with a low electrical permittivity, the two lone electron pairs on the oxygen atom make the molecule basic, so it can coordinate with, say, a lithium ion, which in this instance exhibits acidic properties (it is a Lewis acid).

The literature includes some broad and thorough analyses [10–15] of solvent properties. Many authors have divided non-aqueous solvents into protic and aprotic [7, 10, 13]. This classification seems to be the most relevant as regards the application of solvents in non-aqueous batteries. We would now like to reconsider and complement this classification. Protic solvents are those whose molecules contain mobile and therefore reactive protons which participate in proton transfer, exchange reactions, deuterium substitution (H/D exchange) or in acid-base reactions. They include alcohols, organic and inorganic acids, amines, and also water. Aprotic solvents contain non-reactive non-exchangeable protons. They include alkanes, alkenes, aliphatic and alicyclic ethers (e.g. tetrahydrofuran), lactones, organic carbonates (e.g. ethylene carbonate), compounds containing alkyl groups, e.g. dimethyl sulphoxide, acetonitrile, acetone, and solvents containing no hydrogen at all, e.g. SO_2, $COCl_2$, $SOCl_2$ and $POCl_3$.

Regarding the properties of the latter compounds, classified as aprotic, we propose a slightly different solvent classification into three groups: protonating, non-protonating and aprotic. The first includes all solvents with reactive protons, including water, the second – with non-reactive protons in the sense of Brönsted and Lewis acid-base equilibria. Aprotic solvents comprise those that contain absolutely no protons in their molecules. Table 7.1 sets out some of the physicochemical properties of 38 non-aqueous solvents classified in accordance with our suggestion.

As in all classifications, there are some borderline cases which are hard to assign unequivocally to a particular group. One example are the amines which, according to the suggested classification, should be and have been grouped among the non-protonating solvents. Even so, reactions are known in which amines (including ammonia) behave like acids, e.g. they form amides and imides, but reactions like these do not occur in the systems used in galvanic cells.

Protonating solvents cannot be used in cells which have an alkali metal anode, as the latter are strong bases according to the Lewis [7, 14] and Usanowicz acid-base concepts. There is a strong tendency to transfer electrons from the alkali metal to the functional group of a solvent molecule, which contains mobile hydrogen. A reactive anion is formed from the solvent molecule sometimes followed by dehydrogenation and release of gaseous hydrogen, such as in the reaction of metallic lithium with ethanol, when the reaction products are soluble and corrode the metal, e.g.

$$C_2H_5O\text{–}H + Li^0 \rightarrow C_2H_5O^- + Li^+ + \tfrac{1}{2}2H_2(\uparrow)$$

This type of reaction is the reason why protonating solvents (including water) are not used in high-energy cells. Water itself reacts analogously, i.e. soluble products are formed and H_2 is released.

Our third group, the aprotic solvents, also comprises many compounds which react spontaneously with alkali metals – mainly lithium. However, in this reaction layers of insoluble products are formed that protect the metal against further corrosion. Their reaction with lithium means that some aprotic solvents even can be used as depolarizers (cathode materials) as their free enthalpy with lithium is highly negative. Hence, compounds like thionyl, sulphuryl and phosphoryl chlorides can play a dual role, i.e. that of solvent and of depolarizer. A further advantage of these solvents is that no gaseous products are evolved (or only minimal amounts, as with $SOCl_2$), and that the composition of the electrolyte solutions in such solvents remains virtually unchanged because the reaction products are precipitated and do not pass into solution.

Non-protonating solvents are those most frequently applied in high-energy cells, but few of them can act as a cathode material. Many, if not most, form sparingly soluble compounds with metals of the first and second

Table 7.1 The physicochemical properties of selected protonating, non-protonatinig and aprotic non-aqueous solvents at 25°C

Solvent	Chemical formula	Melting point, °C	Boiling point, °C	Relative electrical permittivity, ε	Dipole moment, $\mu_D \times 10^{30}$, C m (μD [D])	Viscosity, η [mPa s]	Density, ρ, g cm^{-3}	Donor number DN [12, 17, 117, 118] kcal mol^{-1}
1	2	3	4	5	6	7	8	9
Protonating solvents*								
Methanol [12, 23, 102, 104]	CH_3OH	–97.68	64.52	32.63	5.6 (1.68)	0.552	0.7865	19.1
Formamide [103, 104, 119]	$H_2N-C(=O)-H$	2.45	218	109.5		3.30	1.29	24
Methyl formate [75, 104]	$H-C(=O)-OCH_3$	–99	31.5	8.5 (20°C)	5.9 (1.77)	0.328	0.974 (20°C)	
Water [18, 21, 35, 35a, 102]	H_2O	0	100	78.54	6.2 (1.85)	0.8903	0.99707	18.0
Non-protonating solvents								
Acetonce [18, 21, 35, 104, 112, 120]	$(H_3C)_2C=O$	–94.7	56.3	20.56	11.5 (3.44)	0.303	0.704	17.0

* Examples of protonating solvents which cannot be used in non-aqueous lithium cells. Ethanol, other alcohols, ethylene glycol, phenol, carboxylic acids, sulphonic acids and phenylacetylene can be added to them.

1	2	3	4	5	6	7	8	9
Dimethylsulphoxide [104, 105, 106]	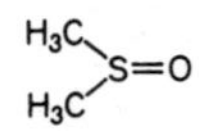	18.54	189.0	46.5	14.4 (4.3)	1.992	1.095	29.8
Tetrahydrothiophene-1,1-dioxide (sulpholane) [75, 104]	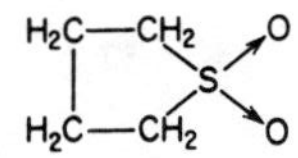	28.45	287.3	43.29 (30°C)		10.28 (30°C)	1.262 (30°C)	14.8
1,2-Propylene carbonate [2, 12, 30a, 40, 48, 61, 104, 120]	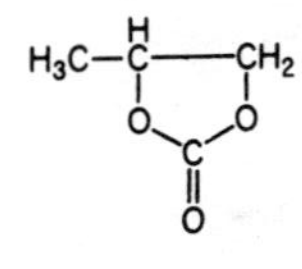	–54.53	241.7	64.92	16.6 (4.98)	2.552	1.198	15.1
Ethylene carbonate [12, 104, 107]	$(CH_3O)_2CHCH_3$	36.4	238	89.6 (40°C)		1.85 (40°C)	1.323 (40°C)	16.4
γ-Butyrolactone [12, 51, 75, 104, 108]	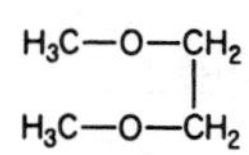	–43.5	204	39.1	13.8 (4.12)	1.7315	1.1242	18 [122]
1,2-Dimethoxyethane [12, 75, 104, 108, 121]	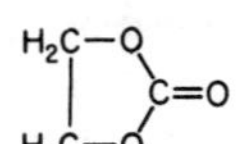	–58	85	7.07	5.7 (1.71)	0.407	0.8612	24 [122, 123]
1,1-Dimethoxyethane [51, 104, 108]	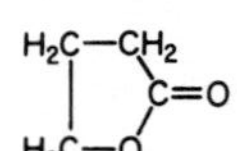	–	64	4.3		0.376	0.8446	

1	2	3	4	5	6	7	8	9
Tetrahydrofuran [12, 53, 104]	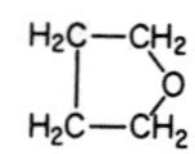	–108.6	65.965	7.43	5.7 (1.71)	0.459	0.8819	20.0 20 [122]
2-Methyltetrahydrofuran [114, 115]	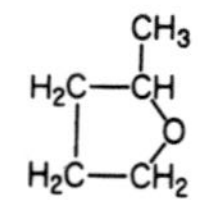	–	78–79	6.24		0.457	0.848	18 [122]
Acetyl chloride [18, 21, 35]	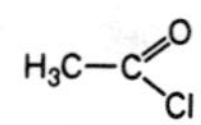	–112	51	15.8 (22°C)	5.7 (1.71)		1.104 (21°C)	0.7
Acetonitrile [12, 75, 104, 107]	$H_3C{-}C{\equiv}N$	–48.83	81.60	35.95	11.5 (3.44) (20°C)	0.341	0.7767	14.1
Propionitrile [18, 21, 35, 104]	$H_3C{-}CH_2{-}C{\equiv}N$	–92.78	97.35	27.2 (20°C)	11.9 (3.57)	0.389	0.7768	16.1
Benzonitrile [12, 104]	C_6H_5CN	–12.75	191.10	25.2	–	1.237	1.0006	11.9
Butyronitrile [104]	$H_3CCH_2CH_2CN$	–111.9	117.94	20.3 (21°C)	11.9 (3.57) (20°C)	0.624 (15°C)	0.795 (15°C)	
Acetic anhydride [12, 18, 21, 35, 104]	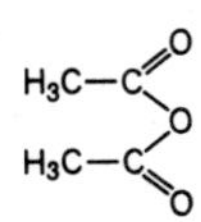	–73.1	140	20.5 (20°C)		0.851	1.082 (20°C)	10.5

1	2	3	4	5	6	7	8	9
Ammonia [7, 10, 40]	NH_3	–77.7	–33.4	22 (–34°C)	4.9 (1.47) (–34°C)	0.265 (–33.5°C)	0.683 (–33.4°C)	
N-Methylformamide [12, 104]	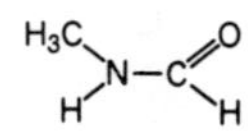	–5.4	185	186.9	12.9 (3.86)	1.65	0.998	
N,N-Dimethylformamide [12, 104]	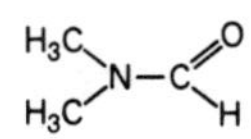	–60.43	153.0	36.71	12.9 (3.86)	0.793	0.944	26.6
N-Methylacetamide [104, 109]	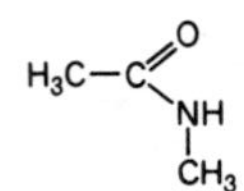	29.8	206	165.5 (40°C)	–	3.0 (40°C)	0.942 (40°C)	
N,N-Dimethylacetamide [12, 104]	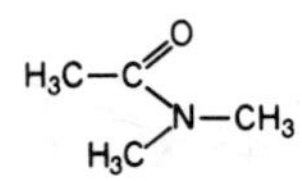	–20	166.1	37.8	12.4 (3.72)	0.927	0.935	27.8
Hexamethylphosphoric triamide (HMPT) [12, 104, 110]	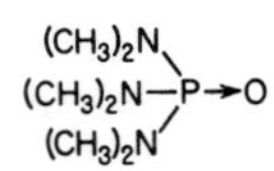	7.2	233	29.6	14.4 (4.31)	3.22	1.020	38.8
Hydrazine [104, 111]	$H_2N{-}NH_2$	1.8	113.5	51.7	6.1 (1.83)		0.995	
Nitrobenzene [12, 40, 104]	$C_6H_5NO_2$	5.76	210.8	34.82	13.4 (4.0)	1.818	1.198	4.4

1	2	3	4	5	6	7	8	9
Dimethyl sulphide [104]	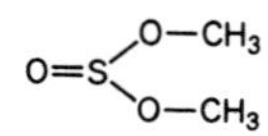	–141	126	22.5 (23.3°C)		0.771	1.207 (24°C)	
Pyridine [10, 12, 104]	C_5H_5N	–41.55	115.25	12.4		0.884	0.978	33.1
N-Methyl-2-pyrrolidone [75]	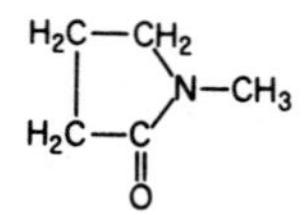	–24	204	32.0		1.666	1.028	
Aprotic solvents								
Phosphoryl chloride [10, 107]	$POCl_3$	1.25	105.8	13.7		1.065	1.645	11.7
Sulphuryl chloride [10, 18, 21, 35]	SO_2Cl_2	–54.1	69.4	9.15 (22°C)		0.674	1.657	0.1
Thionyl chloride [10, 18, 21, 35]	$SOCl_2$	–104.5	77	9.05 (22°C)		0.603	1.629	0.4
Phosphoryl fluorobromide [7]	$POFBr_2$	–117.2	110.1					
Thiophosphoryl chloride [7]	$PSCl_3$	–35	125					
Pyrosulphuryl chloride [7]	$S_2O_5Cl_2$	–37.5	152.5					
Sulphur dioxide [7,107]	SO_2	–72.7	–10.08	12.0 (–20°C)	5.4 (1.61)	0.26 (–20°C)	1.46 (–10°C)	acceptor

groups of the periodic table that cover the anode material with thin passivation layers, protecting them against corrosion [1, 2, 15, 16].

The basic function of non-protonating and aprotic solvents is to secure sufficient solubility of the electrolyte, and hence to obtain as high a conductivity as possible of the solution, of the magnitude of at least $10^{-3}\ \Omega^{-1}\text{cm}^{-1}$ [12]. The higher the absolute value of the Gibbs solvation energy ΔG_{solv}, the better an electrolyte dissolves in the solvent. If ΔG_{solv} is larger than the lattice energy of crystal formation ΔG_{cryst}: $|\Delta G_{\text{solv}}| > |\Delta G_{\text{cryst}}|$ then good salt solubility in a solvent S is achieved.

The ionizing properties of a solvent can be characterized by its relative electrical permittivity – a macroscopic parameter – and by molecular quantities such as the dipole and quadrupole moments, and the donor-acceptor properties of the molecule as a whole or of its functional groups, commonly described by the general term "polarity". Steric factors determining the mode of contact between solvent molecules and dissolved ions also affect the solvation energy. The donor-acceptor abilities are crucial to solvation and association processes in solution [17, 18]. In almost all solvents, ions begin to associate with increasing electrolyte concentration, first into pairs and then into higher solvated aggregates, or into contact ion-pairs and clusters [19, 20]. Any electrically-neutral species that are found do not carry current. Some ionization properties of solvents are indicated in Table 7.1. It is these which determine an electrolyte's solubility, degree of dissociation and thermodynamic dissociation constant. The electrical conductivity of a solution, which is of great importance when assessing its suitability for use in cells, depends on the properties of the solvent and electrolyte, and on the solution viscosity, which affects considerably the mobility of the ions.

There are several "scales of polarity" which enable the donor-acceptor and acid-base properties of solvents to be compared [18, 21, 22]. All such scales are based on some reference compound or solvent, the other solvents being ordered against that standard [23]. A by no means trivial matter is the use of the appropriate polarity scale, especially when one takes into account the fact that the very term "polarity" is ambiguous [24]. The real potentials of electrode materials depend on those properties of a solvent which determine the solvation energy [25]. The lattice energies of crystal formation are well-known and have been published. Determining the Gibbs solvation energies of non-aqueous solvents causes more problems, and hence these are less-known and tabulated. An approximate value of ΔG_{solv} can be found semi-empirically enabling application of the Born equations [26]. An accurate quantum-mechanical computation of the interactions between an ion and its adjacent solvent molecules is extremely difficult. If, however, there is no chemical reaction between them, then coulombic integrals dominate in the expression for quantum-mechanical energy [27]. In view of this, calculations can be restricted only to electrostatic terms. According to Born, the difference between the work done when charging a sphere of radius equal to the crystallographic radius of an ion r_i in a vacuum, and that done when charging a similar sphere in a dielectric of relative permittivity ε_s, is equal to the Gibbs solvation energy. For one mole of ions at infinite dilution he obtained the expression

$$\Delta G_{solv} = -\frac{N_A z_i^2 e^2}{8\pi\varepsilon_0} \cdot \frac{1}{r_i}\left(1 - \frac{1}{\varepsilon_s}\right) \tag{7.1}$$

where z_i is the charge number, e the elementary charge, ε_0 the permittivity of a vacuum, r_i the crystallographic radius of the ion, N_A Avogadro's number and ε_s is the relative electrical permittivity of the solvent. Thus according to equation (7.1), the Gibbs solvation energy is proportional to the reciprocals of the relative electrical permittivity and the ionic radius. A more sophisticated form of this equation takes into consideration the "dielectric saturation" of the ion's environment: if the ionic radius is extended a little so that it is now approximately equal to double the radius of a solvent molecule r_s, it may be assumed that at a distance of $(r_i + 2r_s)$, equal to the radius of the first solvation shell surrounding the ion, the electrical permittivity is equivalent to the relative macroscopic value of ε_s of the pure solvent

$$\Delta G_{solv} = -\frac{N_A z_i^2 e^2}{8\pi\varepsilon_0 (r_i + 2r_s)}\left(1 - \frac{1}{\varepsilon_s}\right) \tag{7.2}$$

It is clear from equations (7.1) and (7.2) that a linear relationship should exist between ΔG_{solv} and $1/\varepsilon$. However, experimental values do not always confirm this linearity, particularly in mixed solvents in which selective solvation of ions by the molecules of one of the components in the mixture is usually observed. It should therefore be assumed that the averaged electrical properties of a mixed solvent cannot be expressed solely by the electrical permittivity. The local charge distribution in the solvent molecules surrounding the ions is of far greater importance; this determines the nature of the ion-solvent interactions, and hence the solvation energy. Therefore, the energy of an ion-dipole interaction is better expressed by the Bernal–Fowler equation [28], which takes into account the dipole moment of the solvent molecule μ_D, the distance r between the charge centres of the ion and ligand dipole (the solvent molecule) and the coordination number of the ion n, equal to the number of ligands is the nearest solvation layer

$$\Delta G_{\mu_D} = -N_A n \frac{z_i e \mu_D}{8\pi\varepsilon_0 r^2}\left(1 - \frac{1}{\varepsilon_s}\right) \tag{7.3}$$

Equation (7.3) describes ion-dipole interaction at the molecular level. In order to express the solvation energy more accurately, one should add to equation (7.3) a term accounting for the energy of interaction between the oriented solvent dipoles in the solvation layer of the ion (i.e. ligand-ligand interactions), and another term describing the energy of interaction between the ion and the bulk of the solution treated as a continuum of average electrical permittivity ε. Ion-quadrupole interactions should also be taken into account as they reflect the differences in cation and anion solvation with respect to the same polar solvent, which is particularly evident when comparing cations and anions of the same size, although the acidic properties of cations and the basic properties (according to Lewis' theory) of anions are of prime importance (see below).

A second source of difference in the solvation energies of anions and cations of the same radii is the different accessibility of the ion to the oriented dipole. In molecules like dimethyl sulphoxide, the partial negative charge is concentrated on the oxygen atom and the partial positive charge is on the sulphur atom. Since the oxygen atom is on the outside of the molecule, the distance between a cation and the negative end of the dipole is shorter than the distance to which an anion can approach the sulphur atom, situated near the centre of the molecule. With small monoatomic anions such as fluoride, and when there is no rotation, the solvation energies can be expected to be similar (Fig. 7.1). The problem of ion solvation in solutions has been discussed by Buckingham [29], Mishchenko [30], Marcus [31], Krestov [32] and Blum [33].

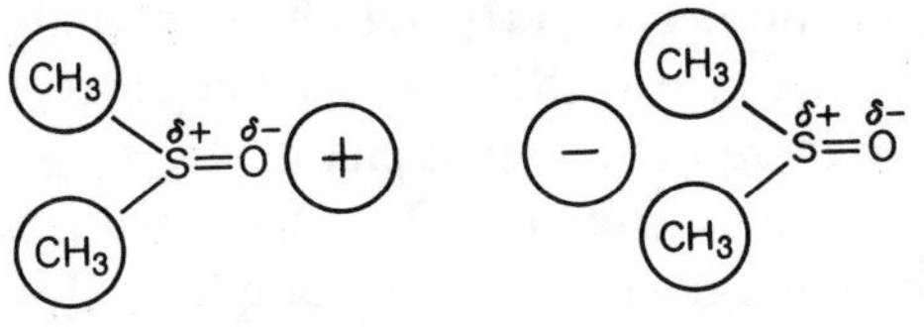

Fig. 7.1 Solvation of cation and anion by a dipolar solvent molecule (dimethyl sulphoxide).

It has already been mentioned that the donor-acceptor properties of a solvent are important as regards ion-dipole interactions. From this point of view, Gutmann [17, 18] has divided coordinating solvents into two groups: (1) donor solvents, which tend to react with electron-pair acceptors, and (2) acceptor solvents, which react with electron-pair donors. Once in contact with electrolytes, donor solvents interact more strongly with cations, and acceptor solvents with anions. All non-ionic substances are classifiable in the same manner. Furthermore, protonating solvents containing heteroatoms with free electron pairs can fulfil both functions – they are donors, as they have a lone pair on oxygen, for example, and acceptors because of their tendency to form a strong hydrogen bonds by means of the positive charge accumulated on the hydrogen atoms. Gutmann classifies alcohols, nitriles, amines (with ammonia), and compounds with carbonyl C = O, phosphoryl P = O, sulphoxy S = O and similar groups as donor solvents. Acceptor solvents include strong Lewis acids, covalently-bonded halides like BBr_3 and $AlCl_3$, $SOCl_2$, SO_2Cl_2, $POCl_3$ [10], and weak Brönsted acids like phenols [7, 10]. The reaction between a dimethyl sulphoxide molecule and a lithium ion is a typical example of a donor solvent reacting with a cation – a Lewis acid:

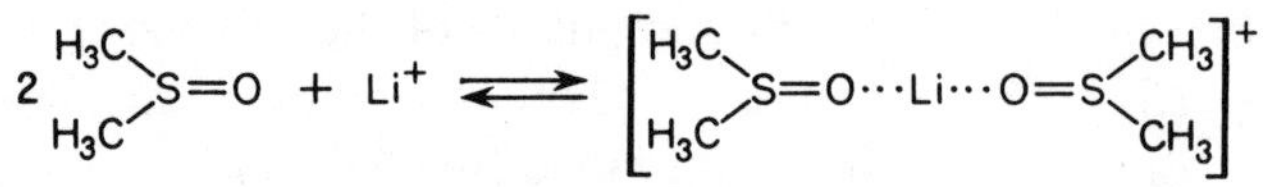

The reaction of thionyl chloride with a chloride ion is typical of reactions between donor-acceptor solvents and anions – a Lewis base:

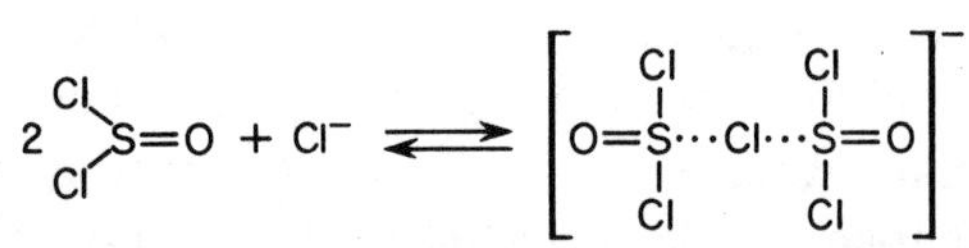

The latter solvation reaction illustrates the ambiguity inherent in any classification of solvents. Since thionyl chloride contains the S = O group, it ought to be classified as a donor solvent, but according to the Lewis theory it is an acid and therefore an acceptor solvent. Thus, in Gutmann's classification, such solvents are distinctive in that they have a dual donor-acceptor function. The fairly high boiling point of $SOCl_2$ (77°C) as compared with that of, say, SO_2 (−10°C) suggests that the former has an associated structure in the liquid phase.

Donor solvents can therefore react with acceptor solvents forming strong molecular donor-acceptor complexes

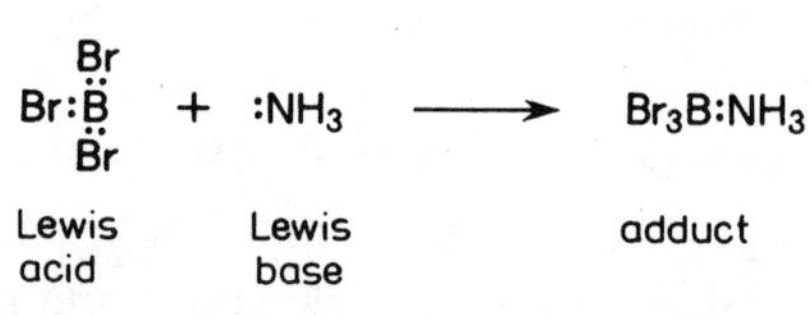

Solvents differ in their ability to provide electron pairs i.e. their "donor strength", varies. A quantitative measure of this ability is the donor number DN introduced by Gutmann [18]. Since antimony pentachloride $SbCl_5$ forms 1:1 complexes with solvent molecules under standard conditions, it has been chosen as the reference against which the donor number of other solvents is compared.

The negative value of the enthalpy change [34] $-\Delta H$ for the formation of an $SbCl_5$–1,2-dichloroethane complex in dilute solution is the reference point and the donor strength in such solutions is taken to be zero – this is the basis of the Gutmann donor strength scale. The $-\Delta H$ value [kcal mol^{-1}] of the reactions of $SbCl_5$ with other donors D is considered to be its donor number DN

$$D + SbCl_5 \rightleftharpoons D\text{–}SbCl_5, \qquad -\Delta H_{D\text{–}SbCl_5} = DN$$

Hence the thermodynamic equilibrium constant $K_{D\text{–}SbCl_5}$ takes the form

$$K_{D\text{–}SbCl_5} = \frac{a_{D\text{–}SbCl_5}}{a_D a_{SbCl_5}} \tag{7.4}$$

As the logarithm of the equilibrium constant is proportional to the donor number, both log $K_{D\text{–}SbCl_5}$ and its negative logarithm p$K_{D\text{–}SbCl_5}$ can be used to measure the donor strength of a substance.

The point in Fig. 7.2 in combination with the donor number can be interpreted as follows. A solvent whose strong donor properties are ex-

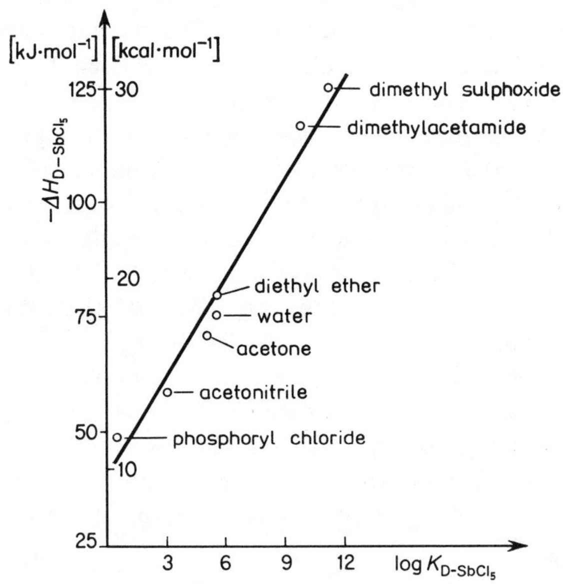

Fig. 7.2 Change in reaction enthalpy $-\Delta H_{D\text{–}SbCl_5}$ vs. logarithm of the equilibrium constant, log $K_{D\text{–}SbCl_5}$, as a measure of the donor strength of solvents ($-\Delta H_{D\text{–}SbCl_5}$ expressed in kcal mol^{-1} is identical with the donor number DN).

pressed by a high donor number, e.g. dimethyl sulphoxide (DN 30), will react with a low-DN solvent, e.g. phosphoryl chloride (DN = ca. 12), classified by Gutmann as an acceptor solvent, yielding molecular complexes of the type

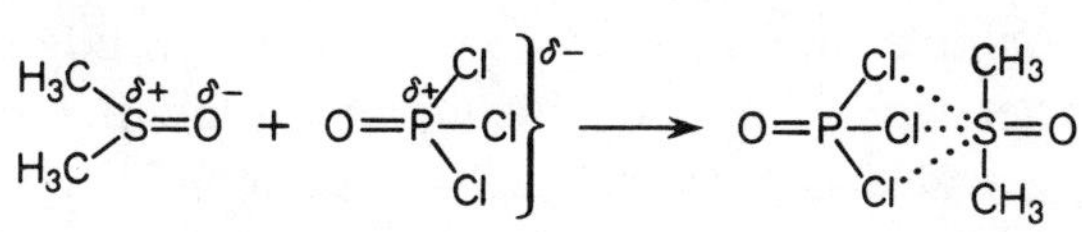

Thus, the donor number is a value approximately expressing the ability of the free electron pairs in a substance to react with another substance present in the system as a substrate; it is useful in interpretating or predicting intermolecular or ion-molecule interactions.

The increasing use of solvents with different properties has prompted much research effort in establishing a more general theory of solvent-solute interactions based on semi-empirical or empirical and semi-theoretical correlation [21, 22, 35].

At present the interactions between a solvent and dissolved substance are divided into two basic contributions. The first is the interaction between a solvated solute molecule and the medium (solvent) treated as a homogeneous isotropic continuum and described by macroscopic parameters – the relative electrical permittivity and the refractive index. The forces due to these interactions are called long-range forces, or non-specific or universal interactions. The second contribution is the interaction between a solute molecule and the adjacent solvent layer described in molecular terms, e.g. the dimensions of the molecule and its dipole and quadrupole moments, which involves short-range forces. Their nature is manifested by weak chemical bonds linking solvent molecules and the solute molecule or ion. Such interactions result in association or solvation complexes, whose molecules are held together by localized and spatially-oriented donor-acceptor bonds and which affect the properties of the solution, e.g. its viscosity [36]. One kind of close-range interaction, called electrophilic solvation (E), is that between a Lewis-acid solvent and a basic solute. In the opposite case of nucleophilic solvation (B), the solvent is basic and the solute acidic.

By taking specific and non-specific interactions into account, it has been possible to formulate a general semi-empirical equation comprising four terms describing the overall interaction Q between a solvent and a solute [35]

$$\delta Q = pf(n^2) + Yf(\varepsilon) + eE + bB \tag{7.5}$$

In this equation, the first term on the right-hand side is a function of the refractive index (n) and describes the polarizability; the second is a function of the relative electrical permittivity, i.e. a measure of the medium polarity; the third and fourth terms describe the electrophilic and nucleophilic solvations respectively; p, Y, e and b are constants.

Using equation (7.5) to make an approximate assessment of the overall solvent-solute interaction requires a knowledge of E and B. These parameters are not defined unequivocally, but in terms of a variety of comparative scales such as the Lewis scale of acidity and basicity, Gutmann's scale of donor-acceptor numbers, and a whole range of other scales which, though less commonly used, are constantly being corrected and extended in order to enhance their usefulness [32, 37, 38].

In principle, a donor strength scale should not be identified with the scales of acidity or basicity, since the latter refer to proton transfer [21, 22]. The donor strength is the total free energy of interaction between a donor and $SbCl_5$, including the electrostatic interactions in very dilute solutions of 1,2-dichloroethane [18]. However, there is a correlation between the donor strength scale and Trémillon's thermodynamic absolute acidity scale [35, 39], which is based on measurements of free energy changes in reactions of the type

$$SbF_5 + HF \rightleftharpoons SbF_6^- + H^+_{solv}$$

These scales can be interrelated if the free solvation energies of $SbCl_5$ and SbF_5 with the same chemical species are measured in the same solvents.

7.2 IONIZATION AND DISSOCIATION

Electrolyte solutions are conventionally divided into strong and weak. Strong electrolytes are completely dissociated into independently moving ions which are separate chemical species in the solution. In solutions of weak electrolytes the following forms of substances can exist in equilibrium: an undissociated molecule, an ionized but undissociated ion pair, and dissociated ions. Apart from simple solvated ions there may also be other charged forms such as ion triplets or larger ion clusters. By definition, thermodynamic dissociation constants K_D describe solutions and the properties of weak electrolytes, as they are the quotient of the activities of these forms: $K_D = a_{A^+} \cdot a_{B^-}/a_{AB}$. Hence, the ionic strength of an electrolyte solution depends on the composition of the solution and the kind of electrolyte. Salts that are completely ionized in the solid state, e.g. NaCl, LiCl, are solvated when dissolved in polar solvents with a high relative electrical permittivity (more generally – in ionizing solvents); each kind of ion is solvated independently. So when such a substance dissolves, its degree of ionization remains unchanged; only its crystal lattice breaks up and ionic aggregates dissociate at the expense of the solvation energy yielding a solution of a strong electrolyte. However, even in the case of very soluble salts, electrically neutral ion pairs or higher, charged ion associates appear in a homogeneous solution as the salt concentration increases or if a non-ionizing solvent is added. In non-polar solvents even completely ionized electrolytes, if soluble, remain mostly in the form of electrically neutral, undissociated ion pairs or clusters which do not participate in electrolytic conduction, which means that in such solvents they function as weak electrolytes. Such systems can be described by means of the dissociation constants of ion pairs K_D or ion triplets K_T (see Section 4.4). Instead of dissociation constants, association constants (which are their reciprocals) $K_A = 1/K_D$, $1/K_T$ etc. are often used. In this way it is emphasized that a system is a strong electrolyte in less concentrated solutions or in a strongly ionizing solvent, while the electrolyte in the solid state is an ionic crystal.

The ionization of neutral, covalently-bonded molecules, e.g. weak acids, is described by the ionization constant K_i

$$AB + xR \rightleftharpoons A^+R_xB^-, \qquad K_i = \frac{a_{A^+R_xB^-}}{a_{AB}} \tag{7.6}$$

where R is a solvent molecule. Ion pairs can dissociate into ions

$$A^+R_xB^- + yR \rightleftharpoons A^+(\mathrm{solv}) + B^-(\mathrm{solv})$$

Only the latter process is described by dissociation constants.

Dissociation constants can be determined by electrochemical methods, e.g. conductometrically, but ionization constants are better determined spectroscopically.

7.3 SOLUBILITY OF IONIC COMPOUNDS

The solubility and dissolution of salts or electrolytes which are completely ionized in the solid state (i.e. the ions are located at lattice nodes) depend on solvation. If the solvation energy is roughly the same as or greater than the lattice energy, the electrolyte dissolves into free, solvated ions. Both the electrical permittivity and the donor number of the solvent are then sufficiently high [40]. A detailed description of dilute and not-too-concentrated solutions is given by the Debye–Hückel–Onsager–Fuoss theory [19] and the Falkenhagen theory [41] (see Section 4.5). As the electrical permittivity of the solvent decreases, so does its capacity to solvate separate ions out of lattice nodes. This is reflected by the parallel increase of differences between the solvation energy and the lattice energy of an electrolyte. Solutions of weak electrolytes obtained in this way are described by the Arrhenius and Fuoss–Accascina theories [19].

Organic solvents such as acetyl chloride (DN $\approx$ 0.7, $\varepsilon = 15.8$), whose donor number [42] and electrical permittivity are both low, dissolve electrolytes only slightly. As the number of kinetic species (ions) capable of carrying charge is too small in such solutions, these are unsuitable for use in electrochemical cells. However, such solvents as the example given above, acetyl chloride, are sometimes used as an additive to a polar solvent, mainly to reduce the dynamic viscosity. In general, the ability of a solvent to form complexes with ions decreases with decreasing donor number, thereby lowering salt solubility.

The poor solubility of salts such as CuF is why some half-cells, e.g.

Cu/CuF, are irreversible. It is due to a high lattice energy resulting from the small size of both cation and anion. Replacing F^- by the large complex ion PF_6^- improves the solubility of the copper(I) ion because the electrostatic energy of interaction between Cu^+ and PF_6^- is much lower, so the lattice energy of the salt decreases [43].

These correlations allow one to predict which salts will dissolve well and in which organic solvents. Only those salts with low lattice energies in the solid state will dissolve well in non-aqueous solvents.

Salts containing large, barely polarizable ions such as ClO_4^-, BF_4^-, AsF_6^-, PF_6^- are readily soluble [10, 44]. In these ions, the central atom with a high formal valency, e.g. Cl(VII), B(III), As(V), P(V), attracts the electron pairs of anionic ligands like F^-, O^{2-}, forming strong covalent co-ordination bonds. The multiatomic complex anion formed in this way has a low surface charge density and a low polarizability because of the "counterpolarization" of the ligands by the central ion. Such anions interact only weakly with cations and belong to the "hard" class of ions [45]. It has already been stated that the lattice energy of compounds containing such ions is relatively low. The salts formed by small, monatomic cations and anions are usually sparingly soluble because of their fairly high surface charge density hence strong coulombic interactions. As monatomic, monovalent anions become larger, the surface charge density falls and the solubility of salts increases, e.g. in the series LiF, LiCl, LiBr, LiI. An increase in anionic charge reduces the solubility, so Li_2S is sparingly soluble, but $LiClO_4$ and KPF_6 are very soluble.

The same holds for cations, which can also be divided into small, barely polarizable or of high valency, the "hard" Li^+, Be^{2+}, Al^{3+}, Si^{4+}, and the larger, more easily polarizable low-valency cations, the "soft" Cs^+, Cu^+, Tl^+, Hg^{2+}. Ions like Cu^{2+}, Zn^{2+}, Pb^{2+}, Fe^{2+} are sometimes classified as intermediate.

7.4 SPECIFIC CONDUCTIVITY OF SOLUTIONS IN NON-AQUEOUS SOLVENTS

Electrolyte solutions to be used in non-aqueous cells must have a high specific conductivity [19], which is a measure of the ability to transport electrical charge per unit volume of electrolyte solution. The specific conductivity κ of a solution is the density of current i carried by 1 cm^3 of solution down unit gradient of an external electrical field $X = 1\ V cm^{-1}$

$$\kappa = \frac{i}{X} = \frac{\sum n_i z_i e v_i}{X} = \sum n_i z_i e u_i \tag{7.7}$$

The specific conductivity therefore depends on the summed product of the number of ions n_i per cm^3 of solution, their valencies z_i, the elementary charge e and the mobility of the ions in the medium u_i. The velocity of the ions is υ_i.

The flow of ions under stationary conditions is defined by two forces balancing one another: the external field force F_{ext} expressed by the product of the field gradient X (the potential gradient in a closed cell) and the ionic charge $z_i e$

$$F_{ext} = z_i e X$$

and the retarding force of the ions F_{ret}, which is the product of the ionic velocity v_i and the resistance coefficient ρ of the electrolyte solution

$$F_{ret} = \rho v_i = 6\pi r \eta v_i$$

Hence, under stationary conditions

$$z_i e X = 6\pi r \eta v_i$$

where η is the dynamic viscosity coefficient of the solvent. This coefficient is the principal macroscopic parameter, which includes and averages out those molecular properties of the solvent influencing the velocity of ions in the medium. The viscosity coefficient is a measure of the internal friction which an ion must overcome during its passage through the solution.

The specific conductivity is a complex, non-linear function which may attain a maximum value at a system-characteristic electrolyte concentration. A huge amount of experiment data has been accumulated on the variations in specific conductivity as a function of electrolyte concentration and temperature, which makes it possible to select systems of maximum conductivity [12, 46–50, 57–62]. The specific conductivities of electrolytes in various solvents are set out in Table 7.2, while Figs. 7.3 and 7.4 illustrate the specific conductivity-concentration relationships of some electrolytes.

Many systems are known in which at a given temperature there is no maximum on the specific conductivity vs. concentration curve [51]. Generally, these are solutions in which the solubility of the salt is poor. Figure 7.5 shows the specific conductivities for solutions of $LiClO_4$ in solvents with widely differing electrical permittivities and donor numbers [52, 53]. In 1,2-dimethoxyethane (1,2-DME) and in tetrahydrofuran (THF) lithium perchlorate behaves as a weak electrolyte, e.g. in pure dry THF the dissociation constant of $LiClO_4$ [54] is ca. 4×10^{-8}. Although ε is ca. 7 for both THF and 1,2-DME, these two solvents differ considerably in their ability to dissolve $LiClO_4$. The maximum solubility of $LiClO_4$ in 1,2-DME

Table 7.2 Specific conductivity of electrolyte solutions in aprotic and non-protonating solvents at 25°C

Solvent	Electrolyte	Concentration, mol dm^{-3}	Specific conductiviy, κ, $\Omega^{-1}cm^{-1}$	References
1	2	3	4	5
Aprotic solvents				
$POCl_3$	$LiPF_6$	0.4	3.7×10^{-3}	[113]
$POCl_3$	$LiAlCl_4$	0.9	7.8×10^{-3}	[113]
$POCl_3$	$LiBCl_4$	1.2	8.7×10^{-3}	[113]
$POCl_3$	$LiBF_4$	1.5	1.1×10^{-3}	[113]
$POCl_3$	$LiSbCl_6$	1.0	2.2×10^{-3}	[113]
$SOCl_2$	$LiAlCl_4$	1.0	1.46×10^{-2}	[113]
$SOCl_2$	$LiAlCl_4$	2.0	2.04×10^{-2}	[113]
$SOCl_2$	$LiSbCl_6$	1.0	1.25×10^{-2}	[113]
SO_2Cl_2	$LiAlCl_4$	1.0	7.4×10^{-3}	[113]
Non-protonating solvents				
Acetonitrile	$LiClO_4$	about 1	3.6×10^{-2}	[5]
	KPF_6	0.93	4.8×10^{-2}	[5]
	KSCN	1.6	2.2×10^{-2}	[5]
	LiCl	2.2	4.3×10^{-4}	[5]
Propylene carbonate	$LiAlCl_4$	0.6	6.6×10^{-3}	[5]
(PC)	$LiClO_4$	0.66 mol kg^{-1}	5.4×10^{-3}	[2]
	KPF_6	0.5	6.8×10^{-3}	[5]
	$LiPF_6$	0.86 mol kg^{-1}	5.41×10^{-3}	[2]
	KSCN	1.097 mol kg^{-1}	7.0×10^{-3}	[47]
	LiCl	1	3.4×10^{-4}	[2]
	$LiBF_4$	1	4.0×10^{-3}	[2]
	LiBr	1.15 mol kg^{-1}	1.8×10^{-3}	[2]
PC (42.13% by mass) + 1,2-DME	$LiClO_4$	1.39 mol kg^{-1}	1.46×10^{-2}	[47]
Ethylene carbonate	$LiClO_4$	1.0	7.9×10^{-3}	[5]
γ-Butyrolactone (BL)	$LiClO_4$	1.1	7.6×10^{-3}	[52]
γ-BL + 1,1-DME (mole fraction 0.496)	$LiClO_4$	0.75	4.5×10^{-3}	[51]
γ-BL + 1,2-DME (mole fraction 0.49)	$LiClO_4$	1.1	1.3×10^{-2}	[52]
Tetrahydrofuran	$LiClO_4$	1.49	4.3×10^{-3}	[53]
	LiCl	0.16	1.6×10^{-4}	[2]
	$LiAlCl_4$	1.6	1.7×10^{-2}	[5]
	$LiBF_4$	2.6	4.0×10^{-3}	[5]

1	2	3	4	5
Dimethyl sulphoxide	$LiClO_4$	1.50	1.0×10^{-2}	[5]
	KPF_6	2.50 (sat.)	7.92×10^{-3}	[47]
	LiCl	2.13 mol kg^{-1}	2.92×10^{-3}	[47]
	NH_4PF_6	1.5 (sat.)	1.37×10^{-2}	[47]
Dimethylformamide	$LiClO_4$	1.16	2.2×10^{-2}	[2]
	LiCl	1 mol kg^{-1}	8.3×10^{-3}	[2]
	KPF_6	about 1	2.4×10^{-2}	[5]
Methyl formate	$LiClO_4$	about 3	2.8×10^{-2}	[5]
Propionitrile	$LiClO_4$	about 1	1.6×10^{-2}	[5]
Water (for comparison)	NaCl	0.5	4.7×10^{-2}	[5]
2-Me-THF	$LiAsF_6$	1.5	4.0×10^{-3}	[2]

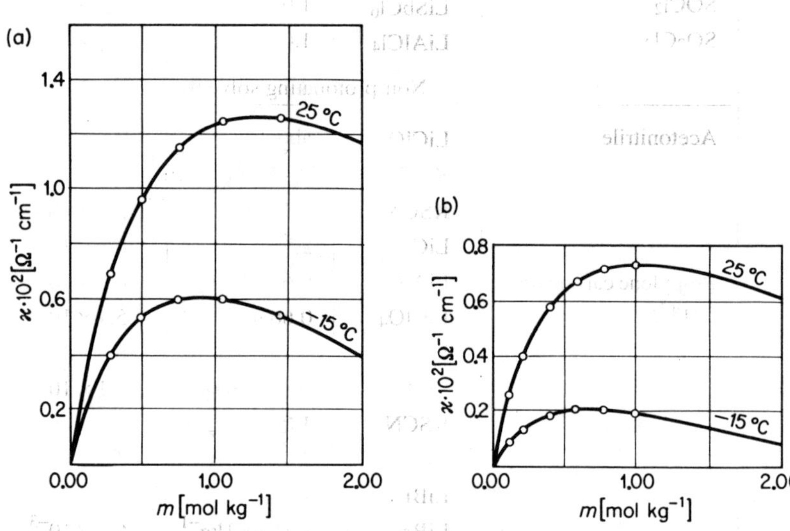

Fig. 7.3 Specific conductivity [46] of $LiClO_4$ solutions in: (a) a mixture of propylene carbonate (42.13% by mass) and 1,2-DME; (b) neat propylene carbonate.

is not even 0.4 mol dm^{-3}, whereas concentrations of 1.3 mol dm^{-3} can be achieved in THF. Nevertheless, neither of the specific conductivity curves of $LiClO_4$ solutions in these solvents attain a maximum (Fig. 7.5) – curves *1* and *2* rise monotonically. The specific conductivities of almost saturated solutions of $LiClO_4$ in THF are high enough for them to be applicable in non-aqueous cells. Distinct κ maxima can be seen on the specific conductivity curves of $LiClO_4$ (Fig. 7.5) for γ-butyrolactone (γ-BL) (curve *3*) and for an equimolar mixture of γ-BL and 1,2-DME (curve *4*). At the same time, there are considerable differences between the conductivity of $LiClO_4$ in pure γ-BL and in its mixtures with 1,2-DME [55, 56] (Figs. 7.6

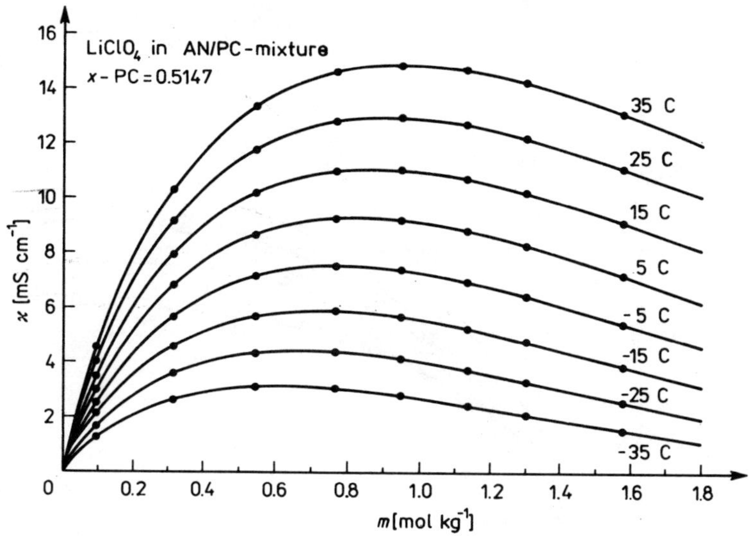

Fig. 7.4 Specific conductance of $LiClO_4$ solutions in a mixture of propylene carbonate and acetonitrile at –35 + 35°C, x – PC = 0.5147 mole fraction [48].

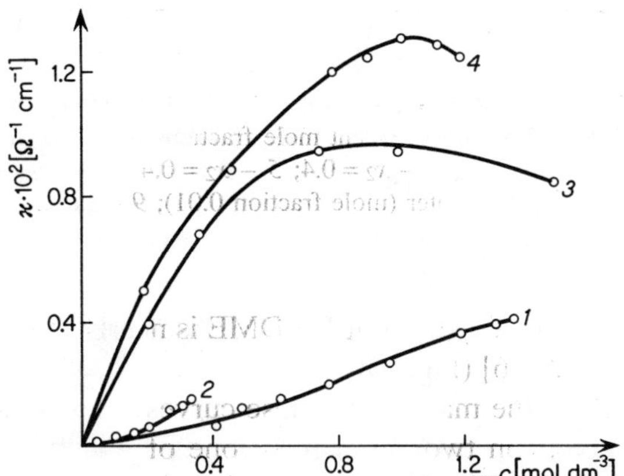

Fig. 7.5 Specific conductivity [52, 56] of $LiClO_4$ solutions as a function of concentration at 25°C; *1* – solutions in tetrahydrofuran; *2* – solutions in 1,2-dimethoxyethane; *3* – solutions in γ-BL; *4* – solutions in an equimolar mixture of γ-BL and 1,2-DME.

and 7.7). Values of κ_{LiClO_4} are higher for the mixed solvent [49, 50, 56, 58]. The κ_{max} conductivity of $LiClO_4$ was found to be highest in an equimolar mixture of γ-BL and 1,2-DME at 25°C (Table 7.3).

The reason for the observed increase in the maximum conductivity of lithium perchlorate in these mixtures is the effect of solvent viscosity on ionic mobility, which is generally underestimated. The viscosity of a 1:1

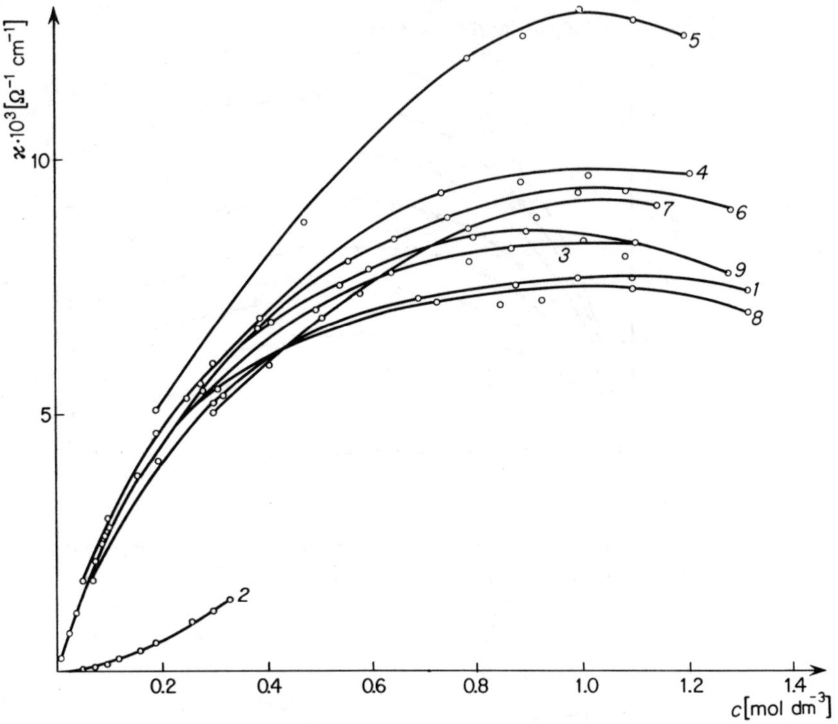

Fig. 7.6 Specific conductivity [55] of $LiClO_4$ solutions in mixtures of γ-BL and 1,2-DME at 25°C for different mole fractions of 1,2-DME (x_2): *1* – $x_2 = 0$; *2* – $x_2 = 1$; *3* – $x_2 = 0.1$; *4* – $x_2 = 0.4$; *5* – $x_2 = 0.49$; *6* – $x_2 = 0.6$; *7* – $x_2 = 0.7$; *8* – $LiClO_4$ in γ-BL + water (mole fraction 0.01); *9* – $LiClO_4$ in γ-BL + water (mol. fraction 0.07).

molar mixture of γ-BL and 1,2-DME is nearly three times lower than that of pure γ-BL [56] (Fig. 7.8).

Generally, the maxima on these curves occur when the quantity measured depends on two parameters, one of which is increasing, while the other is falling. The Molenat equation derivative [63, 64], which defines the specific conductivity of an electrolyte solution

$$\kappa = \Lambda \rho \tag{7.8}$$

where Λ is the molar conductivity and ρ is the electrolyte concentration expressed in mol cm^{-3}, results in

$$d\kappa = \Lambda d\rho + \rho d\Lambda \tag{7.8a}$$

If ρ is assumed to increase monotonically, changes in $d\Lambda$ may either be equal to zero, or be more or less than zero.

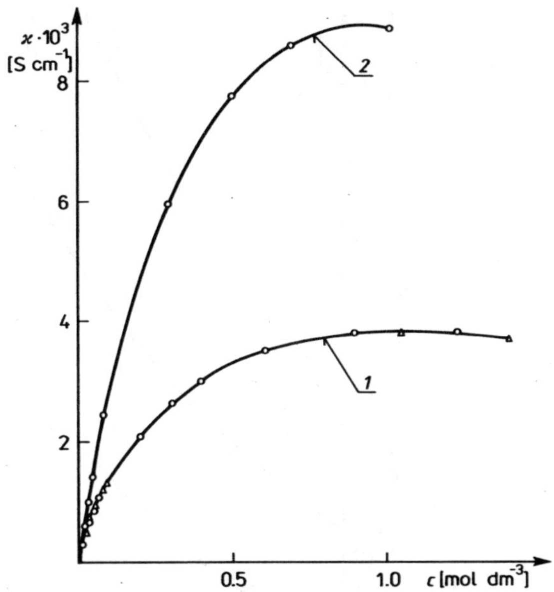

Fig. 7.7 Specific conductance vs. concentration c plot for $LiBF_4$ (*1*) and $LiAsF_6$ (*2*) in γ-BL and 2-methyl-THF ($x_2 = 0.5$ mol. fraction) mixtures at 25°C [50, 62].

It is known from the nature of the relationship between the molar conductivity and the concentration, that changes in $d\Lambda$ with increasing concentration are negative and the curve is then a decreasing one. In some systems in which the electrical permittivity is low, this curve decreases to a minimum [53] (Fig. 7.9) and subsequently begins to rise as $d\Lambda$ passes from negative to positive values. This is probably due to the appearance of charged ion triplets. Specific conductivity maxima κ_{max} ($d\kappa = 0$) are possible only in the former case, that is, in systems in which $d\Lambda < 0$, and $d\rho > 0$. This is the result of two opposing trends: the increase in the num-

Table 7.3 Maximum specific conductivity κ_{max} of lithium perchlorate solutions in mixtures of γ-butyrolactone and 1,2-dimethoxyethane at 25°C and −30°C [55, 56]

Mole fraction of 1,2-DME in the mixture, x_2	25°C		−30°C	
	$\kappa_{max} \times 10^3$, $\Omega^{-1}cm^{-1}$	c_{max}, $mol dm^{-3}$	$\kappa_{max} \times 10^3$, $\Omega^{-1}cm^{-1}$	c_{max}, $mol dm^{-3}$
0	6.3	0.95	2.50	0.80
0.103	8.55	1.03	3.00	0.80
0.498	10.4	1.25	4.30	0.90
0.499	13.00	1.2	–	–
0.596	9.45	1.06	4.40	0.90
0.698	9.35	1.00	4.50	1.00

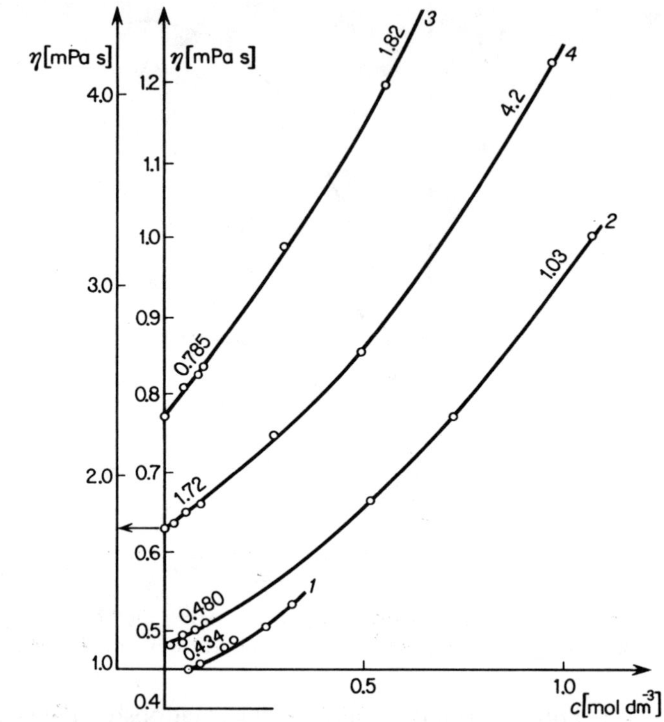

Fig. 7.8 Coefficient of dynamic viscosity η as a function of $LiClO_4$ concentration at 25°C [56]: *1* – solutions in 1,2-DME; *2* – solutions in THF; *3* – solutions in an equimolar mixture of γ-BL and 1,2-DME; *4* – solutions in γ-BL. The figures on the curves are the values of η for extreme concentrations of $LiClO_4$, i.e. 0, and ca. 1.1 mol^{-1}.

ber of charged species per unit volume and the concurrent reduction in their mobility as the electrolyte concentration increases. This is in agreement with the theory of conductivity described in Section 4.5.

Casteel and Amis [64] have put forward a semi-empirical equation linking the specific conductivity κ with its maximum value κ_{max} and the electrolyte concentration m at a given temperature:

$$\kappa = \kappa_{max}\left(\frac{m}{m_{max}}\right)^{a}\exp\left[b(m-m_{max})^2-\frac{a}{m_{max}}(m-m_{max})\right] \qquad (7.9)$$

The constants a and b are calculated from experimental data by the least squares method, m is the electrolyte concentration in moles per kg of solvent and m_{max} is the concentration at which κ reaches a maximum (κ_{max}) for the system. Equation (7.9) satisfactorily describes the experimental

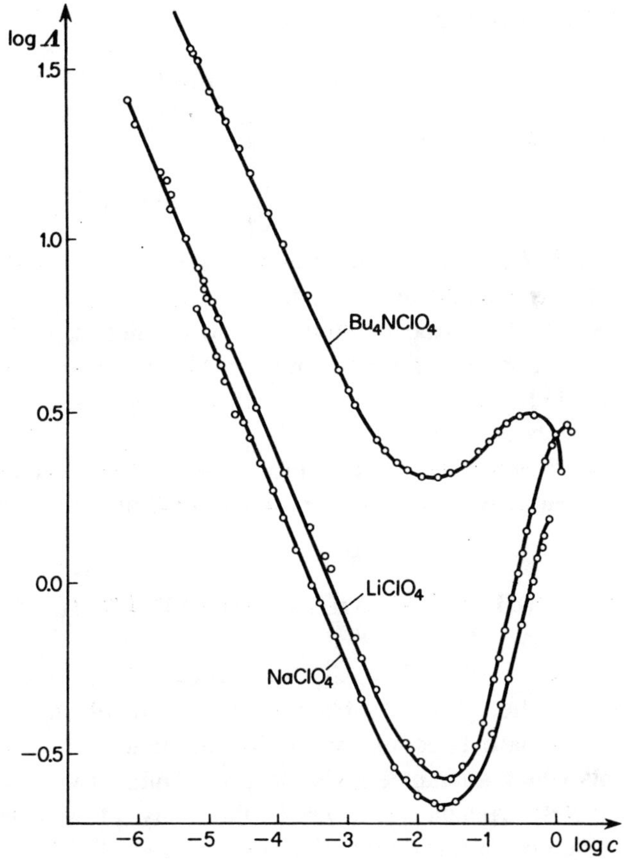

Fig. 7.9 Molar conductivity $\Lambda\,[\Omega^{-1}\text{cm}^2\text{mol}^{-1}]^{-1}$ of $LiClO_4$, $NaClO_4$ and Bu_4NClO_4 solutions in tetrahydrofuran at 25°C [53] as a function of salt concentration expressed in mol dm^{-3}.

values of κ as a function of m for electrolytes in mixed solvents, e.g. propylene carbonate and 1,2-DME [46], and propylene carbonate and acetonitrile [48] (see Fig. 7.3).

Some predictions can be made with equation (7.9). The specific conductivity maximum in a system at constant temperature and fixed solvent composition may occur only when the electrolyte is sufficiently soluble, otherwise the curve of conductivity vs. electrolyte concentration rises monotonically without reaching a maximum. In the systems where it is possible to obtain κ_{max}, the derivative m_{max} with respect to temperature is always greater than zero $(\partial m_{max}/\partial T) > 0$, which implies that the conductivity maximum appears at ever-higher electrolyte concentrations as the temperature rises.

The general empirical equation defining the conductivity-temperature relationship [65], which holds for electrolyte solutions of both high and medium concentrations, is

$$\ln \kappa(m,T) = a_0 + a_1 T^{-1} + a_2 T^{-2} \ldots \quad (7.10)$$

Differentiating yields

$$\frac{\mathrm{d}\ln \kappa(m,T)}{\mathrm{d}(1/T)} = a_1 + 2a_2 T^{-1} \quad (7.11)$$

The plot from this equation is a straight line from which the temperature coefficient can be determined.

So far, it has been impossible to predict κ_{max} of an electrolyte with respect to the composition of a mixed solvent from the appropriate equations. In this case, the system with the largest κ_{max} can only be found experimentally.

Much research work is in progress on concentrated electrolyte solutions and their possible application in non-aqueous cells [59–62].

7.5 STABILITY OF SOLVENTS AND THEIR SOLUTIONS

The factors which intensify solvent-electrolyte interactions also tend to increase the reactivity of a solvent towards electrode materials and may cause its partial decomposition. In particular, most of the non-aqueous solvents which guarantee good electrical conductivity react chemically with active electronegative materials, thus only a few of them, e.g. liquid ammonia, do not react with alkali metals (see Chapter 3). Alkali metals dissolve in liquid ammonia, with solvates of these metals (which are still formally in the zero oxidation state) and solvated electrons appearing in the solutions. Until recently [67] it was believed that lithium does not react with chain ethers, but the latest spectroscopic studies by Aurbach *et al.* [16] have indicated that lithium alcoholates, e.g. $LiOCH_3$ or $LiOCH_2CH_2OLi$, might be formed on the metal surface in 1,2-DME solutions. Lithium alkyl carbonates RCO_3Li may form in the presence of CO_2 in air.

During the reaction an electron is transferred from a lithium atom to a 1,2-DME molecule, forming a radical anion stabilized by the Li^+ cations. The radical subsequently decomposes. This two-step reaction yields lithium alcoholate and ethylene. It should be added that Aurbach *et al.* [16] did not recover the reaction products – they only observed the spectra of the lithium surface.

Spectroscopic studies have also revealed that inorganic compounds of lithium containing oxygen and nitrogen are present in the films passivating lithium in 1,2-DME solutions. If solutions of an electrolyte, e.g. $LiClO_4$ in

0.1 M 1,2-DME contain a certain amount of water (> 0.005 M), the principal reaction product in the passivating film is partially hydrated lithium hydroxide

$$Li_2O \xrightarrow{H_2O} 2LiOH \xrightarrow{2\,H_2O} 2LiOH \cdot H_2O$$

In previous chapters, attention was drawn to the fact that in most organic solvents, the reaction products are deposited as passivating films on active metals like lithium, which are therefore suitable for use in cells, since the quantity of products of the reaction with the anode material is relatively small – the passivating film coats the metal surface and impedes further chemical reaction [68].

Appropriately-equipped laboratories are now studying the nature of the products of the reaction between solvents and metals (mainly lithium) [69–72]. As yet, only a few of the reaction mechanisms giving rise to identified products have been fully confirmed. Among the earliest was trimethylene carbonate (1,3-propylene carbonate), whose reactions with alkali metals are quite well known [75]. Trimethylene carbonate reacts irreversibly with lithium to form propene and lithium carbonate

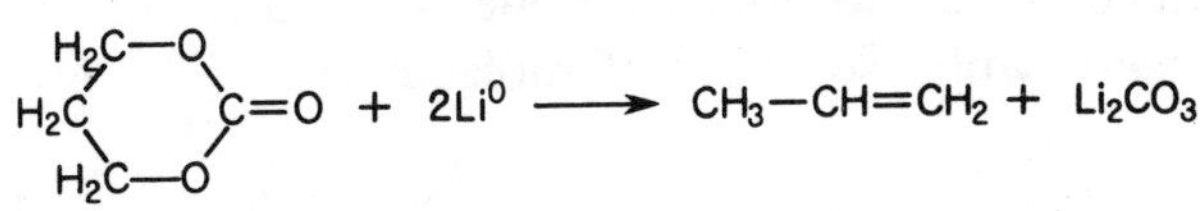

This irreversibly breaks down the solvent, and lithium carbonate passivates the metal. If the trimethylene carbonate contains traces of water, hydrolysis may occur, induced by the lithium hydroxide formed during the decomposition of water by lithium

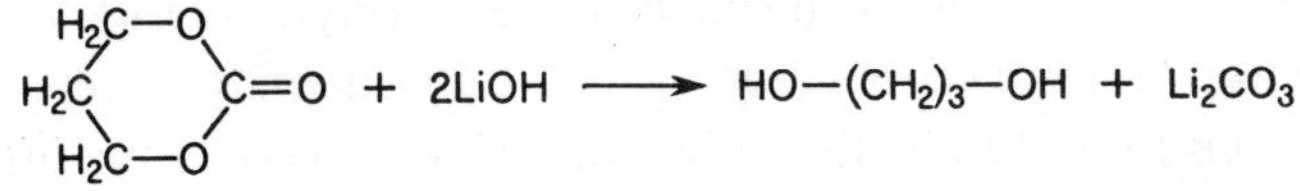

Thus the reaction products are lithium carbonate and 1,3-propanediol (propylene-1,3-glycol). Propylene carbonate reacts very rapidly in a similar way [69–71].

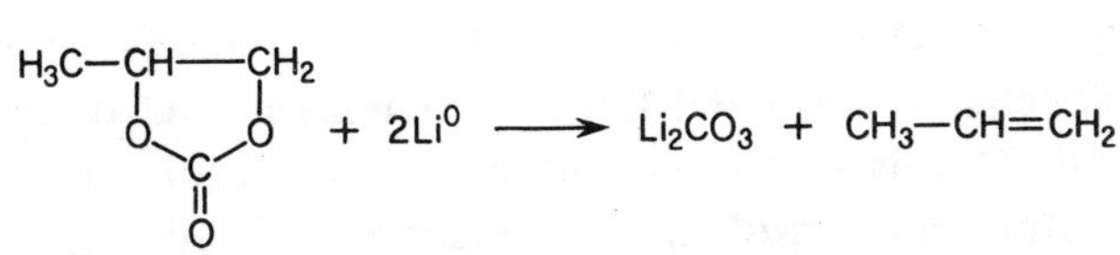

Koch's studies [15] of the tetrahydrofuran-lithium reaction have shown

that the principal reaction products are lithium butenolate and gaseous hydrogen

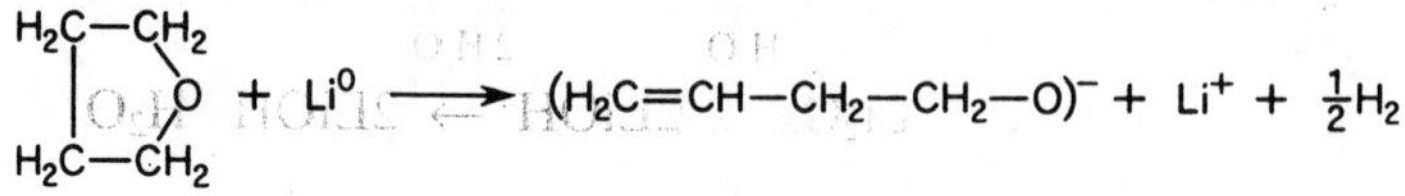

Dimerization leading to the 1,8-diolate without the evolution of gaseous hydrogen also probably occurs

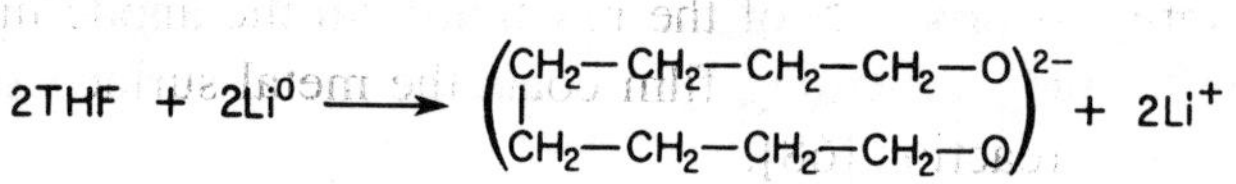

This is the "reductive splitting" of a cyclic ether

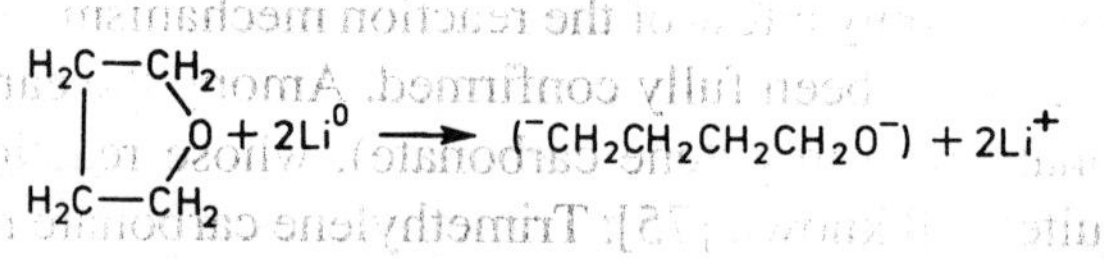

Further reaction products include the acetaldehyde enolate anion and ethylene, which are formed in the consecutive reaction of the butenolate carbanion with a second THF molecule

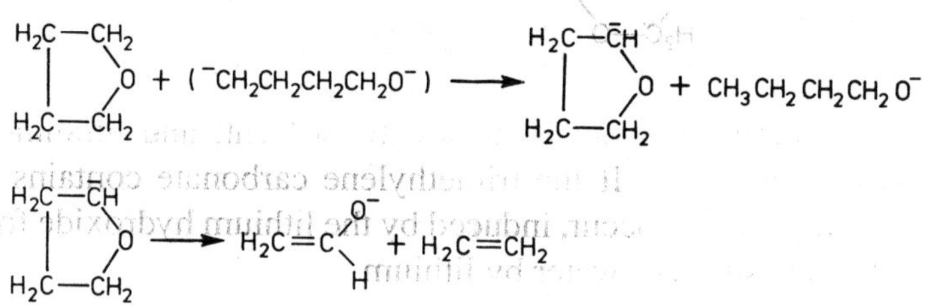

Lithium metal can also reduce the electrolyte. If $LiAsF_6$ is used, the AsF_6^- anion is reduced to AsF_3 which subsequently reacts with the organic anions produced by the reduction of THF to cover the lithium surface with an insoluble, brown, passivation film consisting of LiF and the polymerized products of this reaction. The presence of this layer inhibits further decomposition of the electrolyte solution.

2-Methyltetrahydrofuran is a solvent rather less reactive towards lithium [76]. The substitution of a methyl group in the 2-position with respect to the oxygen atom raises the energy of the lowest unfilled molecular orbital (LUMO) and reduces its reactivity towards the alkali metal. This is tantamount to an increase in the activation energy of an electron transferred in the reaction of lithium with 2-methyltetrahydrofuran, as compared with its reaction with THF or 3-methyl-THF [77]; Kerr [72] expresses the view that the mechanism of the reduction of

2-methyl-THF by lithium is similar to that of the Li–THF reaction, except that the former reaction proceeds far more slowly.

An example of the instability of a solvent towards an electrolyte is the reaction of methyl chloroformate with $LiAlCl_4$ [75]. This double salt is in equilibrium with $AlCl_3$ which catalyzes a multi-step reaction forming gaseous methyl chloride and carbon dioxide

Thionyl chloride reacts with a number of metals directly [78]. Metals whose chlorides have higher heats of formation than their oxides react preferentially yielding chlorides; this applies to sodium and other Group IA metals, as well as to silver, tin and antimony. Sulphur monoxide is rather unstable and reacts further

$$\begin{matrix} H_3C\text{–}O \\ Cl \end{matrix}\!\!\Big\rangle C=O + AlCl_3 \rightarrow CH_3Cl + CO_2 + AlCl_3$$

$$SOCl_2 + 2M \rightarrow 2MCl + SO$$

$$2SO \rightarrow O{=}S{=}S{=}O \rightarrow SO_2 + S$$

and

$$xSO \rightarrow (SO)_x \rightarrow \text{polymerization}$$

It may also react with metals forming sulphates and sulphides.

Sulphur dioxide reacts with monovalent metals yielding hyposulphites

$$2SO_2 + 2M \rightarrow M_2S_2O_4$$

Metals whose oxides have a higher heat of formation than their chlorides form the corresponding oxides. These are Group 2 and 3 metals, i.e. magnesium, zinc, aluminium, etc.

$$SOCl_2 + M \rightarrow MO + SCl_2$$

Sulphur dichloride reacts further with divalent metals with the formation of sulphides and chlorides

$$2M + SCl_2 \rightarrow MCl_2 + MS$$

The overall reaction is

$$3M + 2SOCl_2 \rightarrow 2MCl_2 + MS + SO_2$$

Sulphur dioxide reacts further

$$2SO_2 + M \rightarrow MS_2O_4$$

Sulphuryl chloride reacts with metals in very much the same way as thionyl chloride: chlorides and SO_2 are formed, which may react further in the above manner, but these reactions are retarded by passivating films, so that SO_2Cl_2 apparently does not attack aluminium and magnesium at room temperature. The presence of moisture (water) accelerates the reactions considerably; this is understandable, seeing that the solubility increases, and other reaction products are formed, e.g. alkali and alkaline earth metal hydroxides of a fairly loose structure. At elevated temperatures all metals react with SO_2Cl_2 to yield chlorides, e.g. the reaction with gold gives gold(III) chloride while that with platinum yields platinum(IV) chloride.

Oxides of metals, e.g. divalent metals, also react with sulphuryl chloride to give metal chlorides and sulphates

$$2MO + SO_2Cl_2 \rightarrow MSO_4 + MCl_2$$

for example

$$2HgO + SO_2Cl_2 \rightarrow HgCl_2 + HgSO_4$$

The reactivity of sulphur dioxide has already been discussed.

One solvent in which lithium salts dissociate well and which is used as a component of mixtures when preparing electrolyte solutions is γ-butyrolactone (γ-BL) [47, 56, 58]. However, γ-BL decomposes at clean surfaces of lithium, sodium, potassium and their amalgams [79] with the evolution of methane and smaller quantities of volatile hydrocarbons and the precipitation of lithium carboxylates. It is also probable that lithium salts of suberic acid $(CH_2)_6–(CO_2)_2Li_2$ are formed at the lithium electrode when γ-BL ring is opened:

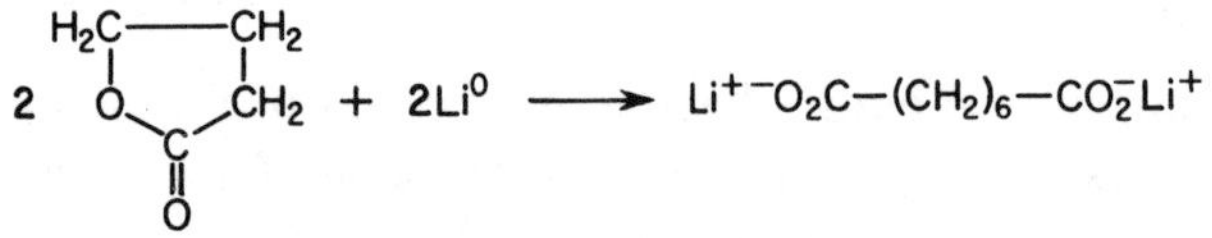

A no-less-important though separate problem is the electrochemical stability of electrolyte solutions at suitably high voltages. This problem usually arises when storage batteries are being charged. The high voltages may then cause the electrolyte to decompose. For example, the perchlorate ion is oxidized at the anode to chloride dioxide ClO_2 and free oxygen at voltages in excess of +3 V [75, 80]

$$ClO_4^- - e \rightarrow ClO_2 + O_2$$

Even trace amounts of water can reduce the stability of organic solvents and their electrolyte solutions, hence the importance of removing it and determining it. The water content is determined *inter alia* by sensitive electrochemical methods [81]. Currently, the automatic titrators based on Karl Fischer's method [16] are the most convenient for quick and accurate water determinations, but gas chromatography and other methods can be used.

7.6 SOLVENTS IN NON-AQUEOUS STORAGE BATTERIES

The principal object of current research is to improve the stability of electrolyte solutions used in high-energy non-aqueous storage batteries. A study of rechargeable lithium cells for space applications has shown that the main factor limiting the cycle life of the cell is the electrolyte [124].

If a solvent is to be used in such batteries, it must satisfy several stringent requirements. As we know it must have a low melting point and high boiling point, a high ability to dissolve electrolytes and to solvate lithium ions, low viscosity, high electrical permittivity and, last but not least, the greatest possible chemical and electrochemical stability [82, 83].

These objectives are achieved in several ways. Stability-increasing functional groups are introduced into solvent molecules, e.g. 2-methyltetrahydrofuran and 4-methyldioxolane are more stable than unsubstituted THF or DOL. Polymerization is inhibited by adding such compounds as hexamethylphosphoramide (HMPA) and 2-methylfuran. More stable solvents like sulpholane (SH) can be added.

Important criteria in the assessment of a solvent's usefulness are its donor and acceptor numbers, which define its ability to solvate anions and cations [84].

Binary mixtures of solvents are used in high-energy storage batteries, e.g. propylene carbonate + sulpholane [85], propylene carbonate + 1,2-DME [86], propylene carbonate + acetonitrile [87], and sulpholane + 1,2-DME [88]. 1,2-DME seems to be the most common additive to other solvents [89]. New solvent systems like substituted aliphatic nitriles [90] or alkoxydioxolanes [91] are continually being introduced. Liquid polymer electrolytes are also investigated [125].

With regard to the reconstitution of lithium anode and of cathode materials, and the efficient cyclic operation of a cell, it is particularly important that passivating films, when formed, are thin and readily removable.

Summarizing, the most efficient cyclic operation with lithium electrodes has been achieved with ether solvents such as methyl-THFs, dialkyl ethers, dioxolane and 1,2-DME.

However, even the most carefully selected mixtures of organic solvents are not always sufficiently stable, especially after a large number of charge/discharge cycles. That is why attention has recently turned to low-melting mixtures of salts, liquid at room temperature, which may be eutectics without the addition of a separate solvent. An example is the ammonium salt-$AlCl_3$ mixture. This melts at $-75°C$, and has a conductivity of the order of 3.4 $mS\,cm^{-1}$ at $+20°C$. Amide-nitrate mixtures are also low-melting [92, 93].

It is precisely these low-melting eutectic mixtures of organic and inorganic salts not containing organic solvents which may turn out to be the best electrolytes for high-energy storage batteries capable of a large number of deep charge/discharge cycles.

7.7 SOLID ELECTROLYTES

The durability required of high-energy cells and the need for miniaturization in microelectronics – principally by reducing cell size, and thus the distance between the electrodes – have prompted the search for solid electrolytes which are fairly rigid and mechanically strong. The use of such electrolytes minimizes self-discharge even at elevated temperatures (up to 100–150°C), and allows the distance between the electrodes to be reduced to a few tenths of a millimetre.

There are two categories of solid electrolyte: salts ionized in the solid state, sometimes with added metal oxides, and solid organic polymers with added ionized salts.

The former type was the first to appear; LiI, and LiI with Al_2O_3, as aluminium oxide (alumina) had already been used for this purpose in high-temperature cells [94]. The small, mobile lithium cations diffusing down a potential gradient conduct the current, but their use is restricted by the low current intensity – up to only 20 μA cm^{-2} at room temperature – because of the high electrical resistance. Nowadays lithium salts or glasses containing $2.5\,LiI \cdot Li_4P_2S_7$ are used in lithium cells which at room temperature supply a temporary current density of up to 1–2 mA cm^{-2} and a continuous current density of 0.1 mA cm^{-2} at a specific conductivity of the order of 10^{-3}–10^{-4} S cm^{-1} [95].

Calcium boro-aluminate glasses have been used in lithium cells and will no doubt also be used in calcium anode cells [96].

The highest conductivity so far obtained, of the order of 10^{-1}–10^{-2} S cm^{-1}, has been with the solid electrolyte of composition $Rb_4Cu_{16}I_6Cl_{14}$ used in special-purpose copper anode cells (see Chapter 8) [97].

Another way of obtaining a solid electrolyte is to use organic polymers containing donor functional groups, with atoms like oxygen or sulphur, acting as Lewis bases. These atoms form complexes with the cations of the anode material, lithium ions in particular. Cation-exchange resins (cationites) are suitable for this purpose.

Poly(ethylene oxide) (PEO) with added $LiClO_4$ or $LiCF_3SO_3$ [98] with a composition of $(PEO)_9LiClO_4$ and $(PEO)_9LiCF_3SO_3$ [99] have already been used. Other PEO-based solid electrolytes with added propylene carbonate, ethylene carbonate and $LiClO_4$ in various proportions [100] have also been prepared. They ensure elevated conductivity and are eminently suitable for high-energy storage batteries.

A polymer of the type $(PEO)_{20}LiClO_4$ has been tested in a storage battery with polypyrrole, polydithiophene and polydithienethiophene electrodes [101, 126].

REFERENCES

[1] R. Jasiński, *Electrochem. Technol.*, **6**, 28 (1968).
[2] W.R. Fawcett, Electrochemical Symposium (Polish Chemical Society and Polish Academy of Sciences), Jachranka near Warsaw (1981).
[3] J.P. Gabano (ed.), *Lithium Batteries*, Academic Press Inc., New York (1983).
[4] Y. Marcus and A.S. Kertes, *Ion Exchange and Solvent Extraction of Metal Complexes*, Wiley-Interscience, New York (1969).
[5] L. Werblan and A. Cisak, *Wiadomości Chem.*, **27**, 843 (1973).
[6] Y.Y. Fialkov, A.N. Zhitomirskii and Y.A. Tarasenko, *The Physical Chemistry of Non-Aqueous Solutions* (translated from Russian), PWN, Warszawa (1983).
[7] T. Moeller, *Inorganic Chemistry*, Chs. 9 and 10, PWN, Warszawa (1959).
[8] I.M. Kolthoff, *Non-Aqueous Electrochemistry* (J.C. Marchon, ed.), Butterworths, London (1971), p. 312.
[9] A.F.O. Germann and C.R. Timpany, *J. Am. Chem. Soc.*, **47**, 2275 (1925).
[10] B.E. Conway, Electrochemical Symposium (Polish Chemical Society and Polish Academy of Sciences), Jachranka near Warsaw (1981).
[11] M.R.J. Dack (ed.), *Techniques of Chemistry*, Vol. 8, Part I, Wiley, New York (1976).
[12] J. Barthel, H.J. Gores, G. Schmeer and R. Wachter, in: *Physical and Inorganic Chemistry*, Akademie-Verlag, Berlin (1983), p. 33.
[13] J. Burgess, *Metal Ions in Solutions*, Ellis Horwood, Chichester (1978).
[14] W.B. Jensen, *The Lewis Acid-Base Concepts*, Wiley, New York (1980).
[15] V.R. Koch, *J. Electrochem. Soc.*, **126**, 181 (1979).
[16] D. Aurbach, M.L. Daroux, P.W. Faguy and E. Yeager, *J. Electrochem. Soc.*, **135**, 1863 (1988).
[17] V. Gutmann, *The Donor-Acceptor Approach to Molecular Interactions*, Plenum Press, New York (1978).
[18] Topics in Current Chemistry, Vol. 27 (V. Gutmann, ed.), Springer-Verlag, Berlin (1972), p. 59.
[19] R.M. Fuoss and F. Accascina, *Electrolytic Conductance*, Interscience Publ., New York (1959).
[20] R. Cookson, *Chem. Rev.*, **74**, 5 (1974).
[21] T.M. Krygowski, E. Milczarek and P.K. Wrona, *J. Chem. Soc., Perkin Trans. II*, **1980**, 1563.
[22] J.S. Jaworski and M.K. Kalinowski, Similarity models in organic electrochemistry, in: *Similarity Models in Organic Chemistry, Biochemistry and Related Fields* (R.J. Zalewski, T.M. Krygowski and L. Shorter, eds.), Ch. 8, Elsevier, Amsterdam (1991).
[23] C. Reichardt, *Solvent Effects in Organic Chemistry*, Verlag Chemie, Weinheim, New York (1979).
[24] W. Pająk and A. Cisak, *Bull. Pol. Acad. Sci. Chem.*, **34**, 359 (1986).
[25] G.W. Heise and N.C. Cahoon, *The Primary Battery*, Vol. 1, Wiley and sons, New York (1971), p. 357.
[26] M. Born, *Z. Phys.*, **1**, 45 (1920).
[27] W. Kołos, *Quantum Chemistry* (in Polish), Ch. 18, PWN, Warszawa (1986), p. 266.

[28] J.D. Bernal and R.H. Fowler, *J. Chem. Phys.*, **1**, 515 (1933).
[29] A.D. Buckingham, *Quart. Rev. Chem. Soc., London,* **13**, 183 (1959); *Disc. Faraday Soc. on Electrolytes*, **24**, 151 (1957).
[30] K.P. Mishchenko and G.M. Poltoratski, *Thermodynamic and Structure of Aqueous and Nonaqueous Electrolyte Solutions* (in Russian), Chemistry, Leningrad (1976).
[30a] M. Salomon and E.J. Plichta, *Electrochim. Acta*, **28**, 1681 (1983).
[31] Y. Marcus, *Ion Solvation,* Ellis Horwood , Chichester, J. Wiley (1985), p. 306.
[32] G.A. Krestov (ed.), *Ionic Solvation* (in Russian), Nauka, Moskva (1987), Chs. I and II.
[33] L. Blum, *J. Phys. Chem.*, **92**, 2969 (1988).
[34] V. Gutmann and E. Wychera, *Inorg. Nucl. Chem. Lett.*, **2**, 257 (1966).
[35] I.A. Koppel and V.A. Palm, *Advances in Linear Free Energy Relationships*, Ch. 5, Plenum Press, London–New York (1972).
[35a] R.H. Stokes and R. Mills, *Viscosity of Electrolytes and Related Properties*, Vol. 3, Pergamon Press Ltd. (1965), p. 74.
[36] L. Werblan, in: *The Electrochemistry of Interfaces and Ionogenic Media* (in Polish) (W. Libuś and E. Dutkiewicz, eds.), PWN, Warszawa (1984), p. 37.
[37] M.J. Kamlet and R.W. Taft, *J. Chem. Soc., Perkin Trans. 2*, **1979**, 337.
[38] E.M. Kosower, *An Introduction to Physical Organic Chemistry*, J. Wiley, New York (1968).
[39] P.-L. Fabre, J. Devynck and B. Tremillon, *Chem. Rev.*, **82**, 591 (1982).
[40] B. Kratochvil and H.L. Yeager, in: *Topics in Current Chemistry* (V. Gutmann, ed.), Vol. 27, Springer-Verlag, Heidelberg–New York (1972), p. 1.
[41] H. Falkenhagen and W. Ebeling, *Theorie der Elektrolyte*, S. Hirzel Verlag, Leipzig (1971).
[42] V. Gutmann and K.H. Wegleitner, *Z. Electrochem.*, **77**, 77 (1972).
[43] G. Clinckspoor, H. Schmidt and W. Vielstich, *Chemie-Ing.-Techn.*, **42**, 206 (1970).
[44] G. Eichinger and J.O. Besenhard, *J. Electroanal. Chem.*, **72**, 1 (1976).
[45] R.G. Pearson, *J. Am. Chem. Soc.*, **85**, 3533 (1963).
[46] J. Barthel, H.J. Gores and G. Schmeer, *Ber. Bunsenges. Phys. Chem.*, **83**, 911 (1979).
[47] J. Barthel, R. Wachter and H.J. Gores, in: *Modern Aspects of Electrochemistry* (B.E. Conway and J.O'M. Bockris, eds.), No. 13, Plenum Press, New York (1979), p. 1.
[48] J. Barthel, R. Neueder, M. Poxleitner, J. Seitz-Beywl and L. Werblan, *J. Electroanal. Chem.,* 1992, in press.
[49] A. Bałkowska, G. Szymański and L. Werblan, *J. Electroanal. Chem.*, **287**, 229 (1990).
[50] A. Bałkowska, J. Lesiński, G. Szymański and L. Werblan, *J. Electroanal. Chem.,* **289**, 171 (1990).
[51] L. Werblan, A. Suzdorf and E. Lin, *Pol. J. Chem.*, **56**, 1071 (1982).
[52] L. Werblan and J. Lesiński, *Pol. J. Chem.*, **53**, 2571 (1979).
[53] L. Werblan, A. Suzdorf and J. Lesiński, *Bull. Acad. Polon. Sci., ser. sci. chim.*, **28**, 627 (1980).
[54] L. Werblan, A. Suzdorf and J. Lesiński, *Roczniki Chem.,* **51**, 2233 (1977).
[55] L. Werblan, A. Suzdorf, E. Lin, G. Szymański and J. Lesiński, *Bull. Acad. Polon. Sci., ser. sci. chim.*, **33**, 285 (1985).
[56] L. Werblan, J. Lesiński, A. Suzdorf, J. Izydorek and R. Drachal, *J. Power Sources*, **9**, 409 (1983).
[57] M. Salomon and E.J. Plichta, *Electrochim Acta*, **29**, 731 (1984).
[58] M. Salomon and E.J. Plichta, *Electrochim. Acta*, **30**, 113 (1985).
[59] Y. Matsuda and M. Morita, *J. Power Sources*, **20**, 273 (1987).

[60] M.Uchiyama, S. Slane, E. Plichta and M. Salomon, *J. Power Sources*, **20**, 279 (1987).
[61] J. Barthel, H.-J. Gores, P. Carlier, F. Feuerlein and M. Utz, *Ber. Bunsenges. Phys. Chem.*, **87**, 436 (1983).
[62] L. Werblan, A. Bałkowska, G. Szymański and J. Lesiński, *International Conference on Solution Chemistry*, Jerusalem, Israel, August 6–11, 1989, Program, p. 35.
[63] J. Molenat, *J. Chim. Phys.*, **66**, 825 (1969).
[64] J.F. Casteel and E.S. Amis, *J. Chem. Eng. Data*, **17**, 55 (1972).
[65] J.F. Casteel, J.R. Angel and H.B. McNeely, *J. Electrochem. Soc. Electrochem. Sci. Technol.*, **122**, 319 (1975).
[66] V.M. Valashko and A.A. Ivanov, *Zh. neorg. khim.*, **19**, 2978 (1974).
[67] B.K. Makarenko, Y.M. Povarov, P.A. Sereda and N.Y. Bobrova, *Elektrokhimiya*, **9**, 705 (1973).
[68] G.E. Blomgren, Ch. 3, E. Peled, Ch. 4, M.B. Clark, Ch. 6, H. Ikeda, Ch. 8, in: *Lithium Batteries* (J.P. Gabano, ed.), Academic Press, Inc., New York (1983).
[69] D. Aurbach, M.L. Daroux, P.W. Faguy and E. Yeager, *J. Electrochem. Soc.*, **134**, 1611 (1987).
[70] Y. Geronov, P. Schwager and R.H. Muller, *J. Electrochem. Soc.*, **129**, 1422 (1982); *ibid.*, **132**, 285 (1985).
[71] H. Yamin and E. Peled, Paper 46, *Proceedings of the 48th Meeting of the Israel Chemistry Society*, Beersheva, Israel (1981).
[72] J.B. Kerr, *J. Electrochem. Soc.*, **132**, 2839 (1985).
[73] R. Selim and P. Bro, *J. Electrochem. Soc.*, **121**, 1467 (1974).
[74] V.R. Koch, J.L. Goldman, C.J. Mattos and M. Mulvaney, *J. Electrochem. Soc.*, **129**, 1 (1982).
[75] J.O. Besenhard and G. Eichinger, *J. Electroanal. Chem.*, **68**, 1 (1976).
[76] J.L. Goldman, R.M. Monk, J.H. Young and W.R. Koch, *J. Electrochem. Soc.*, **127**, 1461 (1980).
[77] K.M. Abraham and J.L. Goldman, International Meeting on Lithium Batteries. Roma (1982), Extended Abstracts, No. 4.
[78] P. Pascal, *Nouveau traite de chimie minerale*, Vol. XIII/2, Masson et Cie, Paris (1962), p. 1522.
[79] (a) F.P. Dousek and J. Jansta, *J. Electroanal. Chem.*, **74**, 195 (1976).
(b) D. Aurbach and H. Gottlieb, *Electrochim. Acta*, No. 3, 141 (1989).
[80] F. Beck, *Electroorganische Electrochemie*, Verlag Chemie, Weinheim (1974), p. 110.
[81] J. Riha, *Electrochemical Power Sources,* 28th Meeting of the International Society of Electrochemistry, Varna (1977), Extended Abstracts, p. 569.
[82] Y. Matsuda, *Progr. Batt. Solar Cells*, **6**, 17 (1987).
[83] Y. Matsuda and M. Morita, *J. Power Sources*, **20**, 273 (1987).
[84] M. Yoshido, H. Nakamura and K. Holzleitner, *Progr. Batt. Solar Cells*, **7**, 377 (1988).
[85] S. Furukawa, K. Nishio and M. Fujimoto, Pat. Jap., 63,102,174 (07.05.1988).
[86] R. Shishikura, Pat. Jap., 63,102,162 (07.05.1988).
[87] S. Ooishi and M. Nagamine, Pat. Jap., 63,102,163 (07.05.1988).
[88] T. Nagai, K. Kaijta and T. Manaba, Pat. Jap., 63,102,173 (07.05.1988).
[89] S. Guibert, M. Carion and J. Simonet, *Bull. Soc. Chem. Fr.*, **1988**, 924.
[90] S. Guibert and J. Simonet, *Bull. Soc. Chim. Fr.*, **1988**, 919.
[91] K. Cedzyńska, A.J. Parker and P. Singh, *J. Power Sources,* **10**, 13 (1983).
[92] S.D. Jones and G.E. Blomgren, Pat. USA, 4,764,440 (05.05.1987).
[93] G.E. McManis, A.N. Platcher, D.E. Bliss and M.H. Miles, *Electrochim. Acta*, **31**,

1271 (1986).
[94] J.P. Malugani, B. Fahys, R. Mercier, G. Robert, J.P. Duchange, S. Baudry, M. Broussely and J.P. Gabano, *Solid State Ionics*, **1983,** 653.
[95] J.R. Akridge and H. Vourlis, *Progr. Batt. Solar Cells*, **6**, 81 (1987).
[96] R.D. Watkins, *Chem. Abstr.*, **108**: 59384e.
[97] T. Sotomura, T. Kambara and S. Sekido, *Progr. Batteries Solar Cells*, **6**, 25 (1987).
[98] B. Scrosati, A. Selvaggi, F. Grose and W. Gang, *J. Power Sources*, **24**, 287 (1988).
[99] C.C. Hunter, D.C. Sinclair, A.R. West and A. Hooper, *J. Power Sources*, **24**, 157 (1988).
[100] J.M. North, Eur. Pat. 279, 554 (*Chem. Abstr.*, **109**: 173606).
[101] C. Arbizzani, M. Mastragostino, S. Panero, P. Prosperi and B. Scrosati, *Synth. Metals*, **28** (1–2), C663 (1989).
[102] L. Werblan, *Bull. Acad. Polon. Sci., ser. sci. chim.*, **27**, 873 (1979).
[103] J.N. Butler, *Adv. Electrochem.*, **7**, 77 (1970).
[104] J.A. Riddick and W.B. Bunger, in: *Organic Solvents*, Vol. 2, (A. Weissberger, ed.), 3rd ed., J. Wiley, New York (1970).
[105] L. Werblan and J. Lesiński, *Pol. J. Chem.*, **52**, 1211 (1978).
[106] R.T.M. Bicknell, K.G. Lawrence, M.-A. Seeley, D. Feakins and L. Werblan, *J. Chem. Soc. Faraday Trans.*I , **72**, 307 (1976).
[107] Y.M. Povarov, *Electrochemical Power Sources*, 28th Meeting of the International Society of Electrochemistry, Varna (1977), Extended Abstracts, No. 137.
[108] L. Werblan and J. Lesiński, *Pol. J. Chem.*, **54**, 507 (1980).
[109] G. Jander, H. Spandau and C.C. Addison, *Chemie in nichtwässerigen ionisierenden Lösungsmitteln*, Vieweg, Braunschweig (1963).
[110] H. Normant, *Angew. Chem., Int. Ed.*, **6**, 1046 (1967).
[111] V. Gutmann, *Coordination Chemistry in Non-Aqueous Solutions*, Springer Verlag, Wien (1968); *Chemie Lexikon* (H. Rompp, ed.), Franckh'sche Verlag, Stuttgard (1966).
[112] M.K. Wong, W.J. MacKinney and A.I. Popov, *J. Phys. Chem.*, **75**, 56 (1971).
[113] J.J. Auborn, K.W. French, S.I. Lieberman, V.K. Shah and A. Heller, *J. Electrochem. Soc.*, **120**, 1613 (1973).
[114] D. Nicholls, C. Sutphen and M. Szwarc, *J. Phys. Chem.*, **72**, 1021 (1968).
[115] K.M. Abraham, J.L. Goldman and M.D. Dempsey, *J. Electrochem. Soc.*, **128**, 2493 (1981).
[116] S. Taniewska-Osińska and A. Kacperska, *Pol. J. Chem.*, **53**, 1351 (1979).
[117] D.F.C. Morris, *Struct. Bonding, Berlin*, **4**, 63 (1968).
[118] G. Nemethy and H.A. Scheraga, *J. Chem. Phys.*, **36**, 3382 (1962).
[119] M. Zief (ed.), *Purification of Inorganic and Organic Materials*, Dekker, New York (1969); M. Zief and R. Speights (eds.), *Ultrapurity Methods and Techniques*, Dekker, New York (1972).
[120] A.K. Covington and T. Dickinson (eds.), *Physical Chemistry of Organic Solvent Systems*, Plenum Press, London (1973).
[121] E. Renard and J.-C. Justice, *J. Solution Chem.*, **3**, 633 (1974).
[122] Yu.A. Karapetian and V.N. Eichis, *Physicochemical Properties of Electrolytic Non-Aqueous Solutions* (in Russian), Khimia, Moskva (1989), p. 29.
[123] R. Schmid and V.N. Sapunov, *Non-formal Kinetics*, Verlag Chemie, Weinheim, (1982), Ch. 7.
[124] J.M. Bodet, J. Labat and J. Goualard, *Proc. 26th Intersoc. Energy Convers. Engl. Conf.*, **1991**, 411.
[125] J.L. Harvie, *Diss. Abstr. Int. B*, **52**, 273 (1991).
[126] M. Liu, S.J. Steven and L.C. De Jonghe, *J. Electrochem. Soc.*, **138**, 1891 (1991).

CHAPTER 8

Non-Aqueous Cells and Storage Batteries

Complete cells at the development stage or already in production are discussed in this chapter.

8.1 CELL CHARACTERISTICS

The usefulness of cells or storage batteries is assessed by means of a number of parameters. These include (1) the specific energy expressed in watt hours per kilogram of electrode materials used in the electrochemical system or per kilogram of the complete cell [$W\,h\,kg^{-1}$], and also in watt hours per litre [$W\,h\,l^{-1}$] (the term "energy density" is also used); (2) the working voltage U in volts; (3) the maximum current density I in milliamperes per cm^2 of the electrode's geometrical surface area [$mA\,cm^{-2}$]; (4) the charge capacity in [mA h] or [A h]; (5) the energy capacity in watt hours [W h]; (6) the power in watts [W] or kilowatts [kW]; (7) the power density in watts per kilogram [$W\,kg^{-1}$] or in watts per litre [$W\,l^{-1}$]; (8) the energy efficiency; (9) the coulombic efficiency of the cell. The cell's operating temperature, shelf life and mechanical strength, i.e. its ability to withstand shock and vibration, are also important.

1. The specific energy should be considered from two angles: as a theoretical value referring exclusively to a particular electrode system and calculated from thermodynamic data [1–8], and as the practical value obtainable from the cell, dependent on the electrode system, the properties of the electrolyte solution used and the cell construction. The conditions under which the cell operates must also be considered, i.e. its voltage and the working current drawn from it.

The theoretical specific energy $\mathcal{E}_w$ [W h kg^{-1}] is calculated from the free energy change under standard conditions ΔG^0:

$$\mathcal{E}_w = -\Delta G^0 / \sum m_i \tag{8.1}$$

where $\sum m_i$ denotes the total mass of electrode materials and ΔG^0 is the free energy change when one mole of reactants is transformed into products during the cell reaction [1]

$$\Delta G^0 = \Delta G^0 \text{ (products)} - \Delta G^0 \text{ (reactants)}$$

Nowadays, ΔG^0 is expressed in [kJ mol^{-1}], and the theoretical specific energy in watt-hours per kilogram of pure electrode materials.

The e.m.f. of the cell, i.e. the open-circuit voltage of the cell, is evaluated from ΔG^0, and is measurable in a compensating system under zero current conditions. The e.m.f. is often called the maximum voltage of the electrode system used in a cell or open circuit voltage (OCV)

$$E^0 = -\Delta G^0 / nF \tag{8.2}$$

where n is the number of electrons transferred in the unit electrode reaction and F is Faraday's constant. The theoretical specific energy $\mathcal{E}_w$ can also be found from the cell e.m.f., the maximum charge q [C] obtained during the reaction, and the total mass of electrode materials

$$\mathcal{E}_w = Eq / 3600 \sum m_i \tag{8.3}$$

For constructional reasons, practical specific energies are several times lower than the calculated values of $\mathcal{E}_w$.

2. If corrosion of electrode materials is neglected, then in principle all of the free energy $-\Delta G$ which results in the reaction appears as electric energy $\mathcal{E}_{el} = -\Delta G = nFE = qE$. As stated before E = e.m.f. is a thermodynamic constant $E = -\Delta G / nF$ characteristic for the electrochemical reaction and does not depend on the conditions under which the galvanic cell is working.

During the cell's operation the current i is flowing in the external circuit represented by the external resistance r_{ext}, but also in the galvanic cell, inside which the circuit is closed via the internal resistances r_{int}. The working voltage U is the voltage obtainable from a cell under a particular set of operating conditions (under load) and is determined by the external resistance r_{ext} of the current receivers (load resistor): $U = i r_{ext}$. Within the cell an internal voltage drop occurs: $U_{int} = i r_{int}$.

Hence

$$E = U + U_{\text{int}}, \quad \text{or} \quad -\Delta G = nFU + nFU_{\text{int}} = \mathcal{E}_{\text{ext}} + \mathcal{E}_{\text{int}} \tag{8.4}$$

One can see that $\mathcal{E}_{\text{el}}$ must be split into two components: one is electric energy available in the external circuit $\mathcal{E}_{\text{ext}}$ which can do useful work – in a particular cell it is equivalent to the energy capacity $Q_{\mathcal{E}}$ (cf. further down point 5). Another is the energy dissipated as heat inside the cell, $\mathcal{E}_{\text{int}}$; the latter is lost for all practical purposes, unless one wishes to use the working cell as a heating element (cf. point 6 below). The working voltage U is thus the difference between the open cell voltage (OCV) or E and internal voltage drop: $U = E - U_{\text{int}}$ or $U = E - ir_{\text{int}}$. U_{int} may be important also in low-internal-resistance cells, such as lithium/thionyl cells, the heating of which under operating conditions may occasionally lead to explosions so that special safeguards must be used (cf. Section 8.6.4.). $U \rightarrow$ e.m.f. when $i \rightarrow 0$, but for a given cell and at constant temperature only r_{ext} is an independent adjustable variable.

3. The current density I is expressed in amperes or milliamperes per unit electrode surface area [A cm^{-2}]. Note that this area is the calculated geometrical or macroscopic surface area of the electrode; only exceptionally is the real or microscopic area (taking account of the electrode's porous structure) required.

4. The charge capacity Q_q is measured in ampere hours [A h]

$$Q_q = it_d, \quad i = \text{const}$$

where t_d is the discharge time, or

$$Q_q = \frac{1}{r}\int_0^{t_d} U\,dt, \quad r = \text{const}$$

where r is the load resistor.

5. The energy capacity $Q_{\mathcal{E}}$ is expressed in watt hours [Wh]

$$Q_{\mathcal{E}} = \int_0^q U\,dq$$

Both the charge capacity and energy capacity are of great practical importance. When non-stoichiometric electrode materials participate in electrode processes, theoretical charge capacities are expressed in electrons (e) per atom of reacting element, e.g. per atom of vanadium in the compound [9] $(\text{Fe,Cr})_{0.13}\text{V}_{0.87}\text{O}_{2.17}$. The unit is then e/atom V, but faradays per mole [F mol^{-1}] could be used.

6. The power P is given in W or kW

$$P = i(U + U_{int}) = i(ir_{ext} + ir_{int}) = i^2 r_{ext} + i^2 r_{int} \tag{8.5}$$

where i is the current flowing and the associated cell voltage U determined principally by the external load resistance. The first term $i^2 r_{ext}$ is the power output P_U, the second $i^2 r_{int}$ is the power lost (usually as heat) in overcoming the cell's internal resistances (cf. point 2).

7. The power density is expressed in watts per kilogram of electrode materials or of the total cell mass.

8. The practical energy efficiency of a cell w is given by the negative quotient of the amount of energy actually supplied by the cell and the theoretical energy, equivalent to the free energy change of the electrochemical reaction in the cell. It is a dimensionless quantity always less than unity

$$w = -\frac{\int_0^q U \mathrm{d}q}{\Delta G} = \frac{Q_E}{\Delta G} \qquad (q \le nF \text{ and } U < E) \tag{8.6}$$

where q is the total charge [C] transferred by the cell's external circuit. Its theoretical maximum value is nF. The real value of q is that which results from the cell's properties and the conditions under which it operates. Deviations of q from nF are due to self-discharge, corrosion of electrode materials and unconsumed electrode materials, because a cell hardly ever discharges completely.

9. The coulombic efficiency is similarly defined as the charge obtained in practice divided by the theoretical charge nF due to the quantity of electrode materials and the reaction stoichiometry

$$\Phi = q/nF \tag{8.7}$$

This too is a dimensionless quantity. Φ expresses the coulombic efficiency not only of a whole cell but also of a half-cell or a single electrode. Then, q is the number of coulombs of charge obtained at a cathode or an anode from one mole of substrates in an electrode reaction, and n is the number of electrons participating in the reaction.

The above parameters enable cells of different types and sizes to be compared. The many other parameters and formulae describing the operation of storage batteries and their charge/discharge characteristics are discussed in the literature [10, 11].

8.2 THERMODYNAMIC AND FUNCTIONAL CHARACTERISTICS OF CELLS

The working properties of a cell or storage battery are described by appropriate functions, usually in graphical form. One basic property is the relationship of working voltage U vs. time t during which a cell operates under various external resistances r_{ext}

$$U = f(t)_{r_{ext}} \tag{8.8}$$

The shape of this characteristic can vary a lot, depending on the types of reactant and product participating in the electrochemical reaction. The working voltages U of some cells are very close to the open-cell voltage, the e.m.f., and remain practically unchanged during discharge until one of the electrode materials is exhausted. This type of cell has a flat discharge curve, the voltage dropping abruptly when one of the materials has reacted completely. An example is the classic Zn/HgO cell. The reactants and products in such cells are separate, pure phases. During discharge, the activity of each component therefore remains constant ($a = 1$), and the working voltage U is practically unaltered, any changes being of the order of a few hundredths of a volt (Fig. 8.1).

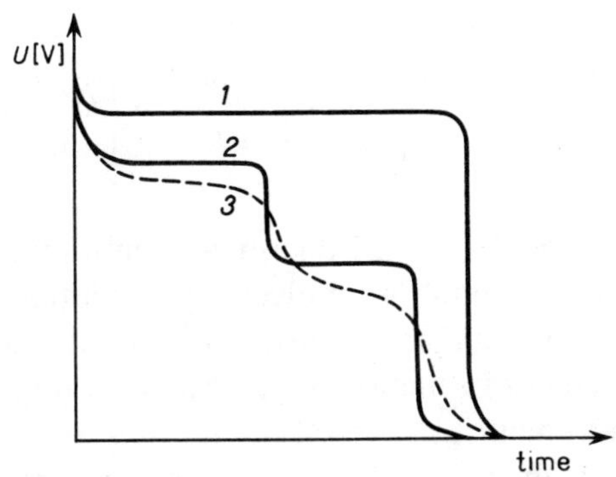

Fig. 8.1 Discharge curves (cell voltage vs. time at low current densities) of cells containing different cathode materials: *1* – HgO cathode material, $HgO \rightarrow Hg^0$, $a = 1$, one-stage reduction; *2* – AgO cathode material, $AgO \rightarrow Ag_2O \rightarrow Ag^0$, $a = 1$, two-stage reduction; *3* – MnO_2 cathode material, Mn(IV) → Mn(III) → Mn(II), activity variable, two-stage reduction.

When almost saturated electrolyte solutions are used, drops in the working voltage during operation can be kept to a minimum. However, it is impossible to obtain flat discharge characteristics for cell reactions involving large changes in the concentrations of the ionic substrates. The use of metal oxide cathode materials can lead to thermodynamically more complex situations. If the metal passes through different oxidation states on being reduced, two cases may occur; if the reaction products are strictly stoichiometric and form separate solid phases, a mixture of pure crystals appears. A number of flat sections corresponding to the several values of the working voltage are then seen on the discharge curve (Fig. 8.1, curve *2*). If the reaction products are mixed crystals, i.e. solid solutions consisting of a single phase, the formation of non-stoichiometric compounds is favoured. The activity of the components will not then be unity; the several sections of the discharge curve pass more gently from one to another: the changes in working voltage with time are gradual (Fig. 8.1, curve *3*), and can be used to measure the extent of cell discharge.

8.2.1 *Temperature characteristics of a cell*

The temperature coefficient of the open-cell e.m.f., E, and of the standard e.m.f., E^0, can be calculated from the Gibbs–Helmholtz equations

$$\left(\frac{\partial E}{\partial T}\right)_p = \frac{\Delta S}{nF} = \frac{1}{T}\left(\frac{\Delta H}{nF} + E\right) \tag{8.9}$$

and

$$\left(\frac{\partial E^0}{\partial T}\right)_p = \frac{\Delta S^0}{nF} = \frac{1}{T}\left(\frac{\Delta H^0}{nf} + E^0\right) \tag{8.10}$$

where ΔH and ΔS are the energy and entropy changes of the cell reaction at a given concentration of electrolyte solution, and ΔH^0 and ΔS^0 are the corresponding standard values. ΔH^0 is easily calculated by subtracting the standard heat of formation of the products from that of the substrates (published data can be used [1–8]).

Knowing the temperature coefficient, it is possible to evaluate the open-cell voltage for any temperature, if the cell e.m.f. at one temperature, e.g. 298 K, is known. The temperature coefficient may be positive or negative and is usually expressed in [$\mu V\,K^{-1}$]; it ranges from 0 to ca. 1 $mV\,K^{-1}$.

8.3 THERMAL EFFECTS IN A CELL

When a charge q flows through a cell, a quantity of heat Q is evolved which can be calculated from

$$Q = \frac{T\Delta S}{n} = F(E - U) = \frac{\Delta H}{n} + FU \tag{8.11}$$

in which E is the cell e.m.f., U the working voltage and F is Faraday's constant. Q is negative, which means that heat is being given out while the cell is operating. The change in entropy ΔS may be positive or negative, although for most cell systems it is positive during discharge. Even if the working voltage is almost the same as the cell e.m.f. (which is the case in cells and storage batteries with very low internal resistances), a quantity of heat equal to $T\Delta S$ is evolved. The entropy of some cells, particularly those in which gases are released during operation, may increase considerably and the cell then give out large quantities of heat. The equation also takes account of heat effects due to ohmic loss within the cell, both in the electrolyte solution and at the electrodes. When the electrodes are shorted, i.e. when the working voltage is nearly zero, the heat evolved by the cell is equal to the enthalpy change of the redox reaction: $Q = \Delta H$.

8.4 ENERGY LOSSES IN CELLS AND STORAGE BATTERIES

Whenever a current flows through a cell, the voltage measured at the cell terminals deviates from the thermodynamically calculated value owing to [11]:

(1) ohmic losses V_{ir} in the bulk of homogeneous phases, i.e. in the bulk of the electrolyte solution as well as within the electrode materials and the conductors taking current to the appliance, and
(2) voltage drops, called the overvoltage η, at the interfaces where the cathode and anode reactions occur.

1. Ohmic losses within phases, especially electrolyte solutions are fairly simple to calculate provided the cell geometry ensures an even distribution of current density throughout the phase. For cells whose electrode surfaces are parallel to one another (Fig. 8.2), the ohmic loss V_{ir} is given by

$$V_{ir} = ir = i(l/\kappa a)$$

where l is the depth of the phase (e.g. the electrolyte solution), κ the specific conductivity, a the cross-sectional area, r the internal resistance of the phase, and i the current.

If the current distribution is not homogeneous, calculating the effective value of V_{ir} becomes much more complicated because the ohmic loss is a complex function of electrode polarization, specific conductivity and cell

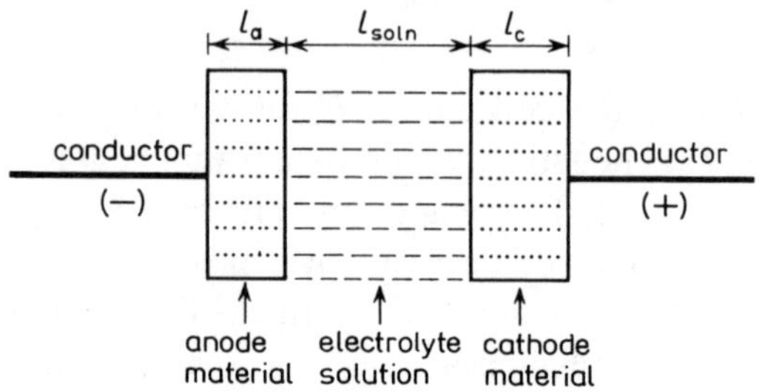

Fig. 8.2 Cell with parallel electrodes.

geometry. In cases where electrode polarization and ohmic losses in electrode materials are negligible, the above formula is still applicable.

Ohmic losses are large in semiconductor electrode materials but negligibly small in metals.

In homogeneous phases, particularly in electrolyte solutions of fixed composition, ohmic losses are temperature-dependent since the specific conductivity κ is a function of temperature and decreases considerably with temperature fall in electrolyte solutions and semiconductor electrode materials. The effect of temperature in metals is the opposite: electron conductivity increases as the temperature falls, but only to a relatively insignificant extent within useful temperature ranges.

A fall in temperature leads to an increase in the viscosity of electrolyte solutions and a consequent decrease in the diffusion coefficients and ionic mobilities in the solution. The influence of temperature on properties determining the conductivity of solutions and diffusion coefficients is discussed in Chapter 4.

2. Voltage drops also occur at interfaces. They cause electrode potentials V to deviate from the thermodynamic values E and are defined as the overvoltage η

$$\eta = E - V$$

The overvoltage can be obtained directly by measuring the potential difference between an electrode operating under non-equilibrium conditions and another electrode of the same type at equilibrium (see Chapter 3).

It comprises the concentration overvoltage η_c, the activation overvoltage η_a and the ohmic overvoltage η_{ir}. The concentration overvoltage is due to the equilibrium shift in the direction of product formation, leading to an increase in product concentration in the electrode space, as well as to the decrease in concentration of those substrates which must diffuse from the bulk of the solution; the concentration of substrates which are simultaneously electrode materials can be regarded as constant at the electrode surface. For reversible electrode systems, the concentration overvoltage can be evaluated from the Nernst equation, as long as the activation overvoltage is zero or negligibly small (the application and range of applicability of the Nernst equation is discussed in Chapter 2)

$$E = E^0 + \frac{RT}{nF} \ln \frac{a_{\text{ox}}}{a_{\text{red}}} = E^0 + |\eta_c| \tag{8.12}$$

For instance, at a temperature of +30°C, a tenfold change in the ratio of product to substrate activities at the electrode surface in a one-electron process changes the potential of that electrode by ca. ± 60 mV; this change is equivalent to the concentration overvoltage η_c. In the majority of cells, the transport of substrates to or from the electrode surface is primarily controlled by diffusion processes.

Activation overvoltages η_a appear when any step in an electrode reaction is kinetically limited; in this case, the overvoltage cannot be calculated from the Nernst equation and more general equations derived from the theory of irreversible processes must be applied. Eyring's theory of absolute reaction rates [12] is applied in the simplest cases: accordingly, the rate constants of heterogeneous processes, including those at electrode surfaces, are finally determined by the activation energy $\mathcal{E}^{\neq}$ of these processes (see Chapter 3). An increase in the activation energy causes the current i flowing at a given voltage to decrease; restoring i to the value corresponding to that of a reversible process requires an additional voltage which, for this reason, is distinguished as the activation overvoltage.

The surface ohmic overvoltage η_{ir} is due mainly to ohmic losses in the passivating films covering the electrode materials, although by definition it covers all possible voltage losses not included in the concentration and activation term or in voltage drops within the electrodes. Any conductivity anomaly in any phase adjacent to an electrode surface causes an ohmic overvoltage. Passivating films, i.e. oxides, solvent decomposition products, etc., above all determine the magnitude of the ohmic overvoltage, because their resistance is high. Sometimes it is this very overvoltage which determines the efficiency of a cell.

If passivating films are not more than a few molecular layers thick, distinguishing between ohmic and activation overvoltages becomes difficult and even artificial because experimentally the two are measured together.

Distinguishing the voltage drops in the bulk of a phase and at the electrode surfaces of an operating cell is possible using suitable measurements and plots. In the simplest case, when the voltage drops in the bulk of the electrode material are minimal and the electrode overvoltage low, the main voltage drops occur in the electrolyte solution.

8.5 MAXIMUM CELL POWER

The greatest power W_{max} of a cell or storage battery in which ohmic voltage drops prevail, i.e. are of the V_{ir} type, is available when the external load is large enough to lower the cell voltage to half its e.m.f. or E

$$W_{max} = \frac{E}{2} i = E^2/4V_{ir} \quad (i = E/2V_{ir}) \tag{8.13}$$

but at the same time, the energy efficiency of the cell also decreases by 50% because $Q_{\mathcal{E}} = nFU = nFE/2$. So for $q = nF$, $w = \mathcal{E}_w/2$ (see Section 8.1 and equations (8.1) and (8.6)). In this case i is the current when the voltage is equal to half the e.m.f. ($E/2$) and V_{ir} is the cell's internal resistance. Unfortunately, when overvoltages appear at electrode surfaces, i.e. voltage drops of the second type are significant, equation (8.13) no longer applies. Then the maximum cell power is found experimentally by determining the voltammetric characteristics of the system, which in turn enables such voltages and currents to be found whereby their product is equal to the maximum power of the cell.

8.6 EXPERIMENTAL AND COMMERCIAL NON-AQUEOUS CELLS AND STORAGE BATTERIES

In recent years, considerable advances have been made in the design and technology both of disposable cells and non-aqueous storage batteries. The effort put into the development of non-aqueous sources of electrical energy was stimulated principally by the fact that it was technically no longer feasible to increase yet further the energy capacity or unit efficiency and power of traditional Leclanché-type cells (Zn/MnO_2), lead-acid accumulators, and the (by then) classic alkali accumulators. In the modern world, it is always the case that when the practical limits of a technology have been reached it is replaced by another more efficient one on which laboratories and industrial centres active in this field then concentrate. The Zn/AgO or Zn/HgO systems in large cells are being replaced by the Zn/air system or the more modern lithium-thionyl chloride system whose practical energy capacity is many times higher than that of the traditional systems. The $Li/SOCl_2$ system has not yet come into general use because the problems concerning its absolutely safe utilization have not been finally solved. In military applications, however, it is one of the most favoured systems because of its excellent operational characteristics. It is suitable for use not only in button cells but also in cylindrical cells in which exceptionally high energy capacities are attainable. Clearly, the application of lithium was a great step forward in the field of electrochemical energy sources, but the choice of cathode material depends in practice on the operational parameters required of the cell in question. Lithium cells can therefore be divided into two groups: those supplying current to low-power electronic devices over a long period (perhaps several years), and those operating at high current densities, i.e. under heavy loads, but for relatively short periods (measured in minutes).

The problem is more complex where storage batteries are concerned; systems that have reached a stage of development where they can replace the classic aqueous acid or base accumulators are not yet in large-scale production. However, low-temperature storage batteries containing lithium-aluminium alloys and intercalation sulphide cathode materials used in conjugation with organic solvents, and high-temperature storage batteries with molten salts and iron sulphide as the cathode material are showing promise. The LiAl/FeS system is already being used to power experimental vehicles.

The first low-temperature non-aqueous storage battery available on the market was a rechargeable cell manufactured by Exxon. Later EIC (USA) and Moli Energy (Canada) launched further models of such batteries. Now Hitachi and Matsushita (Japan) have added their versions [137]. More information on the cells made by these firms will be found in Sections 8.6.1 and 8.6.2.

Selected cells and storage batteries arranged according to the cathode materials used (see Chapter 6) will now be discussed.

8.6.1 *Cells with sulphide or dichalcogen cathode materials and a lithium anode*

These belong to the earliest systems studied, and some which have attained the final stages of development have been manufactured and put on the market.

The first contemporary papers on the use of sulphides in cells were published in the 1950s [13,14]. FeS, FeS_2, Cu_2S, PbS and ZnS were tested. However, this was mainly theoretical work aimed at determining the e.m.f.s of these systems with respect to a standard hydrogen electrode in aqueous electrolyte solutions. These sulphides are only sparingly soluble in organic solvents.

At the end of the sixties, nickel sulphides were used with lithium in non-aqueous cells, with $LiClO_4$ – propylene carbonate solutions [15, 16]. The discharge of cells containing different nickel sulphides – Ni_3S_2, NiS, NiS_2 – was studied. Coulombic efficiencies of over 50% were attained at current densities of ca. 1 mA cm^{-2}. To improve conductivity, the prototype cells had 10% graphite added to the NiS and NiS_2, which are almost insulators. Ni_3S_2 conduct electrons and does not require the addition of graphite. The current efficiency achieved with a drain of 1–10 mA cm^{-2} was > 60%. The discharge curves were strongly dependent on the method of preparing the cathode material and its final structure. There were single plateaus on the discharge curves in the voltage range 1.8–1.6 V.

Both the Ni^{2+} and S_2^{2-} ions in NiS_2 are reduced; this is why the Li/NiS_2 system has a higher theoretical specific energy than the Li/NiS system (Table 8.1). The lower specific energy of the Li/Ni_3S_2 system is probably due to 1/3 of the nickel atoms in this compound being in the reduced (metallic) state.

Both copper(II) sulphide CuS and copper(I) sulphide Cu_2S are quite commonly employed in non-aqueous cells. Copper(II) sulphide is a semiconductor which is very sparingly soluble in organic solvents. Gabano *et al.* [25, 26], Dampier [27] and Dechenaux *et al.* [28] were the earliest workers to study this system. Probably the first sulphide cell contained isopropylamine as solvent [28], and produced a practical specific energy of ca. 250 W h kg^{-1}. Gabano and his co-workers continued their researches into this system using tetrahydrofuran (THF) with added 1,2-dimethoxyethane (1,2-DME) as solvent which yielded a higher practical specific energy of ca. 300 W h kg^{-1}. It turned out, however, that THF reacts with lithium (see Chapter 7).

Dampier [27] continued work on this system using *inter alia* propylene carbonate (PC) as solvent in solutions with $LiClO_4$; however, their viscosity decreased the ionic mobility.

Li/CuS cells discharge in two steps since CuS is reduced in two steps (seen Table 8.1 and Fig. 8.3). The lithium surface is passivated in these

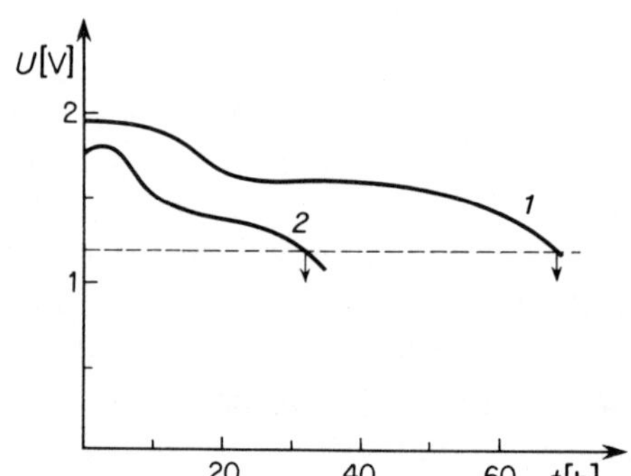

Fig. 8.3 Discharge curve of a Li/CuS cell ($LiClO_4$–PC electrolyte) under various loads *R*: *1* – $R = 10\ \Omega$; *2* – $R = 5\ \Omega$; the straight dashed line corresponds to the final cell voltage of ca. 1.2 V.

solvents. Although the film prevents the lithium anode from corroding and protects the solution from further decomposition, it also retards the elec-

Table 8.1 Theoretical values of the e.m.f. E^0, specific energy $\mathcal{E}_W$ and temperature coefficient dE^0/dT calculated from thermodynamic data for lithium electrode systems with various cathode materials

Electrode system	Overall reaction	E^0, V	$\mathcal{E}_W$, W h kg^{-1} (298 K)	dE^0/dT, μV K^{-1}	References
1	2	3	4	5	6
Lithium with sulphide cathode material					
Li/Ni_3S_2	$Ni(II) \rightarrow Ni(I) \rightarrow Ni(0)$ $4Li^0 + Ni_3S_2 \rightarrow 2Li_2S + 3Ni^0$	1.8	776	–	[15]
Li/NiS	$Ni(II) \rightarrow Ni(0)$ $2Li^0 + NiS \rightarrow Li_2S + Ni^0$	2.02	1040	–106	[1]
Li/NiS_2	$Ni(II) \rightarrow Ni(0)$ and $S_2^{2-} \rightarrow 2S^{2-}$ $4Li^0 + NiS_2 \rightarrow Ni^0 + 2Li_2S$	1.8	1407	–	[15]
Li/CuS	$Cu(II) \rightarrow Cu(0)$ $CuS + 2Li^0 \rightarrow Cu^0 + Li_2S$	2.23	1090	–159	[1]
Li/CuS	$Cu(II) \rightarrow Cu(I)$ $2CuS + 2Li^0 \rightarrow Cu_2S + Li_2S$	2.42	634	–50	[1]
Li/Cu_2S	$Cu(I) \rightarrow (Cu_0)$ $2Li^0 + Cu_2S \rightarrow Li_2S + 2Cu^0$	2.04	632	–267	[1]
Li/S	$2Li^0 + S^0 \rightarrow Li_2S$	2.48	2900	–	[1]
Li/FeS	$Fe(II) \rightarrow Fe(0)$ $2Li^0 + FeS \rightarrow Li_2S + Fe^0$	1.98	1030	–157	[1]

1	2	3	4	5	6
Li/FeS_2	Fe(II) → Fe(0) and $S_2^{2-} \rightarrow 2S^{2-}$ $FeS_2 + 4Li^0 \rightarrow Fe^0 + 2Li_2S$	2.83	1231	–	[17]
Li/TiS_2	Ti(IV) → Ti(III) $xLi^0 + TiS_2 \rightarrow Li_xTiS_2$ (intercalation compound)	2.47 for $x = 0$ 1.87 for $x = 1$	356 for $x = 1$	–	[18] [19]
Li/TiS_2	Ti(IV) → Ti(III) $2Li^0 + 2TiS_2 \rightarrow Ti_2S_3 + Li_2S$	2.24	505	–170	[1]
Li/Ti_2S_3	Ti(III) → Ti(II) $2Li^0 + Ti_2S_3 \rightarrow 2TiS + Li_2S$	1.50	389	–52	[1]
Li/TiS	Ti(II) → Ti(0) $2Li^0 + TiS \rightarrow Li_2S + Ti^0$	0.89	510	–109	[1]
Li/TiS_2	Ti(IV) → Ti(0) $4Li^0 + TiS_2 \rightarrow 2Li_2S + Ti^0$	1.38	1060	–110	[1]
Li/TiS_3	Ti(IV) → Ti(0) $6Li^0 + TiS_3 \rightarrow 3Li_2S + Ti^0$				
Li/MnS_2	Mn(IV) → Mn(II) $MnS_2 + 2Li^0 \rightarrow MnS + Li_2S$	2.64	1070	–14	[1]
Li/MnS	Mn(II) → Mn(0) $MnS + 2Li^0 \rightarrow Mn^0 + Li_2S$	1.36	721	–225	[1]
Li/MnS_2	Mn(IV) → Mn(0) $MnS_2 + 4Li^0 \rightarrow Mn^0 + 2Li_2S$	2.00	1460	–120	[1]

1	2	3	4	5	6
Li/V_2S_3	$V(III) \rightarrow V(II)$ $V_2S_3 + 2Li^0 \rightarrow 2VS + Li_2S$	1.79	453	–74	[1]
Li/V_2S_3	$V(III) \rightarrow V(0)$ $V_2S_3 + 6Li^0 \rightarrow 2V^0 + 3Li_2S$	1.59	1070	–112	[1]
Li/VS	$V(II) \rightarrow V(0)$ $VS + 2Li^0 \rightarrow V^0 + Li_2S$	1.49	826	–131	[1]
Li/MoS_2	$Mo(IV) \rightarrow Mo(III)$ $2MoS_2 + 2Li^0 \rightarrow Mo_2S_3 + Li_2S$	1.68	270	–16	[1]
Li/Mo_2S_3	$Mo(III) \rightarrow Mo(0)$ $Mo_2S_3 + 6Li^0 \rightarrow 2Mo^0 + 3Li_2S$	1.83	892	–97	[1]
Li/MoS_2	$Mo(IV) \rightarrow Mo(0)$ $MoS_2 + 4Li^0 \rightarrow Mo^0 + 2Li_2S$	1.79	1020	–77	[1]
Li/ZrS_2	$Zr(IV) \rightarrow Zr(II)$ $ZrS_2 + 2Li^0 \rightarrow ZrS + Li_2S$	0.97	309	–128	[1]
Li/ZrS	$Zr(II) \rightarrow Zr(0)$ $ZrS + 2Li^0 \rightarrow Zr^0 + Li_2S$	1.08	423	–54	[1]
Li/ZrS_2	$Zr(IV) \rightarrow Zr(0)$ $ZrS_2 + 4Li^0 \rightarrow Zr^0 + 2Li_2S$	1.03	602	–91	[1]
Li/CoS	$Co(II) \rightarrow Co(0)$ $CoS + 2Li^0 \rightarrow Co^0 + Li_2S$	1.97	1000	–47	[1]
Li/TaS_2	$Ta(IV) \rightarrow Ta(0)$ $TaS_2 + 4Li^0 \rightarrow Ta^0 + 2Li_2S$	1.59	625	–74	[1]

1	2	3	4	5	6
Li/Cr_2S_3	Cr(III) → Cr(II) $Cr_2S_3 + 2Li^0 \rightarrow 2CrS + Li_2S$	2.82	707	+284	[1]
Li/Cr_2S_3	Cr(III) → Cr(0) $Cr_2S_3 + 6Li^0 \rightarrow 2Cr^0 + 3Li_2S$	2.05	1360	–38	[1]
Li/CrS	Cr(II) → Cr(0) $CrS + 2Li^0 \rightarrow Cr^0 + Li_2S$	1.66	909	–199	[1]
Li/PbS	Pb(II) → Pb(0) $PbS + 2Li^0 \rightarrow Pb^0 + Li_2S$	1.98	420	–122	[1]
Li/Ag_2S	Ag(I) → Ag(0) $Ag_2S + 2Li^0 \rightarrow 2Ag^o + Li_2S$	2.28	467	–288	[1]
Li/SiS_2	Si(IV) → Si(II) $SiS_2 + 2Li^0 \rightarrow SiS + Li_2S$	2.27	1144	–	[17]
Li/NbS_2	Nb(IV) → Nb(II) $NbS_2 + 2Li^0 \rightarrow NbS + Li_2S$	1.76	550	–62	[1]
Li/NbS	Nb(II) → Nb(0) $NbS + 2Li^0 \rightarrow Nb^0 + Li_2S$	1.42	548	–90	[1]
Li/NbS_2	Nb(IV) → Nb(0) $NbS_2 + 4Li^0 \rightarrow Nb^0 + 2Li_2S$	1.59	921	–76	[1]
Li/WS_2	W(IV) → W(0) $WS_2 + 4Li^0 \rightarrow W^0 + 2Li_2S$	1.87	729	–73	[1]
Li/ReS_2	Re(IV) → Re(0) $ReS_2 + 4Li^0 \rightarrow Re^0 + 2Li_2S$	2.05	788	–72	[1]

1	2	3	4	5	6
Li/SnS_2	$Sn(IV) \rightarrow Sn(II)$ $SnS_2 + 2Li^0 \rightarrow SnS + Li_2S$	2.30	626	–40	[1]
Li/SnS_2	$Sn(IV) \rightarrow Sn(0)$ $SnS_2 + 4Li^0 \rightarrow Sn^0 + 2Li_2S$	2.12	1080	–79	[1]
Lithium with polyacetylene cathode materials					
$Li/[(CH)^{y+}(ClO_4^-)_y]$	$[(CH)^{0.06} + (ClO_4^-)_{0.06}]_x + 0.06xLi \rightarrow$ $\rightarrow (CH)_x + 0.06xLiClO_4$	3.7	176 (practical!)	–	[20]
Lithium with sulphoxy cathode materials					
$Li/SOCl_2$	$3SOCl_2 + 8Li^0 \rightarrow 6LiCl + Li_2SO_3 + 2S$	3.61	1876	–	[21, 22]
Li/SO_2Cl_2	$2SO_2Cl_2 + 6Li^0 \rightarrow 4LiCl + Li_2SO_4 + S$	3.90	2013	–	[22]
Li/SO_2	$2SO_2 + 2Li^0 \rightarrow Li_2S_2O_4$	2.91	290 (practical!)	–	[23, 24]
Lithium with oxide cathode materials					
Li/TiO_2	$Ti(IV) \rightarrow Ti(III)$ $2TiO_2 + 2Li^0 \rightarrow Ti_2O_3 + Li_2O$	1.12	344	–242	[1]
	$Ti(IV) \rightarrow Ti(II)$	0.86	493	–128	[1]
Li/V_2O_5	$V(V) \rightarrow V(III)$ $V_2O_5 + 4Li^0 \rightarrow V_2O_3 + 2Li_2O$	2.18	1110	–139	[1]

1	2	3	4	5	6
Li/VO_2	V(IV) → V(III) $2VO_2 + 2Li^0 \rightarrow V_2O_3 + Li_2O$	1.55	588	–33	[1]
Li/CrO_3	Cr(VI) → Cr(IV) $CrO_3 + 2Li^0 \rightarrow CrO_2 + Li_2O$	3.03	1430	–226	[1]
	Cr(VI) → Cr(III) $2CrO_3 + 6Li^0 \rightarrow Cr_2O_3 + 3Li_2O$	2.97	1980	–449	[1]
Li/MnO_2	Mn(IV) → Mn(III) $2MnO_2 + 2Li^0 \rightarrow Mn_2O_3 + Li_2O$	2.63	752	–71	[1]
	Mn(IV) → Mn(II) $MnO_2 + 2Li^0 \rightarrow MnO + Li_2O$	2.37	1260	–70	[1]
Li/Mn_2O_3	Mn(III) → Mn(II) $Mn_2O_3 + 2Li^0 \rightarrow MnO + Li_2O$	2.10	656	–69	[1]
Li/Co_3O_4	Co_2(III)Co(II) → 3Co(II) $Co_3O_4 + 2Li^0 \rightarrow 3CoO + Li_2O$	2.05	434	–246	[1]
	Co_2(III)Co(II) → 3Co(0) $Co_3O_4 + 8Li^0 \rightarrow 3Co^o + 4Li_2O$	1.86	1350	–201	[1]
Li/CoO	Co(II) → Co(0) $CoO + 2Li^0 \rightarrow Co^0 + Li_2O$	1.80	1080	–186	[1]
Li/NiO	Ni(II) → Ni(0) $NiO + 2Li^0 \rightarrow Ni^0 + Li_2O$	1.80	1090	–195	[1]

1	2	3	4	5	6
Li/MoO_3	Mo(VI) → Mo(IV) $MoO_3 + 2Li^0 \rightarrow MoO_2 + Li_2O$	2.21	752	–126	[1]
	Mo(VI) → Mo(0) $MoO_3 + 6Li^0 \rightarrow Mo^0 + 3Li_2O$	1.75	1510	–196	[1]
Li/MoO_2	Mo(IV) → Mo(0) $MoO_2 + 4Li^0 \rightarrow Mo^0 + 2Li_2O$	1.51	1040	–230	[1]
Li/F_2	F(0) → F(–I) $F_2 + 2Li^0 \rightarrow 2LiF$	6.05	6250	–	[1]

trode processes. Werblan *et al.* [29–32, 33a], strove to find the electrolyte solution yielding the best possible results with Li/CuS electrodes and simultaneously ensuring satisfactory mass and charge transfer. Transfer conditions were best in equimolar mixtures of γ-butyrolactone (γ-BL) and 1,2-DME, with a lithium perchlorate electrolyte [29, 33a]. A better cathode material structure was obtained by selecting a porosity guaranteeing a low resistance and a large surface area of the working CuS electrode. Much work has gone into computer-designing a model of a porous electrode [123]. Flat, two-step discharge characteristics (Fig. 8.4), $U = f(t)$, were obtained with a discharge current of ca. 2 mA cm^{-2} [29], the working voltage being maintained between 2.0 and 1.8 V until 90% of the cathode material had been consumed.

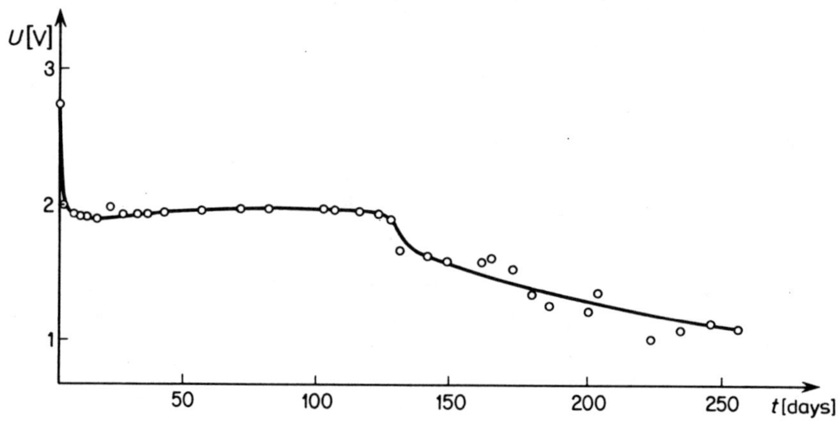

Fig. 8.4 Cell discharge curve of Li | $LiClO_4$ (1 moldm^{-3}) + equimolar mixture of γ-BL and 1,2-DME | CuS; load R = 100 kΩ [29, 30].

The specific energy of these experimental cells was ca. 320 W h kg^{-1}. A series of miniaturized cells with a charge capacity of ca. 100 mA h intended to power electronic devices requiring a low current has been produced in cooperation with the Central Laboratory of Batteries and Cells in Poznań, Poland. The lifetime of such cells is more than 2 years [120–122].

Matsuda *et al.* [34] suggested using liquid ammonia as solvent and ammonium, sodium or potassium thiocyanates as electrolytes with the Li/CuS system. They obtained a one-step, two-electron reduction of CuS: $CuS + 2Li^0 \rightarrow Cu^0 + Li_2S$. Copper(II) sulphide is a good cathode material in this solvent.

Li/CuS cells with mixed organic solvents other than those mentioned so far have been manufactured by the Cordis Corporation for power cardiac pacemakers. Excellent results have been achieved with these cells, and 100 000 of them have already been implanted [35].

Copper(II) sulphide has also found application in non-aqueous storage batteries as a result of the work of Eichinger and Fritz [36, 37]. They demonstrated that a CuS cathode can work reversibly if the sulphide ion concentration in a suitable electrolyte solution is sufficiently high: the slight solubility of both CuS and Li_2S in organic solvents causes this system to be kinetically irreversible. Eichinger and Fritz were able to increase the sulphide ion concentration by shifting the dissolution equilibrium of the reaction product. To do this, they employed macrocyclic crown polyethers to complex alkali metal ions, so improving sulphide solubility and increasing the sulphide ion concentration [36]; a reversible sulphide electrode was thereby obtained. If a semi-permeable or ion-exchange membrane separates the cathode space from the anode space, a Li/CuS cell can be used as a storage battery [37].

This type of cell can be improved by using better separators. Boron nitride, chemically stable in cells even at high temperatures [38], is one of the latest membrane materials.

The high theoretical specific energy of the lithium-sulphur system (2900 $W\,h\,kg^{-1}$) has encouraged much research into cells based on these electrode materials. Two kinds of cells have been studied: low-temperature cells with organic solvents [39] and high-temperature cells with molten salts [40].

Sulphur, an insulator, can be used as a cathode material if graphite is added to it. Solvents include dimethyl sulphoxide, dimethylformamide, PC, as well as various salts such as alkali metal thiocyanates, bromides, iodides and perchlorates. Cell voltages of 2.5–2.0 V have been obtained, but the coulombic efficiency is low due to the formation of polysulphide ions S_x^{2-}; as these are readily soluble, anode material is lost when it reacts with lithium

$$2(x-1)Li^0 + S_x^{2-} + 2Li^+ \rightarrow xLi_2S(\downarrow)$$

This was why such cells were formerly of little practical importance. But according to the reports, a high practical specific energy of 730 $W\,h\,kg^{-1}$ has been obtained from such cells (with respect to the mass of all the cell materials except the can) [41], albeit at low current densities (ca. 0.04 $mA\,cm^{-2}$). Specific energies in excess of 200 $W\,h\,kg^{-1}$ have been attained in the high-temperature version (352°C) with a LiCl–KCl eutectic as the molten electrolyte. However, there are drawbacks in using sulphur alone: it evaporates and dissolves as polysulphides. The activity of sulphur can be reduced by the use of sulphur compounds, e.g. FeS and FeS_2, in both of which iron occurs in the oxidation state (II) combined with sulphide and disulphide ions respectively

$$4Li^0 + FeS_2 \rightarrow 2Li_2S + Fe^0$$

Both Fe^{2+} and disulphide ions are reduced in this overall reaction with lithium. The Li/LiCl–KCl/FeS cell can work as a storage battery, so clearly it is a reversible system. As the research were generally encouraging, *inter alia* because high current densities of the order of 1 A cm^{-2} were obtained, development work on this system continued throughout the 1970s and led to the construction of storage batteries capable of powering vehicles. The cathode material in these batteries was stoichiometric iron sulphide, the electrolyte was the molten LiCl–KCl eutectic and the anode was made of a lithium-aluminium alloy [42]. Storage batteries of this type are capable of over 300 charge/discharge cycles at a reversible operational specific energy of over 100 W h kg^{-1}.

Both the cathode and anode reactions during charge and discharge are highly complex and involve phase transitions. Five separate phases, which appear in electrochemical redox reactions have been identified in FeS alone. This system is still under development. For example, the number of cycles is being raised and has already passed the 500 mark [43]. This research is being done *inter alia* at the Argonne National Laboratory (USA), one of the largest centres of research into electrochemical sources of energy for powering vehicles. A low-temperature cell containing FeS and an organic solvent in the Li/$LiAsF_6$ + THF/FeS system is also being investigated with the aim of constructing a rechargeable cell [17]. However, only a few charge/discharge cycles have been obtained at a current density of 1 mA cm^{-2} and a deep discharge to 1.47 V. Studies have been conducted on a non-aqueous lithium cell which will enable the achievement of a value of the working voltage of 1.5 V, which would be very useful in electronic devices. Both sulphides, FeS [33b,c,d] and FeS_2 [33e], were used as cathode materials. Nelson *et al.* [33e] observed that the FeS cathode gave a slightly lower performance capability than the FeS_2 cathode due to the lower sulphur content and lower cell voltage, but on the other hand, its corrosion was reduced, its lifetime longer and its cost lower. Hence, FeS, especially when porous [33], is considered a superior cathode material. According to the latest reports [176, 177] Werblan *et al.* in the cells, a R-43 type, of Li|equimolar PC + AN + $LiAsF_6$|FeS system, were enable to obtain of practical specific energy reaching 80% of the theoretical value at 20°C. There is also little prospect of MnS_2 cathodes being used in rechargeable cells. Nickel and manganese(II) sulphides are better in this respect. More encouraging results were also obtained with SnS_2 and NiS, while those using CuS and the same electrolyte solution were surprisingly good [17]. Over 40 cycles were attained with the Li/$LiAsF_6$ + THF/CuS cell and a cell capacity of half that theoretically possible. The process is based on the two-electron reduction

$$CuS + 2e + 2Li^+ \rightarrow Cu^0 + Li_2S$$

Perhaps the CuS electrode is reversible here because of the kind of electrolyte used and the formation of intermediate products with the electrolyte anion during charging, as the anode materials can be reconstituted. The number of cycles (ca. 40) is limited because the electrolyte gradually decomposes. The intercalation of lithium ions into the copper sulphide structure is a possible parallel mechanism.

Recent research has shown that many transition metal disulphides are very active electrochemically, and act as reversible cathode materials in systems with alkali metal anodes. This is another way of obtaining reversible cathodes in non-aqueous media. The first way is to construct electrodes reversible with respect to ions from the reacting electrode material passing into solution. In other words, this first way requires the liquid phase of an electrode to be made up in such a way as to enable the solubility of the ionic products of the electrode reaction to be raised. Obviously, the cathode space in such a cell must be separated from the anode space by a membrane, as in the case of the Daniell cell.

The other way of obtaining reversible cathodes in non-aqueous media is to intercalate the cathode material with cations from the solution (Chapter 6).

In order to form three-component phases when intercalated with lithium, a cathode material must

(a) cause a large quantity of free energy to be released during the cell reaction,
(b) need little energy to expand its crystal lattice during intercalation,
(c) have a highly reversible structure, so that permanent deformations of the crystal lattice do not occur during cycling,
(d) guarantee a high mobility of the alkali metal ion in the van der Waals gap of solid-phase intercalated compounds, which in addition ought to retain good electron conductivity.

Table 8.2 compares the ranges of content (x) of lithium and sodium ions and their coefficients of diffusion D in the intercalated sulphides of titanium and tantalum. Averaged values of D are given in this table; note the

Table 8.2 Stoichiometric content ranges (x) of lithium and sodium, and diffusion coefficients (D) in the intercalated sulphides of titanium and tantalum

Compound	D, cm^{-2} s^{-1}	Temperature, K	Range of x
Li_xTiS_2	ca. 6×10^{-9}	303	0–1
Li_xTaS_2	ca. 3×10^{-8}	303	0.25–0.75
Na_xTiS_2	ca. 1×10^{-9}	300	0.25–0.60
Na_xTaS_2	ca. 3×10^{-8}	300	0.3–0.4

paper by Johnson and Worrell [18], who have shown that these coefficients can vary within the useful composition range of these compounds.

An example of this kind of cell is the lithium cell with intercalated titanium sulphide TiS_2 which can also function as a storage battery [44, 141]. The free energy of the reaction $x\mathrm{Li}^0 + \mathrm{TiS}_2 \rightarrow \mathrm{Li}_x\mathrm{TiS}_2$ is quite high (−206 kJ mol^{-1}) and varies only slightly in the range $0 < x < 1$ for lithium ions in Li_xTiS_2 compounds. The lattice parameters vary by only 10%, so the cathode material is not destroyed during multiple charge/discharge cycles.

The electrode reaction is reversible throughout the composition range ($0 < x < 1$). The appropriately high diffusion coefficient of lithium ions in TiS_2 ensures high current densities. The metallic properties of titanium sulphide are reflected in its very good electron conductivity, which does not need to be improved by additives such as graphite. No dissolution of TiS_2 in electrolyte solutions made up in organic solvents was observed.

In this kind of cell, the electrode materials, mainly TiS_2, are used very effectively and currents of the order of 10 mA cm^{-2} are obtained. There is no self-discharge to shorten the cell life. The practical specific energy – though only with respect to the mass of anode material (lithium) – is 455 W h kg^{-1}. At a current density of 10 mA cm^{-2} the charge and discharge curves are fairly flat [44] (Fig. 8.5).

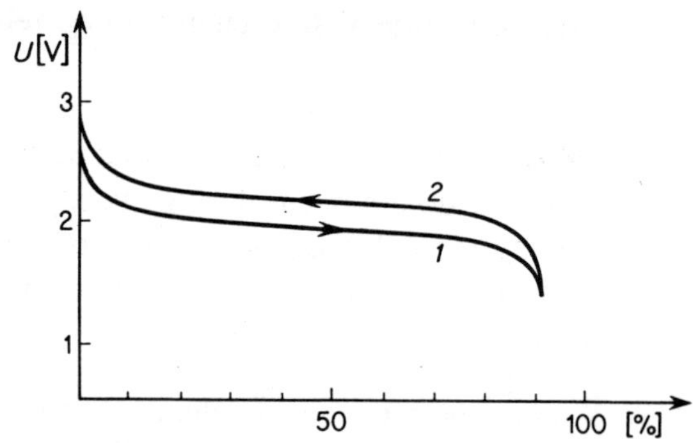

Fig. 8.5 Charge (*1*) and discharge (*2*) curves of a Li | $LiClO_4$ + ether | TiS_2 cell with respect to the % consumption of anode material; charge circuit current density 10 mA cm^{-2}.

Solutions of $LiClO_4$ in organic ethers have been used in low-temperature cells with non-stoichiometric metal sulphides; mixtures of molten salts are used in high-temperature cells.

The lithium content in the intercalation compounds Li_xTiS_2 is usually estimated by coulometric titration, but the results are not always very accurate; if this is the case, a full chemical analysis is carried out. Diffusion coefficients are determined chronoamperometrically with an applied rectangular voltage or by means of relaxation methods using current pulses [18]. TiS_2 is a better cathode material than FeS_2 in high-temperature cells with molten salts. During charge and discharge the structural reconstitution of the intercalation compound is much better in TiS_2 than in pyrite (FeS_2). This is clear from the plot of U vs. x (Fig. 8.6).

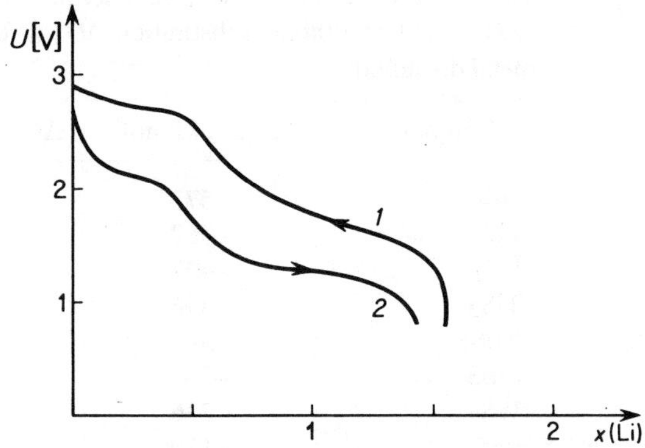

Fig. 8.6 Voltage of a Li/FeS_2 cell as a function of the stoichiometric content of lithium (x) in the cathode material: *1* – charge curve, *2* – discharge curve.

The majority of development work has been done on TiS_2 because it is the most efficient and economical of the sandwich disulphides. But other disulphides, even the less common ones, like VS_2, CrS_2, TaS_2, ZrS_2, MoS_2 have been tested [42, 184]. The free energies of formation $-\Delta G_{form}$, of lithium substitution $-\Delta G_{sub}$ and of intercalation $-\Delta G_{intercal}$ are helpful in understanding the properties of intercalation compounds and in assessing their suitability for use in lithium cells. The more negative the value of $-\Delta G_{form}$, the more readily are three-component phases formed, because the compound's sandwich structure remains stable despite the intercalation of lithium ions. The lower the numerical value of the free energy of lithium substitution $-\Delta G_{sub}$ in reactions of the type

$$\mathrm{Li} + \tfrac{1}{4}\mathrm{MS_2} \rightarrow \tfrac{1}{2}\mathrm{Li_2S} + \tfrac{1}{4}\mathrm{M}$$

in which M is a transition metal, the greater the probability that an intercalation compound will be formed, provided that $-\Delta G_{form}$ of the transition metal sulphide is high. For example, for TiS, $-\Delta G_{form} = 396\ \mathrm{kJ\,mol^{-1}}$, $-\Delta G_{sub} = 120\ \mathrm{kJ\,mol^{-1}}$ and $\Delta G_{intercal} = 206\ \mathrm{kJ\,mol^{-1}}$ for the intercalation $x\mathrm{Li} + \mathrm{TiS_2} \rightarrow \mathrm{Li}_x\mathrm{TiS_2}$. Therefore, the intercalation compound is more stable than the substitution product. Hence, a TiS_2 electrode operating reversibly with respect to the lithium anode is more likely to exist in a system with an intercalation compound than in one where the substitution reaction yields Li_2S and metallic titanium. The free energy changes $-\Delta G_{form}$ and $-\Delta G_{sub}$ for transition metal sulphides [44] are given in Table 8.3. On the other hand, the free energies of intercalation for transition metal disulphides are roughly the same at ca. -200 kJ mol^{-1}.

Table 8.3 Changes in the free energies of formation ΔG_{form} and of lithium substitution ΔG_{sub} for transition metal disulphides

Compound	ΔG_{form}, kJ mol^{-1}	ΔG_{sub}, kJ mol^{-1}
HfS_2	–573	–76
ZrS_2	–562	–78
US_2	–427	–112
TiS_2	–396	–120
NbS_2	–345	–132
MoS_2	–268	–152
WS_2	–236	–160
ReS_2	–170	–176
SnS_2	–142	–183

Free energy lithium intercalation $\Delta G_{intercal} \approx -200$ kJ mol^{-1}.

Another sulphide of titanium, TiS_3 or TiS(S–S), containing a disulphide ion has been tested in lithium cells with a $LiClO_4$–PC solution. This experimental cell yielded a current of 0.02 mA at a working a voltage of 1.5 V and an external resistance of 68 kΩ [45]. The main advantages of using intercalated sulphides are that the electrodes are reversible and the voltages obtained are quite high. As low specific energies are still a problem, higher sulphides like MoS_3, NbS_3, Fe_3S_8, Co_2S_7, Co_2S_9 and Ni_2S_7 have been tried as well as CoS_2, NiS_2 and MnS_2. Voltages of the order of 2.4–2.5 V (open circuit voltage – OCV) and energy densities of ca. 1 kW h kg^{-1} have been obtained with these polysulphides in lithium cells.

Good parameters have been achieved with a reversible Li/CoS_x cell – ca. 0.4 kW h kg^{-1} and high currents of 2–3 A [138]. 1500 charge/discharge cycles were obtained in a cell with a Mo_6S_8 cathode; only 25% of its energy capacity – 2 W h in an AA-size cell – was used up [139]. A mixed sulphide $AgMo_6S_8$ has also been used [140].

Selenides and tellurides have been applied as cathode materials [44]. They include sandwich compounds reversibly intercalating lithium or sodium ions, but for a variety of reasons, like the toxicity of the elements and compounds, the high atomic mass (low specific energy) and low current densities, no practical applications have been found for them so far. Selected parameters of sulphide and selenide cells are given in Table 8.4.

Sandwich sulphides are at present used in commercial, low-temperature

Table 8.4 Parameters of lithium-dichalcogenide cells [44, 46, 47]

Parameter	$V_{0.75}Fe_{0.25}S_2$	$V_{0.25}Cr_{0.75}S_2$	NbS_2	VSe_2	$TiSe_2$	VS_2	$SbSe_2$	TaS_2
$\Delta G_{intercal}$ Li_xMY_2, kJ mol^{-1}	–	–	–	–191	–182	–222	–200	–211
Initial e.m.f., V, extrapolated to $x = 0$	–	–	–	2.0	2.12	2.05	2.32	2.92
Final e.m.f., V, for Li_xMY_2 when $x = 1$	2.2	2.48	–	1.96	1.68	2.01	1.68	1.46
Theoretical specific energy, W h kg^{-1} (Li)	510	510	–	246	243	505	216	233
Current density, mA cm^{-2}	0.2	0.4	0.33	2	0.5	0.2	0.5	1
Percentage consumption of electrode materials at a given current density*, %	83	65	90	97	92	42	85	60
Calculated specific energy*	395	369	307	243	231	230	182	134

* After the first discharge.

non-aqueous rechargeable cells. The Exxon company uses a Li–Al alloy anode and a TiS_2 cathode; the electrolyte is $LiAlF_6$ in 2-methyl THF.

The lithium storage battery manufactured by Moli Energy contains a lithium anode, a MoS_2 cathode and $LiAlF_6$ dissolved in PC. Hitachi have recently put on the market a Li–Al/TiS_2 button cell (2×0.2 cm) but with a different electrolyte – $LiPF_6$ in a 4-methyl-1,3-dioxolane/1,2-DME mixture. Hexamethylphosphoramide (HMPA) has been added to the electrolyte to inhibit polymerization side reactions [137].

8.6.2 *Cells with graphite or acetylene compounds as cathode materials*

Fluorographite cells are at present among the best lithium current-generating systems and are already being manufactured. Their development was initiated by Braeuer [48] at the end of the 1960s, who was the first to discover the high potential of these compounds in organic solvents. Watanabe *et al.* [49] investigated their properties and designed a synthesis of monofluorographite compounds. In lithium cells, the overall reaction

$$(CF_x)_n + (nx)Li^0 \rightarrow nxLiF + C_n$$

takes place, giving rise to amorphous carbon and lithium fluoride precipitating in the pores of the cathode material. Fukuda *et al.* [50, 51] studied the effect of various electrolyte solutions on the working parameters of the cell, in particular, the effect of the electrolyte on the discharge curves at different temperatures and on the specific energy. They used solutions of $LiClO_4$ in PC and $LiBF_4$ in γ-BL and in PC. Gunther [52] also used mixed solvents of PC and 1,2-DME in this type of cell. Fluorographites are non-stoichiometric, and substances of various compositions have been used in cells. Table 8.5 contains data on fluorographites of different compositions and the theoretical specific energies available from their reactions with lithium.

Irreversible changes have been shown to take place in this cathode material during discharge, because amorphous carbon is formed; cells like these can thus be used only once. Their advantages are high specific en-

Table 8.5 Theoretical specific energies and range of e.m.f.'s in lithium-fluorographite Li/$(CF_x)_n$ systems [51]

Cathode material	Capacity, A h g^{-1}	e.m.f., V	Theoretical specific energy, W h kg^{-1}
$(CF_{1.0})_n$	0.864	2.8 – 3.3	2190
$(CF_{0.8})_n$	0.788		1960
$(CF_{0.6})_n$	0.687		1740
$(CF_{0.4})_n$	0.547		1430

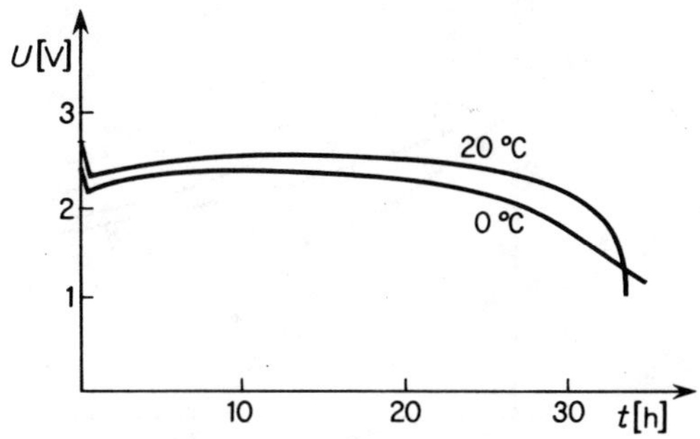

Fig. 8.7 Discharge curves of a LR-2/34-type fluorographite cell at 0°C and 20°C and load $R = 60\ \Omega$.

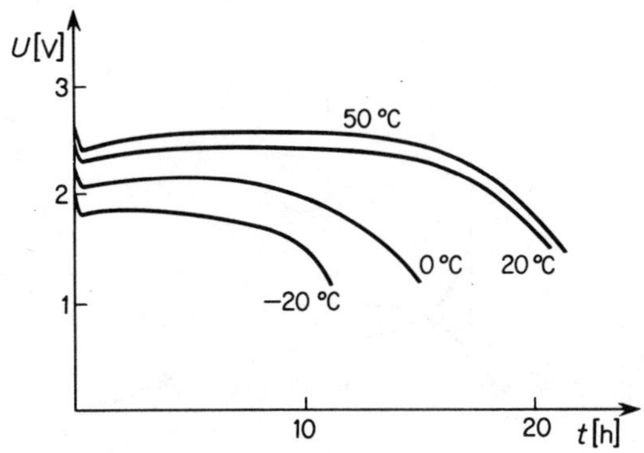

Fig. 8.8 Discharge curves of a fluorographite cell over a temperature range from +50°C to −20°C at a load of 8 Ω.

ergies and exceptionally high working voltages. Their discharge characteristics under various loads are very flat (Fig. 8.7). Figure 8.8 illustrates the effect of temperature on the discharge curve. The apparent changes in the cell's working parameters over a wide range of temperature, from +50°C to −20°C are relatively small, so a wide range of applications is in prospect for such cells. About 80% of the cathode material is consumed at room temperature and a load of 50 Ω; the practical specific energies at this temperature are ca. 300 $\mathrm{W\,h\,kg^{-1}}$. At −20°C they fall to ca. 150 $\mathrm{W\,h\,kg^{-1}}$.

Studies of the influence of the electrolyte solution of current densities have yielded extremely interesting results. Solutions of γ-BL containing $LiClO_4$ or $LiBF_4$ provide higher voltages and current densities. The effect of the electrolyte itself has been studied – solutions of $NaPF_6$ in PC give the worst results. γ-BL with $LiBF_4$ and $LiClO_4$ yielded current densities as high as 35 $\mathrm{mA\,cm^{-2}}$ (Fig. 8.9).

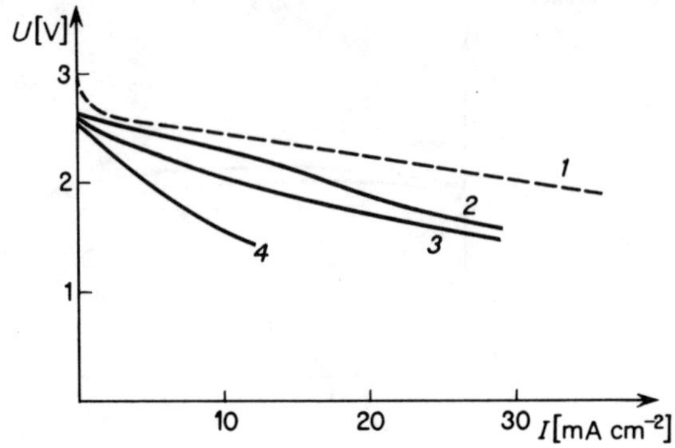

Fig. 8.9 Effect of electrolyte on current density and voltage in Li/$(CF)_n$ cells at 20°C: *1* – 1 M solution of $LiBF_4$ in γ-BL; *2* – 1 M solution of $LiClO_4$ in PC; *3* – 1 M solution of $LiBF_4$ in PC; *4* – saturated solution of $NaPF_6$ in PC.

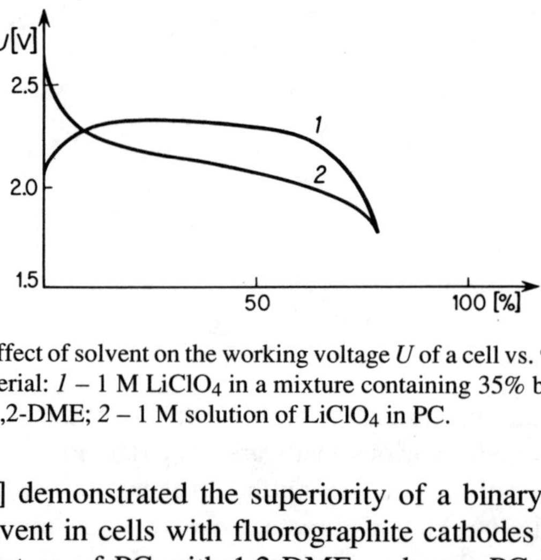

Fig. 8.10 The effect of solvent on the working voltage *U* of a cell vs. % consumption of cathode material: *1* – 1 M $LiClO_4$ in a mixture containing 35% by mass PC and 65% by mass 1,2-DME; *2* – 1 M solution of $LiClO_4$ in PC.

Gunther [52] demonstrated the superiority of a binary mixed solvent over a pure solvent in cells with fluorographite cathodes (Fig. 8.10). He employed a mixture of PC with 1,2-DME and pure PC with $LiClO_4$ as electrolyte. Fluorographite cells are still being improved. At the same time, cheaper halogen-graphite cells have been constructed in which highly reactive fluorine is replaced by the cheaper and less corrosive chlorine [53]. Commercial cells of this kind have a nominal voltage of 3 V, and depending on the model, can have a capacity of 60–1500 mA h and generate currents of 0.2–150 mA [54].

Polyacetylene (PA) is a covalent organic polymer which can be endowed with cathodic or anodic properties, depending on the ion incorporated electrochemically into the polymer structure. The ionic content determines the electronic conductivity, which ranges across 12 orders of magnitude, from that of a typical semiconductor to almost that of a metal. When anions such as perchlorate are incorporated, PA becomes a cathode material which, in combination with lithium in a suitable non-aqueous electrolyte solution, yields an interesting cell with promising characteristics [178].

An open lithium cell with PA + 6% perchlorate ions supplies a voltage of 3.7 V and initial short-circuit current of 25 mA. Such a system can be charged and discharged many times, and is therefore suitable for use in a storage battery. As many as 326 charge/discharge cycles have been performed without the PA changing its structure. The discharge reaction in such a cell is described in Section 6.5. The experimental specific energy was ca. 176 W h kg, but discharge was incomplete and only 60% of the cathode material was used up. The electrochemical processes were carried out in a 0.3 M solution of $LiClO_4$ in PC.

Anode material is formed by the electroreduction of PA, which gives rise to anion groups $(CH)_x^{y-}$ whose charge is neutralized by cations incorporating themselves into the molecular lattice of PA [55–57]. Anode material is most simply produced by placing lithium foil and PA foil in a $LiClO_4$–THF solution and performing internal electrolysis [55]. Then, the overall reaction

$$(CH)_x + yLi^0 \rightarrow Li_y^+(CH)_x^{y-}$$

e.g. $(CH)_{20} + 3Li^0 \rightarrow Li_3^+(CH)_{20}^{3-}$, produces an anode material that can be used in a cell or storage battery.

Other cations, such as tetrabutylammonium, can be incorporated by placing PA foil in solutions of tetrabutylammonium perchlorate and electrolysing. Recently, cathode and anode materials were produced simultaneously by placing large foils of covalent PA in solutions of tetrabutylammonium perchlorate and electrolytically incorporating perchlorate ions into one foil (cathode material) and tetrabutylammonium ions into the other (anode material). This electrolysis is equivalent to charging. In the reverse process, when the external current has been switched off and the circuit closed, a cell is obtained whose characteristics are somewhat inferior to those of lithium anode cells, but which provides a high initial working voltage of 2.5 V at an initial current of 11 mA.

Thus we have two new kinds of non-aqueous cell [20, 55–58]: one based on the lithium anode–PA cathode system $Li/[(CH)^{y+}(ClO_4^-)_y]$, the other on the PA anode–PA cathode system $[(CH)_x^{y-}(Bu_4N)_y^+]/[(CH)^{y+}(ClO_4^-)_y]$.

By incorporating various cations ($Na^+, Li^+, R_4N^+, R_4P^+$) and anions ($ClO_4^-, PF_6^-, AsF_6^-$) in PA film, a whole range of electrode materials and cells or storage batteries of different characteristics can be devised. At present, they are small experimental cells containing only a few milligrams of electrode materials. However, the prospects for this type of cell are considerable: the degree of charging of the electrode materials can be increased threefold, up to 18–20%. A further advantage is the possibility of obtaining high current densities owing to the very loose structure of the PA foil, which consists of intertwined (felted) fibres, 2/3 of whose volume comprises open spaces. When filled with a solution of electrolyte, they offer a several-fold increase in the actual interphase surface area at which the current-generating processes take place. The development of PA cells is continuing [59, 60].

Quite recently, the Matsushita Company of Japan has produced a similar kind of lithium storage battery with an activated charcoal cathode saturated with electrolyte solution which works like an electrolytic condenser. When the battery is charging, electrolyte anions are adsorbed on the positively-charged activated charcoal, which has a very extensive surface area. When it is discharging, the anions are desorbed when the positive charge on the cathode is reduced due to the inflow of electrons released from the anode conduction band during the anodic oxidation of lithium (see also Section 6.5).

8.6.3 *Cells with other organic cathode materials*

Many stoichiometric organic compounds have been tested in recent years. Their properties are reviewed in Section 6.5 and in several papers [21, 61, 62].

Nitro compounds have been the most extensively studied. When reduced completely, they give rise to amines; as 6 protons are consumed by each nitro group, proton donors (usually water) must be included in the system. A four-electron reduction occurs in neutral solution (in which a magnesium electrode can be used) with hydroxylamine as the product; in solution not containing proton donors (where lithium can be used) only solutions yielding low-valency anions are possible. Such materials are therefore not very efficient in lithium cells since only a small number of electrons take part in the overall reactions. Consequently, this line of research has all but been abandoned and few types of lithium cell with organic compounds have been reported. An earlier application of organic compounds is the lithium dichloroisocyanuric acid cell [21]: Li/$LiClO_4$ (2 mol dm^{-3}), methyl formate/dichloroisocyanuric acid. The cell was designed as a thin-layer, activated reserve cell supplying instantaneous currents of up to ca. 200 mA cm^{-2}.

By contrast, storage batteries in which *both* electrode materials are stoichiometric organic compounds have been successful. The anode material is 9,10-anthracenediol (I), the cathode material tetrachlorobenzoquinone (II) (chloranil) [63]:

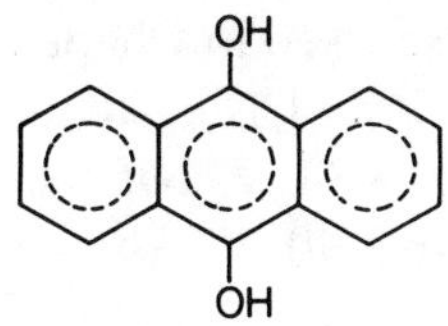

(I)

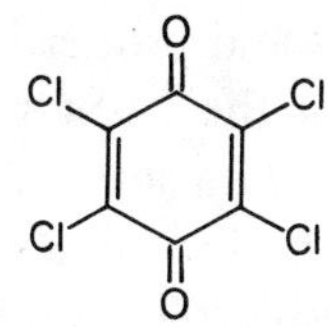

(II)

Such a storage battery is capable of up to 100 charge/discharge cycles with no reduction in capacity. Its e.m.f. is 0.55 V, and the energy density ca. 20 $W\,h\,kg^{-1}$, which is ca. 30% of the theoretical value. Chloranil has been used on its own in a lithium cell with $LiClO_4$ in PC which had an e.m.f. of 3.3 V, falling to 2.6 V after 85% discharge [64].

The standard redox potentials of other quinones range from +0.01 V for 1,4-dimethylanthraquinone to +1.25 V for dichlorodicyano-*p*-benzoquinone. This wide range has made it possible to design cells in which both anode and cathode materials are quinones [65]. The e.m.f. of a cell with a dichlorodicyano-*p*-benzoquinone cathode and a 1,4-dimethylanthraquinone anode has been calculated at 1.24 V. Quinones and hydroquinones are reversible systems, and are therefore applicable in storage batteries [65].

Some aliphatic and aromatic nitro compounds and quinones have been combined with a magnesium electrode, but these are cells with aqueous electrolytes [10].

Certain aliphatic halogen derivatives have been found useful as cathode materials, but only in lithium cells where they double as solvents [143]. 1, 1, 2, 2-tetrachloroethane (m.p. −44°C, b.p. +146°C) with added methyl formate or THF and $LiAsF_6$ as electrolyte in a 2.4 V/1 mA or 2.0 V/5 mA cell has the best properties. Current densities can be high, almost as high as in SO_2 cells. An advantage is that hydrogen ions do not participate in the cathode reaction, so that no proton donor is needed: two halogen atoms are detached as anions and a double bond is formed [143, 191] (see Section 6.5)

$$Cl_2HC{-}CHCl_2 + 2e \rightarrow ClHC{=}CHCl + 2Cl^-$$

Another system has recently been developed: the anode material is a phenyl-vinyl polymer with substituents containing sp^3-hybridized oxygen (e.g. aromatic ethers), so carbonium or oxonium quinoid forms are produced during the anodic oxidation. The cathode is a polymer of similar structure but with sp^2-hybridized oxygen (e.g. ketones), so that reduced carbanionic forms are produced [144].

8.6.4 *Cells with cathode materials having solvent properties*

Cells containing covalent inorganic liquid compounds as cathode materials which are also solvents appeared at the start of the 1970s [21, 66, 179]. They include thionyl, sulphuryl and phosphoryl chlorides, whose reactions and properties are discussed in Chapters 6 and 7. The respective e.m.f.s of the Li/$SOCl_2$ and Li/SO_2Cl_2 systems calculated from thermodynamic data are 3.61 V and 3.84 V, but the experimental values are slightly higher at 3.65 V and 3.90 V. The experimental e.m.f. of the Li/$POCl_3$ system is 2.56 V.

When the experimental e.m.f. exceeds the value calculated from thermodynamic data, it should be assumed that a redox side-reaction accompanies the principal electrochemical reaction. If the redox potentials of these two reactions are approximately the same, one obtains a mixed potential [67] which may be higher than that calculated for the principal reaction, and the experimental e.m.f. exceeds the calculated value.

The above-mentioned oxychlorides should react spontaneously with lithium, and this is what happens at elevated temperatures, but a passivating film prevents the lithium from reacting in this way at room temperature. The theoretical specific energies for Li/$SOCl_2$ and Li/SO_2Cl_2 are 1875 and 2013 $\mathrm{W\,h\,kg^{-1}}$ respectively. In cells with graphite current collectors and a 1–2 M $LiAlCl_4$ electrolyte, practical specific energies reach values of ca. 600 $\mathrm{W\,h\,kg^{-1}}$ and 700 $\mathrm{W\,h\,dm^{-3}}$. At typical current densities of 1 $\mathrm{mA\,cm^{-2}}$ the working voltage is 3.5 V, but when the cell is shorted, maximum currents of 200 $\mathrm{mA\,cm^{-2}}$ can be obtained [22, 68, 69]. The lifetime of a cell is limited because the cathode corrodes and reaction products, mainly lithium chloride, accumulate.

Figure 8.11–8.13 illustrate the discharge characteristics of thionyl and sulphuryl cells. They show that such cells provide not only a high working voltage but also exceptionally high discharge currents, such as are rarely obtainable from galvanic cells. Many research teams throughout the world are working on these cells, elucidating in ever greater detail the mechanisms of their electrode processes [70, 145] and improving their design and operating parameters. Much work has been done on the nature of the film passivating the lithium anode [71]. It has been found, for example, that the structure of the passivating film is complex. It consists of a thin (15–50

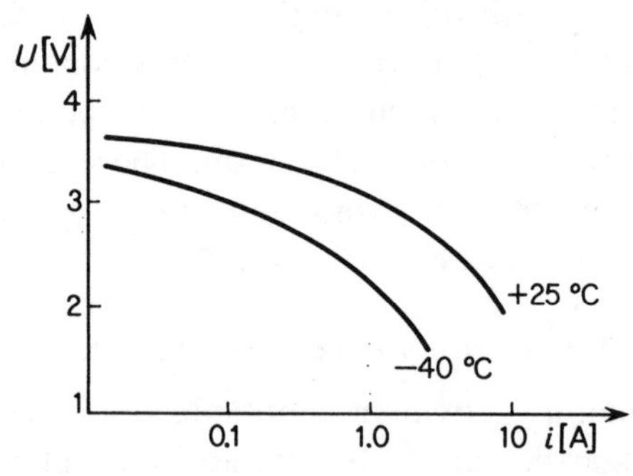

Fig. 8.11 Voltammetric characteristics of a Li/$SOCl_2$ cell at +25°C and −40°C.

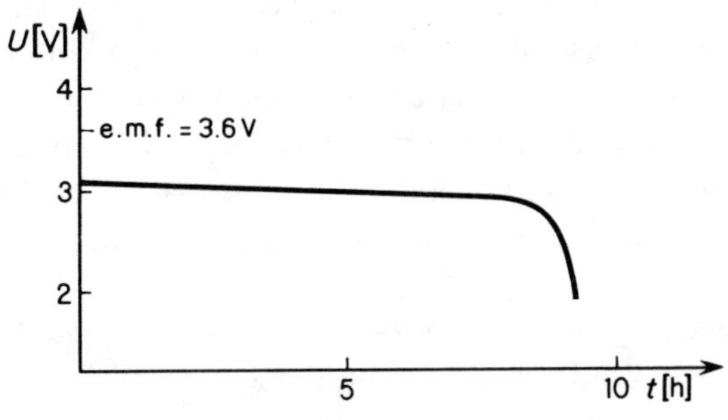

Fig. 8.12 Discharge curve of a $Li/SOCl_2$ cell for a current drain of 1 A (25°C).

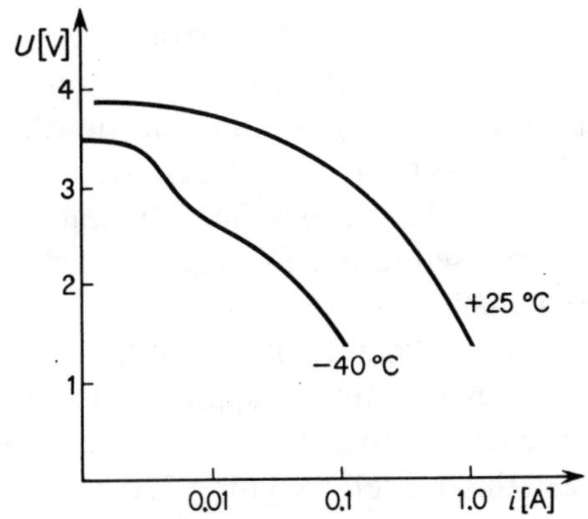

Fig. 8.13 Voltammetric characteristics of a Li/SO_2Cl_2 cell at +25°C and –40°C.

nm) inner layer of LiCl sealing the lithium surface – the passivating film proper – and an outer layer which, though thicker (ca. 2 μm) is too porous to protect the metal. Even though the conductivity and activation energy of the passivating film resemble those of an insulating ionic crystal of LiCl, the thinness of the first layer enables the electrode reaction to proceed and ions to migrate, possibly by means of the tunnel mechanism.

The double-layer structure of the passivating film means that a lithium cell can operate at temperatures up to 150°C, at which its discharge characteristics are good; the cell life at this temperature is also satisfactory [72]. Thionyl batteries have been known to explode [21,73, 181], but this problem can be overcome by using current limiters [146]. The main cause seems to be overheating of the cell, e.g. when too much current is being drawn, and the solvent starts to boil; molten lithium (> 180°C) reacts explosively with oxychlorides.

A safe and no less efficient source of current is a thionyl cell with a calcium metal anode [73, 182] and calcium perchlorate as electrolyte. Calcium is also passivated in $SOCl_2$, and its higher melting point (ca. 840°C) provides an additional safety margin. The higher atomic mass of calcium and the consequent higher cell mass is partially balanced by the two-electron anode reaction of calcium $Ca^0 \rightarrow Ca^{2+} + 2e$. Unfortunately, calcium corrodes in $SOCl_2$, and this diminishes the practical usefulness of this cell [147].

In recent years, miniature, hermetically-sealed Li/$SOCl_2$ cells have been constructed for electronic devices requiring a small current supply. This is no mean achievement, in view of the fact that it was for long impossible to prevent $SOCl_2$ from evaporating and the electrolyte from seeping out of the cell [74]. Lithium-thionyl cells have a shelf life of up to 10 years [75]. They work, though far less efficiently, at temperatures down to below −40°C [23, 148].

Numerous versions of lithium-thionyl cells came to be mass produced in the 1980s, and ranged from very small ones with capacities of the few tenths of an A h used in telecommunications to very large ones with capacities of 10 000 A h and 16 000 A h used to power electrical devices in ballistic missiles and torpedoes. Spirally-coiled lithium anodes with a large surface area are used to obtain larger currents. With appropriate technology and strict quality control these cells are safe, especially when blocking diodes are used as current limiters [149].

Apart from their military applications, thionyl cells are being increasingly used in civilian life, even as cardiac pacemakers [150]. The medium-sized thionyl cell weighing 370 g with a capacity of 45 A h and a nominal voltage of 3.6 V is an example of the practical possibilities of this system. It supplies a current of 35 mA for 1000 hours or 5 mA for 9000 hours at a voltage of 3.5 V [151].

As lithium-sulphuryl cells have extremely good discharge characteristics, work on them is continuing, directed particularly towards reducing the resistance of passivating films [76, 77, 152, 153].

No new information has been discovered in the literature on Li/$POCl_3$ cells, but work is in progress on the use of $POCl_3$ as a solvent in lithium-halogen cells [78]. This state of affairs may be due to the fact that the e.m.f. of lithium-phosphoryl cells is lower than that of lithium-thionyl ones.

Since the beginning of the seventies, much research has been devoted to cells containing liquid sulphur dioxide as both cathode material and solvent [23, 24, 154–157]. The unreactive cathode is carbon deposited on aluminium gauze. An organic solvent such as acetonitrile or PC is added to reduce the vapour pressure of SO_2 (b.pt. −10°C). The electrolyte is usually lithium bromide. The theoretical e.m.f. of the Li/SO_2 system is 2.9 V at room temperature and 2.75 V at −60°C. The practical specific energy obtained at that time was 290 W h kg^{-1}. SO_2 is volatile, so the internal pressure in the cell must be around 0.2–0.4 MPa at room temperature. The lifetime of such cells is estimated at 10 years. Even though the Li/SO_2 system is reversible, it is usually used in disposable (primary) cells. One of their advantages is that they operate well at very low temperatures (< −50°C), thanks to the high specific conductivity of the solution. Both two- and three-component mixtures of solvents have been used, e.g. a mixture of acetonitrile, PC and liquid SO_2 in proportions by volume of 10:3:23 [23]. Apart from LiBr, $LiAlCl_4$ is also used as electrolyte [79]. Lithium-sulphur dioxide cells are still at the development stage and their possibilities have not yet been exhausted. The effect of SO_2 concentration in mixed solvents on the behaviour of the lithium anode has recently been studied. At SO_2 concentrations below 12% by mass, loose gelatinous precipitates of low resistance are formed and the lithium is rapidly corroded. Only when the SO_2 concentration is above 15% by mass do compact layers form, preventing lithium and solvent from reacting with each other [80].

Recent work has concentrated on the possibilities of applying the Li/SO_2 system in reversible cells [158, 159, 179, 192].

8.6.5 *Halogen and halide cells*

One of the earliest was the lithium-chlorine cell (Li/Cl_2) in which chlorine gas was the cathode material [81] and a molten eutectic of lithium and potassium chlorides was the electrolyte. This cell was capable of producing extraordinarily high currents, up to 60 A, at a voltage of 2.5 V, but they were short-lived, lasting only ca. 15–20 minutes. To obtain such a high current, the electrochemical reaction must proceed at a temperature of ca. 620°C. It yields specific energies of ca. 320 W h kg^{-1}. This high-temperature cell, whose unreactive cathode is porous graphite, is used for special purposes [81].

Cells where the lithium anode has been replaced by aluminium [82] operate at 70°C with a molten, three-component electrolyte comprising sodium, potassium and aluminium chlorides. The theoretical specific energy of the Al/Cl_2 system is 1350 W h kg^{-1}, and its e.m.f. is ca. 2.1 V. The cell requires layered graphite compounds to absorb the chlorine and form the chlorine electrode. The practical coulombic energy is 100 A h kg^{-1} but this can be increased to 200 A h kg^{-1} with an appropriate graphite-compound structure [82].

A chlorine cell based on the Li/Cl_2 system with a solution of $LiAlCl_4$ in sulphuryl chloride and an unreactive carbon cathode [83] operating from −30°C to +93°C yielded a current density of 21 mA cm^{-2}. The cell voltage is temperature-dependent: at a resistance of 5 Ω it is ca. 3 V at −22°C and ca. 3.5 V at +93°C. Such cells have flat discharge characteristics. Lithium cells using bromine have not been much studied, although a report was published in 1974 of investigations into the Li/Br_2 system using a mixed LiBr–$LiClO_4$ electrolyte in PC with a polyethylene separator, which only partially prevented self-discharge [84].

Even BrF_3, a liquid at room temperature, has been tested as a solvent and a cathode material [160, 180]. A cell with a lithium anode has been produced; it has a potential of 5.1 V and a current density of up to 20 mA cm^{-2}. However, BrF_3 reacts explosively with water and most organic compounds [161], so its prospects are rather bleak, except perhaps in special areas such as powering remote-controlled electrical equipment. BrF_3 could also be used as a cathode material component in e.g. mixtures with $SOCl_2$.

The lithium-iodine system has been applied in miniature cells supplying the low-current requirements of electronic devices. Their characteristics are such that they are among the best sources of current for cardiac pacemakers as they can operate continuously for up to 10 years. The cathode material in the Li/I_2 system is a charge-transfer complex of iodine and poly(2-vinylpyridine) formed from a polymer of molecular mass ca. 13 000 amu. The iodine-lithium reaction product, precipitating during discharge on the lithium anode as a thin film, acts as the solid conducting electrolyte LiI. The open-cell voltage is 2.80 V. This type of cell operates at a temperature of 37°C under a load of 100 kΩ and maintains an almost constant operating voltage of 2.63 V–2.62 V for 6 years [85]. The calculated specific energy is 1540 W h kg^{-1} [86], but the practical value is 110–148 W h kg^{-1} [85]. Cells of this type have a nominal voltage of 2.8 V and a capacity from 0.03 to 10 A h [86, 87, 171, 185] (Fig. 8.14).

Earlier studies of non-aqueous cells concentrated on systems with copper, silver and nickel fluorides and chlorides [88, 89]. They were followed by numerous reports on the application of other metal halides, from the main groups of the Periodic Table, e.g. $PbCl_2$, SbF_3, AsF_3, as well as al-

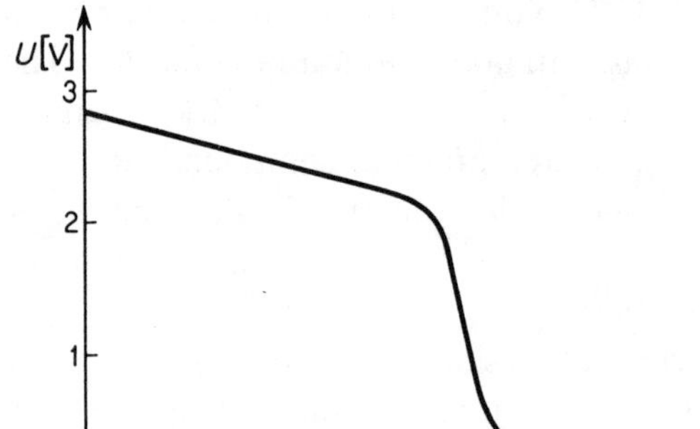

Fig. 8.14 Discharge curve of a Li/I_2 cell for powering cardiac pacemakers; current 20 μA at a temperature of 37°C [171].

most all the subgroups (HgF_2, $CdCl_2$, CrF_3, MnF_3, CoF_3).

A great deal of work has been done on fluorides because of their low atomic mass. The e.m.f. of a Li/NiF_2 cell was 2.83 V and its theoretical specific energy was 1364 $W h kg^{-1}$ [90]. NiF_2 was mixed with graphite to ensure adequate conductivity; the electrolyte was a saturated solution of KPF_6 in PC with polypropylene film separating the cathode space from the rest of the cell. As the electrode reactions in this system could be reversed, such a cell could be discharged and recharged like a storage battery. The fluoride electrode was reversible owing to the presence of PF_6^- ions. The overall reactions are

$$\text{discharge:} \quad NiF_2 + 2Li^0 \rightarrow Ni^0 + 2LiF$$

$$\text{charge:} \quad Ni^0 + 2LiPF_6 \rightarrow NiF_2 + 2Li^0 + 2PF_5$$

$$2PF_5 + 2LiF \rightarrow 2LiPF_6$$

Several tens of charge/discharge cycles were possible. The PF_5 produced is a reactive gas which reacts with insoluble lithium fluoride yielding a soluble salt containing the complex PF_6^- ion. However, these multi-step reactions are not completely reversible.

In theory, the Li/CuF_2 system is very efficient: 1648 $W h kg^{-1}$ at an e.m.f. of 3.53 V [15]. The discharge reaction in a Li/CuF_2 cell with $LiClO_4$ in an aprotic solvent produces insoluble lithium fluoride [91]:

$$CuF_2 + 2Li^0 \rightarrow Cu^0 + 2LiF$$

This leads to voltage drops due to increasing internal resistance. The impaired flow of ions limits mass transfer, so useful current densities are low.

It was not found possible to reconstitute the CuF_2 cathode in the Li/CuF_2, PF_6^- system by recharging [15], so, unlike the nickel fluoride system, it could not be used to construct a storage battery.

However, this was possible with the $Li/CuCl_2$ system [25], in which the electrolyte was $LiAlCl_4$ and methyl formate. The theoretical specific energies of the $Li/CuCl_2$ and Li/CuCl systems are 1100 and 700 $W h kg^{-1}$ respectively. Up to 1000 charge/discharge cycles have been achieved with the $Li/CuCl_2$ system using a methyl formate solvent. Slightly more than 100 cycles were obtained with PC in the same system, whose practical specific energy was ca. 200 $W h kg^{-1}$. With CuCl as the cathode material in the same system, less than 20 cycles were possible; the specific energy was ca. 120 $W h kg^{-1}$ [25] (Fig. 8.15).

The low number of cycles attainable in cells with PC is due to the increasing resistance of the lithium anode as it gradually becomes covered with insoluble products of the solvent–lithium reaction [15].

The reversibility of the $Li/CuCl_2$ system in lithium aluminium tetrachloride–methyl formate solutions is due to the formation of chloro-

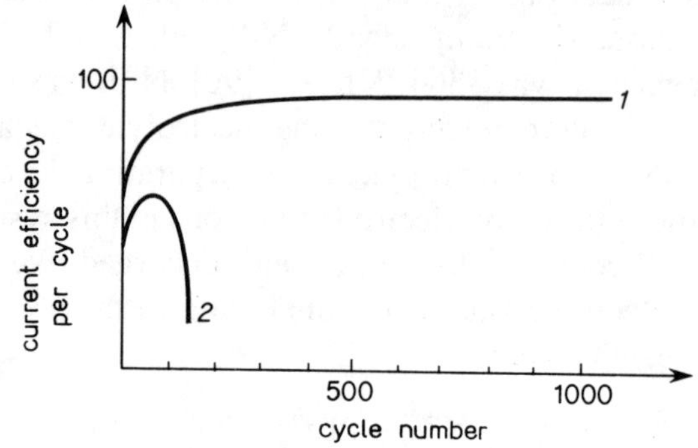

Fig. 8.15 The effect of the electrolyte solution on the number of charge/discharge cycles obtained in a rechargeable Li/$CuCl_2$ cell: *1* – $LiAlCl_4$ (1 mol dm^{-3}) in methyl formate; *2* – $LiAlCl_4$ (1 mol dm^{-3}) in PC.

cuprate(I) ions [21] like $CuCl_2^-$ and perhaps $CuCl_3^-$ in the presence of excess $AlCl_3$ [15] in the charged cell. The charging reactions are

$$3LiAlCl_4 + Cu^0 \rightarrow Cu^{II}Cl_3^- + 3Li^+ + 3AlCl_3 + 2e$$

or

$$2LiAlCl_4 + Cu^0 \rightarrow Cu^{I}Cl_2^- + 2Li^+ + 2AlCl_3 + e$$

These enable the cathode material $CuCl_2$ to be reconstituted while the storage battery is being charged. The same thing happens in $LiAlCl_4$–PC solutions. Charging may proceed as follows:

$$Cu^0 + LiAlCl_4 \rightarrow CuAlCl_4 + Li^+ + e \quad \text{oxidation of } Cu^0 \text{ to Cu(I)}$$

but the equilibrium in the solution

$$CuAlCl_4 + LiCl \rightleftharpoons CuCl + LiAlCl_4$$

causes CuCl to precipitate. The next step may lead to the formation of $CuCl^+$

$$CuCl \rightarrow CuCl^+ + e$$

which is immediately precipitated:

$$CuCl^+ + Cl^- \rightarrow CuCl_2$$

or

$$CuCl^+AlCl_4^- \rightarrow CuCl_2 + AlCl_3$$

It appears that a certain excess of $AlCl_3$ in the charged cell guarantees the reversibility of the reaction [15]. The overall reaction is

$$CuCl_2 + 2AlCl_3 + 2Li^0 \underset{\text{charge}}{\overset{\text{discharge}}{\rightleftharpoons}} Cu^0 + 2LiAlCl_4$$

However, possible complexes like $CuAlCl_4$ may be important intermediate products.

Reports and patents indicate that extensive studies have been undertaken on the possible use of other transition metal halides, e.g. MnF_3, SbF_3, CrF_3, CoF_3 [92], CdF_2, $CdCl_2$ [93], HgF_2 [94], $PbCl_2$ [5], CuBr [96], but the results have not been sufficiently promising to warrant mass production of any of these systems.

8.6.6 *Lithium cells with oxide cathode materials*

The rapid development of lithium cells with oxide cathode materials began with the discovery that the reversible formation of intercalation compounds of lithium and other alkali metals with transition metal oxides was feasible. This property enabled storage batteries to be constructed which were capable of a large number of charge/discharge cycles at satisfactorily high specific energies (higher than those achieved in storage batteries with intercalation sulphides like TiS_2). Many papers have been published recently on crystallography, the diffusion kinetics of lithium ions, and the thermodynamics of the electrochemical processes in lithium-oxide cells. In comparison with chalcogenides, however, there are few oxides with the TiS_2-type sandwich structure. The best known are the stoichiometric MoO_3 and CoO_2 and the non-stoichiometric Mo_9O_{26} (i.e. $MoO_{2.889}$) [97–99]; with lithium they all form intercalation compounds of the Li_xMoO_3 type, and yield high specific energies in lithium cells. A large class of useful transition metal oxides contains MO_6-type octahedra in three-dimensional lattice structures [100] whose corners or edges connect forming channels which can be reversibly intercalated especially by small alkali-metal cations like Li^+. Such oxides include V_2O_5, V_6O_{13} (i.e. the nonstoichiometric $VO_{2.167}$), ReO_3, and rutile – one of the three crystalline forms of TiO_2 [101]. The channels, or voids, may be aligned in one direction only (as in TiO_2), in two directions (e.g. $VO_{2.167}$) or even in three (ReO_3). For structural reasons, reversible lithium-cation penetration is easiest in lattices whose channels are aligned along two directions. Hence, WO_3 is especially useful because, compared with other compounds, its structural voids can accommodate three times as many lithium ions; the γ-MnO_2 structure is also useful.

The high diffusion coefficients ($\leq 10^{-7}\ cm^2\ s^{-1}$) of lithium cations are characteristic of these oxides; hence, they move only slightly less rapidly in these lattices than in liquid electrolyte solutions, in which diffusion coefficients are ca. 10^{-6} cm^2 s^{-1}. The cations in these oxides are so mobile because they do not form covalent bonds with the oxygen atoms at the lattice nodes [102].

The phase transition plots of Li–M_xO_y systems reveal stages indicating that a number of more permanent, separate structural phases in the various lithium content ranges have formed. Therefore, the chemical potential of lithium ions $\Delta\mu_{Li^+}$ also varies, depending on the quantity of lithium intercalated into the oxide structure (Fig. 8.16). However, these differences in $\Delta\mu_{Li^+}$ do not really affect the reversibility of lithium ion intercalation in these structures.

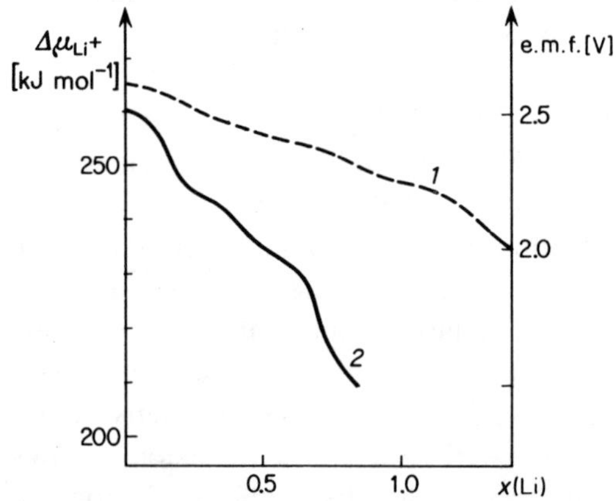

Fig. 8.16 Chemical potential of lithium ions $\Delta\mu_{Li^+}$ vs. the lithium content (x) in Mo_8O_{23} for the intercalation yielding $Li_xMoO_{2.83}$ (curve *2*), and the e.m.f. of the Li/Mo_8O_{23} cell (curve *1*).

The free energy changes due to Li^+ intercalation into oxide structures have been measured [103] and the theoretical specific energies calculated therefrom (Fig. 8.17).

Oxides capable of rapidly intercalating and releasing lithium ions also have high specific energies: for Cr_3O_8 it is 1080 and for V_6O_{13} (i.e. $VO_{2.167}$), 800 W h kg^{-1}. The non-stoichiometric vanadium oxide is a better cathode material in lithium storage batteries than titanium sulphide: when the storage battery was operating under stationary conditions, at 1 mA cm^{-2}, specific energies of 200–350 W h kg^{-1} were obtained from a Li/V_6O_{13} system [104] in which a 1.5 M solution of $LiAsF_6$ in 2-methyl THF and a graphite electrode were used. However, practical studies [104] have shown that the non-stoichiometric vanadium oxides $VO_{2.17}$ and $VO_{2.19}$ still need improving because of their poor electron conductivity. Graphite has been added and some of the vanadium replaced by another transition metal like Cr, Mo and Fe. This has yielded a cathode material of content $M_{0.13}V_{0.87}O_{2.17}$ where M is another transition metal. The initial

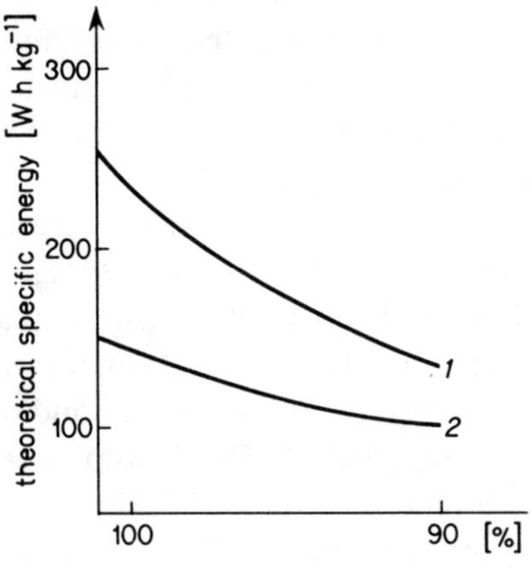

Fig. 8.17 Theoretical specific energies as a function of the percentage of intercalated lithium: *1* – for the Li/V_6O_{13} system; *2* – for the Li/TiS_2 system. The stoichiometric coefficient of lithium in the cathode material is 0.8.

voltage is 3.0 V, but cell polarization is considerable and a voltage plateau occurs at around 2 V. The final discharge voltage is 1.9 V when one lithium atom per transition metal atom is intercalated as in $Li_1(V+M)_1O_{2.17}$ or 1.7 V when the final composition is $Li_{1.1}(V+M)_1O_{2.17}$ (Fig. 8.18).

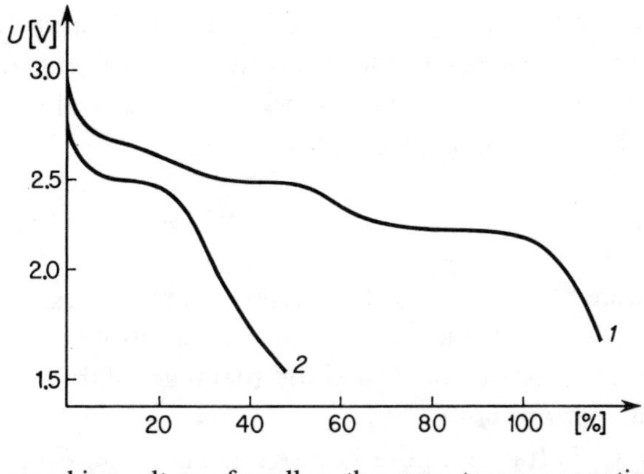

Fig. 8.18 The working voltage of a cell vs. the percentage consumption of its useful capacity in %: *1* – cell with non-stoichiometric vanadium oxide Li/$VO_{2.19}$ referred to formal reduction with the addition of 1.1 electrons per vanadium atom; *2* – cell with stoichiometric vanadium dioxide Li/VO_2, referred to formal reduction with the addition of 0.5 electrons per vanadium atom.

Graphite must be present in vanadium oxide cathode materials. Without it, the capacity is up to 50% lower. A fall in temperature to –16°C also reduced the room-temperature capacity by some 30% [101]. At present, several laboratories are investigating non-stoichiometric vanadium oxides

[105]. Discrepancies have arisen in the results because different methods have been used to synthesize the oxides and prepare the cathode material (leading to different grain sizes); charging and discharging conditions have also varied. Consequently a theoretical mechanism by which lithium is intercalated is being sought [106].

So far, γ-MnO_2 has achieved the greatest practical significance in disposable lithium cells [107], but no storage battery ready for large-scale production has yet been built with this material. MnO_2 is one of those oxides in which both electrons and small cations can move freely about its crystal structure forming a one-phase solid redox system [108]. The latest research has shown that sandwich sulphides like TiS_2 and stoichiometric and non-stoichiometric oxides like MoO_3,$VO_{2.17}$ and $VO_{2.19}$ belong to this category of redox systems [164].

The mechanism of reversible intercalation of small ions like Li^+ applies to electrode materials that are one-phase solid redox systems. Basically, this mechanism involves the simultaneous electroreduction of a multivalent, positive, oxide-forming transition-metal ion to a lower oxidation state by electrons arriving via the external cell circuit, and free, reversible intercalation by diffusion of small cations introduced into the bulk of the oxide in such numbers that the reduction in oxidation state suffered by the transition metal ion is compensated by the charge of the cations from the solution. Thus, the spatial structure of the oxide anion (O^{2-}) cathode material remains undisturbed, since the oxide reaction product does not form a separate phase. In short, oxide anions do not participate in the redox reaction; they merely provide a framework for the crystal structure. The following reversible reaction is an example:

$$Mn^{IV}O_2 + e + Li^{I} \rightleftharpoons Li^{I}Mn^{III}O_2$$

The oxide phase remains homogeneous in the region of variable composition produced by the intercalation of small monovalent cations, because the reduced ions of the transition metal are stabilized only in the presence of small cations (Fig. 8.19).

The potential of a pure homogeneous phase E depends on its composition, in this particular case on the quantity of intercalated lithium, but wherever two constant-composition phases coexist, one of which is the reaction product, the electrode voltage is constant (ignoring polarization effects) so long as some starting material is still present. In the first case, the change in potential of a MnO_2 cathode as Mn(IV) is reduced to Mn(III) can be described by the Nernst equation [109] ($n = 1$):

$$E = E^0 + \frac{RT}{nF} \ln \frac{a_{\text{Mn(IV)(solid phase)}}}{a_{\text{Mn(III)(solid phase)}}}$$

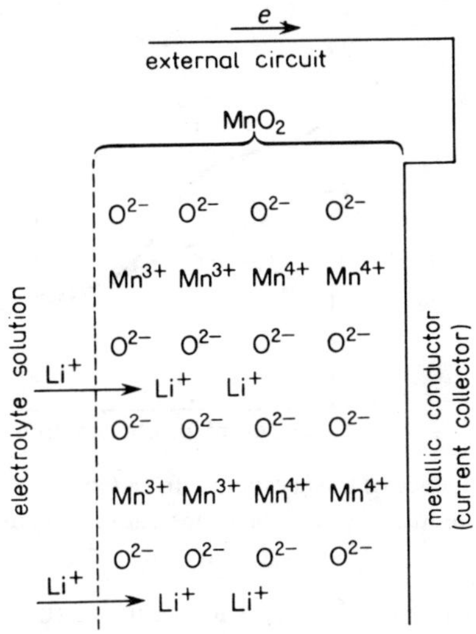

Fig. 8.19 The mechanism of reversible intercalation of small ions such as Li^+ into semiconducting sandwich cathode materials which are constant one-phase redox systems, exemplified by the reduction of Mn^{4+} to Mn^{3+} in MnO_2.

Figure 8.20 shows the plot of MnO_2 electrode potential vs. the degree of reduction of the oxide-forming metal ion Mn(IV) → Mn(III) determined by Li^+ intercalation.

The effect of the electrochemical properties of the γ and β forms of manganese dioxide on the change in electrode potential is shown in Fig. 8.21. The e.m.f. ($E = -\Delta G/nF$) of the open circuit of a Li/MnO_2 cell when discharging to Mn(III) is 2.63 V; the theoretical specific energy is 752 W h kg^{-1}

In many countries, Li/MnO_2 cells are already being manufactured to

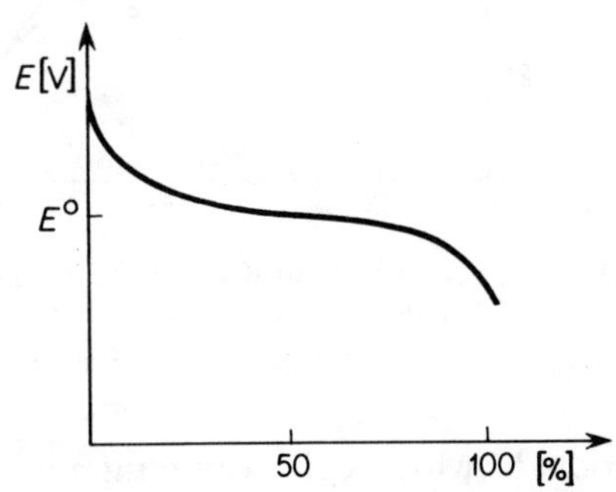

Fig. 8.20 The potential of a semiconducting MnO_2 cathode vs. the degree of reduction of Mn^{4+} to Mn^{3+}: E^0 – standard electrode potential at different activities of both forms $a_{Mn^{4+}} = a_{Mn^{3+}}$.

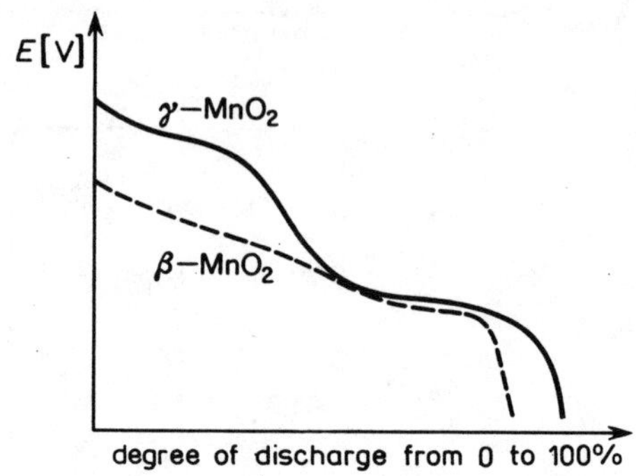

Fig. 8.21 The influence of the crystalline form (γ and β) of manganese dioxide on electrode potential vs. the degree of reduction of the cathode material.

supply very small (≥ 5 μA) and quite large (≤ 0.5 A) currents to all manner of electronic devices [107, 110–115].

Various electrolyte solutions are used, usually mixed organic solvents with an inorganic salt. One coin-type Li/MnO_2 cell contained a mixture of PC + 1,2-DME with 1 M lithium perchlorate [110]. Here, the cathode was mixed γ- and β-MnO_2, produced by heating electrolytic γ-MnO_2 to 350°C. Polypropylene film separated the cathode and anode spaces. The effect of solvent composition on the cell discharge curves is shown in Fig. 8.22.

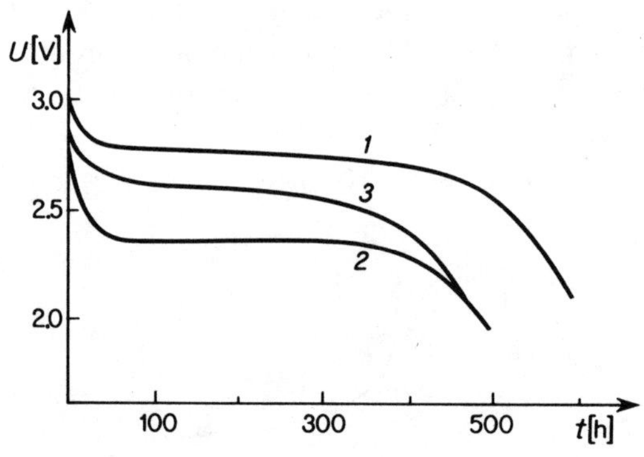

Fig. 8.22 Discharge curves of Li/$LiClO_4$ in PC + 1,2-DME/MnO_2 cells with respect to the composition of the mixed solvent: *1* – 2:1 PC:1,2-DME at 20°C; *2* – 1:1 PC:1,2-DME at –10°C; *3* – 2:1 PC:1,2-DME at –10°C.

A cylindrical Li/MnO_2 cell, weighing 3 g including the can [107], has a capacity of 160 mA h and a practical specific energy (referred to the total cell mass) of 143 W h kg^{-1}. The discharge curves with respect to external resistance are shown in Fig. 8.23.

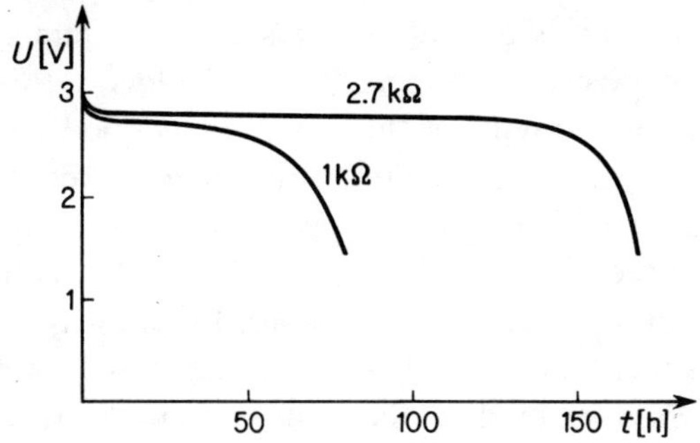

Fig. 8.23 Typical discharge curves of a miniaturized Li/MnO_2 cell of mass 3 g and capacity 0.16 A h under different loads.

The characteristics of a lithium cell with CuS or MnO_2 using the same electrolyte solution, 1 M $LiClO_4$ in 70% THF + 30% DME [115] are compared in Fig. 8.24.

Already being produced on a large scale, lithium-copper(II) oxide cells use a mixed organic solvent containing PC, and a $LiClO_4$ or $LiBF_4$ electrolyte. They have good properties at elevated temperatures up to +150°C [111, 116].

Cells with cobalt and vanadium oxides, capable of reversibly intercalating small cations (Li^+ and Na^+) into the crystal lattice, are increasing in importance [162, 183].

The Allied-Signal company has built small, prototypical AF-size cells of capacity ca. 0.7 A h and working voltage ≤ 3 V, containing cathodes of the intercalation compounds Na_xCoO_2 or $Li_xV_6O_{13}$. The respective anode materials are a sodium-lead alloy Na_xPb where $1 \leq x \leq 3.75$ and a lithium-

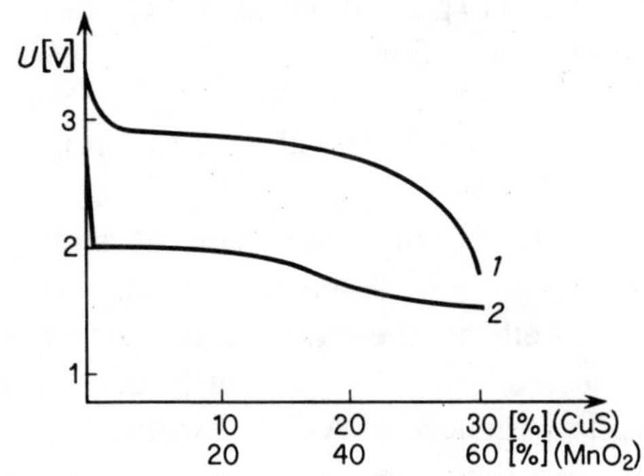

Fig. 8.24 Cell voltage with respect to % cathode-material consumption in: *1* – Li/MnO_2 cells during a one-electron discharge reaction $Mn^{IV}O_2 + Li^+ +$ $+ e \rightarrow LiMn^{III}O_2$; *2* – Li/CuS cells during a two-electron discharge reaction

$$CuS + e + Li^+ \rightarrow \tfrac{1}{2}Cu_2S + \tfrac{1}{2}Li_2S$$

$$\tfrac{1}{2}Cu_2S + e + Li^+ \rightarrow Cu^+ + \tfrac{1}{2}Li_2S$$

aluminium alloy LiAl. The anode structure is complicated: it is a composite in which the finely dispersed anode material, i.e. the alloy of the above composition, has been press-moulded into an organic polymer. The electron conductivity of this polymer is good, and accordingly acts as a current collector: at the same time it can reversibly intercalate or absorb metal cations released by the powdered anode during discharge. When the current direction is reversed, the Na^+ or Li^+ cations return to the alloy, which retains its granular structure. In this way, a cell has been produced which is rechargeable without dendrite formation [162].

The polymer binding the grains of anode material is polyacetylene or polyphenylene (PPP). The electrolyte is 1 M $NaPF_6$ (or $LiPF_6$) in DME. Such cells are highly reversible during extended cycling and supply a practical energy density of ca. 100 $W\,h\,kg^{-1}$.

The Li/Li_xCoO_2 storage battery with a pure lithium anode has also been improved by the use of binary mixtures of organic solvents containing methyl formate and dimethyl carbonate. With a $LiAsF_6$ electrolyte, over 100 deep discharge cycles have been obtained at an average working voltage of 3.7–3.9 V and current densities of $\leq 2\ mA\,cm^{-2}$ as a consequence of the great mobility of Li^+ ions in the Li_xCoO_2 sandwich structure [163].

Lithium-silver oxide cells are made, chiefly to power electronic watches [112]. Also a lithium-nickel oxide cell is described [193].

One of the first systems used to power cardiac pacemakers was a lithium cell with a silver chromate cathode: Li/Ag_2CrO_4. Like other silver compounds, e.g. silver oxide, silver chromate is also reduced yielding separate silver metal and lithium chromate phases. No intercalation compounds are formed. The cell e.m.f. of such a multiphase system is constant [109]. Practical specific energies of 214 $W\,h\,kg^{-1}$ have been obtained in a Li/Ag_2CrO_4 cell [117], at an initial voltage of 3.3 V. The theoretical specific energy of the reaction

$$Ag_2CrO_4 + 2Li^0 \rightarrow Li_2CrO_4 + 2Ag^0$$

is 34 $W\,h\,kg^{-1}$ [118]. In a new cell of similar type, the double salt $AgBi(CrO_4)_2$ is the cathode material [119].

Li/Ag_2CrO_4 cells are the second-most important as regards applications in cardiac pacemakers [35, 112], partly because even after Ag_2CrO_4 has been completely reduced to Ag^0, the voltage (ca. 2.5 V) is still exceptionally high due to a second reduction, i.e. of chromate, containing Cr(VI), to Cr(III) [119].

8.6.7 *Cells with solid electrolytes*

Attempts to obtain high-energy cells providing a very large number of charge/discharge cycles are continually running into difficulties. One is that the organic solvents, which ensure that electrolyte solutions have a sufficiently high conductivity, are insufficiently stable.

Attention has thus turned to inorganic mixtures of ionized salts that form low-melting eutectics or glazes of high ionic conductivity even at fairly low temperatures. Several types of storage battery have already been constructed with such electrolytes [165].

SAFT (France) have manufactured a storage battery with a LiAl-alloy anode, a solid electrolyte (glaze) containing 2.5 LiI, $Li_4P_2S_7$ and a TiS_2 cathode. Current densities of ca. 20 $\mu A\,cm^{-2}$ were obtained. By applying high pressures to enhance contact between anode, electrolyte and cathode, Ackridge and Vourlis [165] obtained a tenfold increase in the current density, $\leq 0.2\ mA\,cm^{-2}$ in class 2016 standard cells. These are button cells 20 mm in diameter and 1.6 mm thick with a volume energy density of ca. 0.2 $W\,h\,cm^{-3}$, so they can be used in electronic watches and calculators.

Computer memory systems have to be powered by very thin storage batteries built into integrated circuits. An ultra-thin solid-state micro-battery with a thickness of 0.01 mm (10 μm) was made for this purpose [194]. A cell with the $Cu/Rb_4Cu_{16}I_6Cl_{14}/Cu_xTiS_2$ system and a mixed rubidium-copper chalcogenide solid electrolyte has been found useful here [166]. The anode is copper metal, so although the working voltage is only ca. 0.5 V, current densities are quite high ($\leq 1\ mA\,cm^{-2}$). This cell can operate within a wide range of temperatures, from −60°C to +60°C, and the cell is housed in plastic film, and not in a metal can. Cross-linked organic polymers are also useful as solid electrolytes. They provide a fairly rigid framework with polar functional groups binding ionized salts, whose ions can move about in the voids of the polymer. They can be used in combination with a lithium anode.

Examples are a lithium cell with the $Li/(PEO)_qLiCF_3SO_3/LiV_3O_8$ system in which a polyether (i.e. poly(ethylene oxide), PEO) is moulded together with the electrolyte [167], or another cell with a $RbCu_4I_{1.5}Cl_{3.5}$ electrolyte and a $Cu_2Mo_6S_8$ cathode [168]. Poly(phosphazene) polymer has also been used instead of PEO [169, 170, 171].

8.7 CONCLUSIONS

Recent research on chemical sources of energy [124–131, 178–190] has been striving to obtain reversible electrodes in non-aqueous media suitable both for reusable cells and storage batteries capable of withstanding even a thousand charge/discharge cycles. This is justified in view of the limited natural resources of lithium, whose technical applications have expanded very rapidly in the last decade, especially in the field of nuclear energy. Hence the high price of lithium and the need to recycle it. Lithium cells must be hermetically sealed and the many problems in this connection have been overcome. Cells of a very high standard are now being manufactured. However, they are expensive, so it is a great waste to discard them – lithium is too valuable to be used in disposable cells.

It seems that this line of development will be continued, and will require ever more detailed studies on the synthesis of compounds and electrode materials, the microscopic and crystallographic structure of both starting materials and products, the reaction mechanisms of lithium and other metals which could replace it. However, the advantages of lithium are unquestioned. A step forward in this direction are lithium alloys.

There is a strengthening tendency towards the production of new types of non-stoichiometric compound by the addition of a second transition metal to sulphides and oxides in order to obtain mixed compounds, often acting like solid solutions. With these, much higher specific energies can be generated in lithium cells.

The problem of preparing reversible electrodes is a very complicated one. Among other things, it involves the production of electrolyte solutions in which an appropriate combination of ligands allows soluble though perhaps unstable complexes to form. When cells with multiphase electrodes discharge, these complexes enable the initial electrode materials, especially the cathode material, to be reconstitued.

Great stress is also laid on reducing the cell's internal resistance so that useful current densities can be increased by choosing suitable mixtures of organic solvents, polymers or inorganic glazes, and the best possible electrolyte.

Polymeric organic compounds such as polyacetylene are highly promising. Through the electrochemical oxidation or reduction of compounds containing intercalated anions or cations, materials with anodic or cathodic properties are obtained.

The construction of very high capacity storage batteries and generating sets for powering electrically-driven vehicles is proceeding. It is helped by computer modelling of batteries for electric vehicles [195] and computer simulation of high-discharge-rate battery systems [196]. This line of research is very important and will surely be continued. Non-aqueous cells and storage batteries are undoubtedly one of those branches of advanced technology requiring strong cooperation between many different scientific disciplines [132–136].

This evaluation of capabilities of non-aqueous lithium cells as written in the Polish edition of this book has been largely confirmed during the next decade as evidenced by a review which appeared recently [172].

One can only add, that since the prospects of electrochemical cells based on cold nuclear fusion are nil [173, 174] cells and batteries based on lithium will continue to be improved, while modern quality and reliability methods described in the new book by Bro and Levy [175] will be implemented.

REFERENCES

[1] J.G. Gibson and J.L. Sudworth, *Specific Energies of Galvanic Reactions and Related Thedmodynamic Data*, Chapman and Hall, London (1973).

[2] W.M. Latimer, *The Oxidation States of the Elements and their Potentials in Aqueous Solutions*, Prentice-Hall, Englewood Cliffs, N.J. (1952).

[3] E. Bichovsky and F. Rossini, *The Thermochemistry of Chemical Substances*, Reinhold Publ. Co., New York (1936).

[4] O. Kubachevski, E.L. Evans and C.B. Alcock, *Metallurgical Thermochemistry*, Pergamon Press, London (1967).

[5] JANAF, *Thermochemical Tables*, The Dow Chemical Company, Midland, Michigan (1965).

[6] W.J. Hamer, M.S. Malmberg and B. Rubin, *J. Electrochem. Soc.*, **112**, 750 (1965).

[7] A. Glassner, *The Thermochemical Properties of the Oxides, Fluorides and Chlorides to 2500 K*, Argonne National Laboratory Publication, ANL-5750, after [1].

[8] K.C. Mills, *Thermodynamic Data for Inorganic Sulphides, Selenides and Tellurides*, Butterworth, London (1974).

[9] K.M. Abraham, J.L. Goldman and M.D. Dempsey, *J. Electrochem. Soc.*, **128**, 2493 (1981).

[10] J. Gomółka, F. Kowalczyk and A. Franke, *Modern Chemical Sources of Current* (in Polish), MON, Warszawa (1977).

[11] G.W. Heise and N.C. Cahoon, *The Primary Battery*, Vol. 1, J. Wiley, New York (1971).

[12] H. Eyring, J. Walter and G.E. Kimball, *Quantum Chemistry*, J. Wiley, VIII ed., New York (1958).

[13] K.E. Wrabetz, *Z. Elektrochem.*, **60**, 722 (1956).

[14] W. Naddack, K.E. Wrabetz and W. Herbst, *Z. Elektrochem.*, **59**, 752 (1955).

[15] R. Jasiński, *Electrochem. Technol.*, **6**, 28 (1968).

[16] R. Jasiński and B. Burrows, *J. Electrochem. Soc.*, **116**, 422 (1969).

[17] F.W. Dampier, *J. Electrochem. Soc.*, **128**, 2501 (1981).

[18] W.B. Johnson and W.L. Worrell, *Synthetic Metals*, **4**, 225 (1984).

[19] M.S. Whittingham, *J. Electrochem. Soc.*, **123**, 315 (1976).

[20] P.J. Nigrey, D. McInnes jr., D.P. Nairns and A.G. McDiarmid, *J. Electrochem. Soc.*, **128**, 1651 (1981).

[21] G. Eichinger and J.O. Besenhard, *J. Electroanal. Chem.*, **72**, 1 (1976).

[22] J.J. Auborn and N. Marincić, *Power Sources 5, Proceedings of the 9th International Symposium, Brighton 1974* (D.H. Collins ed.), Academic Press, London (1975), Abstract No. 43.

[23] A.N. Dey, *Electrochemical Power Sources*, 28th Meeting of the International Society of Electrochemistry, Varna (1977), Extended Abstracts, No. 141.

[24] P. Bro, R. Homes, N. Marincić and H. Taylor, *Power Sources 5, Proceedings of the 9th International Symposium, Brighton 1974* (D.H. Collins ed.), Academic Press, London (1975), Abstract No. 45.

[25] J.P. Gabano, G. Gerbier and J.F. Laurent, *Chemie-Ing.-Techn.*, **42**, 210 (1970).

[26] J.P. Gabano, V. Dechenaux, G. Gerbier and J. Jammet, *J. Electrochem. Soc.*, **119**, 459 (1972).

[27] F.W. Dampier, *J. Electrochem. Soc.*, **121**, 658 (1974).

[28] V. Dechenaux, G. Gerbier and J.F. Laurent, *Entropie*, **10**, 15 (1966).

[29] L. Werblan and J. Lesiński, *Pol. J. Chem.*, **53**, 2571 (1979).
[30] L. Werblan, J. Lesiński, A. Suzdorf, J. Izydorek and P. Drachal, *J. Power Sources*, **9**, 409 (1983).
[31] L. Werblan, A. Suzdorf and J. Lesiński, *Bull. Acad. Polon. Sci., ser. sci. chim.*, **28**, 627 (1980).
[32] L. Werblan, A. Suzdorf and E. Lin, *Pol. J. Chem.*, **56**, 1071 (1982).
[33] J. Lesiński, J. Izydorek and L. Werblan, *J. Power Sources*, **27**, 337 (1989).
(a) L. Werblan, A. Suzdorf, E. Lin, G. Szymański and J. Lesiński, *Bull. Polish Acad. Sci. Chem.*, **33**, 285 (1985).
(b) L. Werblan and J. Izydorek, 8th ISSSSI, University of Regensburg, August 9–14 1987, Abstracts, L5.06., FRG.
(c) Y. Uetani, K. Yokoyama and O. Okamoto, *J. Power Sources*, **5**, 89 (1980).
(d) Ch. Iwakura, N. Isobe and H. Tamura, *Electrochim. Acta*, **28**, 277 (1983).
(e) P.A. Nelson, E.C. Gay and W.J. Walsh, *26th Power Sources Symp., 1984*, The Electrochemical Society, Princeton, N.Y., p. 65.
[34] Y. Matsuda, H. Ueyama and Y. Takasu, *J. Electrochem. Soc.*, **123**, 220 (1976).
[35] B.B. Ovens and K.R. Brennen, *Progr. in Batteries and Solar Cells*, **3**, 38 (1980).
[36] G. Eichinger and H.P. Fritz, *J. Electroanal. Chem. Interfac. Electrochem.*, **58**, 357 (1975).
[37] G. Eichinger and H.P. Fritz, *J. Electroanal. Chem. Interfac. Electrochem.*, **58**, 369 (1975).
[38] R.B. Swaroop and J.E. Battles, *J. Electrochem. Soc.*, **128**, 1873 (1981).
[39] J.R. Coleman and M.W. Bates, *Power Sources 2, Proceedings of the 6th International Symposium, Brighton 1968* (D.H. Collins, ed.), Pergamon Press, Oxford (1968), p. 289.
[40] D.R. Vissers, Z. Tomczuk and R.K. Steunenberg, *J. Electrochem. Soc.*, **121**, 663 (1974).
[41] H. Yamin and E. Peled, International Meeting on Lithium Batteries, Rome (1982), Extended Abstracts, No. 10.
[42] Z. Tomczuk, S.K. Preto and M.F. Roche, *J. Electrochem. Soc.*, **128**, 760 (1981); L. Bartholme, E.C. Gay and H. Shimotaka, Electrochemical Society Meeting, Hollywood (1980), Extended Abstract, p. 298.
[43] J.E. Battles, F.C. Mrazek and N.C. Otto, Report ANL 80-130, Argonne National Laboratory (1980).
[44] M.S. Whittingham, *Progr. Solid State Chem.*, **12**, 41 (1978).
[45] D.W. Murphy and F.A. Trumbore, *J. Electrochem. Soc.*, **123**, 960 (1976).
[46] D.W. Murphy, J.N. Carrides, F.J. Di Salvo, C. Cross and J.W. Waszczak, *Mat. Res. Bull.*, **12**, 825 (1977).
[47] J.N. Carrides and D.W. Murphy, *J. Electrochem. Soc.*, **124**, 1309 (1977).
[48] K. Braeuer, Electrochemical Society Meeting, Boston (1968), Extended Abstracts No. 210, Electrochemical Society Meeting, New York (1969), Extended Abstracts, No. 164.
[49] N. Watanabe and M. Fukuda, FRG Patent No. 919, 394.
[50] M. Fukuda and T. Iijima, Electrochemical Society Meeting, Cleveland (1971), Extended Abstracts, No. 41.
[51] M. Fukuda and T. Iijima, *Power Sources 5, Proceedings of the 9th International Symposium, Brighton 1974* (D.H. Collins, ed.), Academic Press, London (1975), p. 713.
[52] R.G. Gunther, *Power Sources 5, Proceedings of the 9th International Symposium, Brighton 1974* (D.H. Collins, ed.), Academic Press, London (1975).
[53] FRG Patent No. 2 264 057, after G. Eichinger and J.O. Besenhard, *J. Electroanal. Chem.*, **72**, 29 (1976).
[54] Matsushita-cho Moriguchi, *Progr. in Batteries and Solar Cells*, **3**, 45 (1980).

[55] D. McInnes jr., M.A. Druy, P.J. Nigrey, D.P. Nairns, A.G. McDiarmid and A.J. Heeger, *J. Chem Soc., Chem. Commun.*, in press.

[56] C.K. Chiang, M.A. Druy, C.S. Gan, A.J. Heeger, E.J. Louis, A.G. McDiarmid, Y.W. Park and H. Shirkawa, *J. Am. Chem. Soc.*, **100**, 1013 (1973).

[57] T.C. Chung, A. Feldblum, A.J. Heeger and A.G. McDiarmid, *J. Chem. Phys.*, **74**, 5605 (1981).

[58] R. Dagani, *Chem. Ing. News*, **59**, 39 (1981).

[59] A. Proń, D. Billand, I. Kulszewicz, C. Budrowski, J. Przyłuski and J. Suwalski, *Mat. Res. Bull.*, **16**, 1229 (1981).

[60] G.C. Farrington, D. Frydrych and R. Hug, International Meeting on Lithium Batteries, Rome (1982), Extended Abstracts, No. 22.

[61] R. Jasiński, *High Energy Batteries*, Plenum Press, New York (1967).

[62] L. Werblan and A. Cisak, *Wiadomości Chem.*, **27**, 843 (1973).

[63] H. Binder, A. Köhling, G. Walker and G. Sandstede, *Power Sources 6, Proceedings of the 10th International Symposium, Brighton 1976* (D.H. Collins, ed.), Academic Press, Abstract No. 41.

[64] H. Alt, H. Binder, G. Klempert, A. Köhling and G. Sandstede, *Chemie-Ing.-Techn.*, **44**, 195 (1972).

[65] O.S. Ksenzek, W.Z. Basukov, F.E. Dinkewich, W.Z. Moskovski, S.A. Pietrova and E.M. Shembel, *Electrochemical Power Sources,* 28th Meeting of the International Society of Electrochemistry, Varna (1977), Extended Abstract, No. 151.

[66] J.J. Auborn, A. Heller and K.W. French, *Proceedings of the 25th Power Sources Symposium*, 1972, p. 6.

[67] K.J. Vetter, *Electrochemische Kinetik*, Springer-Verlag, Berlin (1961), p. 312.

[68] K.A. Klinedinst, *J. Electrochem. Soc.*, **128**, 2507 (1981).

[69] J.J. Auborn and A. Heller, *Proceedings of the 24th Meeting of the International Society of Electrochemistry, Brighton 1973*, Extended Abstracts, p. 254.

[70] L.A. Beketaeva, *Electrochemical Power Sources*, 28th Meeting of the International Society of Electrochemistry, Varna (1977), Extended Abstracts, No. 150.

[71] R.W. Moshtev, Y. Geronov and B. Puresheva, *J. Electrochem. Soc.*, **128**, 1851 (1981); *Electrochemical Power Sources,* 28th Meeting of the International Society of Electrochemistry, Varna (1977), Extended Abstracts, p. 655.

[72] M. Babai and J. Bineth, International Meeting on Lithium Batteries, Rome (1982), Extended Abstracts, No. 12.

[73] A. Meitav and E. Peled, International Meeting on Lithium Batteries, Rome (1982), Extended Abstracts, No. 35.

[74] N. Marincić and J. Epstein, *Progr. in Batteries and Solar Cells,* **3**, 84 (1980).

[75] M. Babai, *Progr. in Batteries and Solar Cells*, **3**, 110 (1980).

[76] J.P. Gabano and G. Celin, *Power Sources 8, Proceedings of the 12th International Symposium, Brighton 1980* (D.H. Collins, ed.), Academic Press, London (1981), p. 3.

[77] F. Alessandrini, B. Scrosati, M. Lazzari and F. Borino, International Meeting on Lithium Batteries, Rome (1982), Extended Abstracts, No. 11.

[78] J.J. Auborn, K.W. French, S.E. Liebermann, V.K. Shah and A. Heller, *J. Electrochem. Soc.*, **120**, 1613 (1973).

[79] G. Pistoia, *Electrochemical Power Sources*, 28th Meeting of the International Society of Electrochemistry, Varna (1977), Extended Abstract, No. 45.

[80] Y. Geronov, B. Puresheva and R.V. Moshtev, International Meeting on Lithium Batteries, Rome (1982), Extended Abstracts, No. 9.

[81] F.M. Bowers and J.H. Amburs, *Power Sources 5, Proceedings of the 9th International Symposium, Brighton 1974* (D.H. Collins, ed.), Academic Press, London (1975), Abstract No. 36.

[82] S. Maximovitch, G. Bronoel, Nguyen Tan Nghia and J. Sarradin, *Power Sources 6,*

Proceedings of the 10th International Symposium, Brighton 1976 (D.H. Collins, ed.), Academic Press, London (1977), p. 55.
[83] C.C. Liang, M.E. Bolster and R.M. Murphy, *J. Electrochem. Soc.*, **128**, 1631 (1981).
[84] J.L. Weiminger and F.W. Secor, *J. Electrochem. Soc.*, **121**, 315 (1974).
[85] A.A. Schneider, W. Greatbath and R. Mead, *Power Sources 5, Proceedings of the 9th International Symposium, Brighton 1974* (D.H. Collins, ed.), Academic Press, London (1975), Abstract No. 41.
[86] *Progr. in Batteries and Solar Cells*, 3, 43 (1980).
[87] B.B. Owens, International Meeting on Lithium Batteries, Rome (1982), Extended Abstracts, No. 28; J. Broadhead, *Power Sources 4, Proceedings of the 8th International Symposium, Brighton 1972* (D.H. Collins, ed.), Academic Press, London (1973), Abstract No. 30.
[88] P. Ludwig, T. Layloff and R. Adams, *J. Am. Chem. Soc.*, **85**, 4568 (1964).
[89] R. Jasiński, *Adv. Electrochem. Eng.*, **8**, 253 (1971); M. Fukuda and Denki Kagaku, **41**, 593 (1973).
[90] H.N. Seiger, A.E. Lyall and R.C. Shair, *Power Sources 2, Proceedings of the 6th International Symposium, Brighton 1968* (D.H. Collins, ed.), Pergamon Press, London (1968), Abstract No. 17.
[91] R. Keller, *J. Electrochem. Soc.*, **117**, 1266 (1970).
[92] K.H.M. Braeuer and J.A. Harvey, Report AD 654813 (1967).
[93] FRG Patent No. 1813657.
[94] S. Lerner and H.N. Seiger, *J. Electrochem. Soc.*, **117**, 574 (1970).
[95] M.L.B. Rao, *J. Electrochem. Soc.*, **114**, 665 (1967).
[96] French Patent No. 1556247.
[97] J.O. Besenhard and R. Schöllhorn, *J. Power Sources*, **1**, 267 (1976/77).
[98] M. Pasquali, G. Pistoia and F. Rodante, *Solid State Ionics*, **6**, 319 (1982).
[99] K. Mizishima, P.C. Jones, P.J. Wieseman and J.B. Goodenough, *Mat. Res. Bull.*, **15**, 783 (1980).
[100] G. Pistoia, International Meeting on Lithium Batteries, Rome (1982), Extended Abstracts, No. 13; L. Campanella and G. Pistoia, *J. Electrochem. Soc.*, **118**, 1905 (1971).
[101] D.W. Murphy, P.A. Christian, F.J. Di Salvo, J.N. Carrides and J.W. Waszczak, *J. Electrochem. Soc.*, **128**, 2053 (1981).
[102] H.Y.P. Hong, *J. Power Sources*, **5**, 137 (1980).
[103] P.G. Dichens and G.J. Reynolds, *Solid State Ionics*, **5**, 331 (1981).
[104] K.M. Abraham, J.L. Goldman and M.D. Dempsey, *J. Electrochem. Soc.*, **128,** 2493 (1981).
[105] H. Ikeda and S. Narukowa, International Meeting on Lithium Batteries, Rome (1982), Extended Abstracts, No. 16.
[106] F. Delard, D. Deroo and R. Mauger, International Meeting on Lithium Batteries, Rome (1982), Extended Abstracts, No. 14.
[107] H. Ikeda, M. Hara and S. Nakaido (SANYO), *Progr. in Batteries and Solar Cells*, **3**, 44 (1980); *ibid.*, **3**, 64 (1980).
[108] A. Kozawa and J.F. Yeager, *J. Electrochem. Soc.*, **112**, 959 (1965); A. Kozawa and R.A. Powers, *J. Electrochem. Soc.*, **113**, 870 (1966).
[109] A. Kozawa, *Progr. in Batteries and Solar Cells*, **3**, 125 (1980).
[110] J. Watanabe, R. Okazaku, M. Nakai and Y. Toyoguchi, *Progr. in Batteries and Solar Cells*, **3**, 70 (1980).
[111] SAFT, *Progr. in Batteries and Solar Cells*, **3**, 47 (1980).
[112] BEREC, *Progr. in Batteries and Solar Cells*, **3**, 46 (1980).
[113] VARTA, *Progr. in Batteries and Solar Cells*, **5**, 44 (1980).

[114] K. Hampartzumian, A. Mochilov, M. Ilchev and V. Manev, International Meeting on Lithium Batteries, Rome (1982), Extended Abstracts, No. 17.

[115] F.W. Dampier, *J. Electrochem. Soc.*, **121**, 656 (1974).

[116] H. Ogawa, T. Iijima, A. Morita and J. Nishimura, *Progr. in Batteries and Solar Cells*, **3**, 5, 93 (1980).

[117] N. Marincić, *J. Appl. Electrochem.*, **6**, 463 (1976); B. Di Pietro and B. Scrosati, *J. Electrochem. Soc.*, **124**, 161 (1977).

[118] H.P. Siebert, *Elektronik*, **1976**, 48.

[119] M. Broussely, J.P. Rivault and J.P. Gabano, International Meeting on Lithium Batteries, Rome (1982), Extended Abstracts, No. 18.

[120] Implementation "Starting the experimental scale production of high-energy lithium cells" in the Central Laboratory of Cells and Batteries in Poznań, Poland, based on laboratory scale technology, elaborated at the Laboratory of Transport Phenomena in Solution, Department of Chemistry, University of Warsaw. Approved by the Ministry of Machine Industry and by EMA Union on 18. 10. 1979.

[121] L. Werblan, J. Lesiński, B. Pruszkowska-Drachal and A. Suzdorf, "A method of producing a lithium cell", Polish Patent, 131, 310 (30 Nov. 1985).

[122] J. Downarowicz and R. Szłapka, Central Laboratory of Batteries and Cells, Poznań, Poland (1984), report.

[123] J. Lesiński, G. Szymański and L. Werblan, *J. Electroanal. Chem.*, **205**, 91 (1986).

[124] K. Wiesener, J. Garche and W. Schneider, *Electrochemische Stromquellen*, Akademie-Verlag, Berlin (1981).

[125] V.S. Bagotzki, 3th Intern. Symposium Elektrochem. Stromquellen, Dresden (1978), p. 29.

[126] A. Sadkowski, J. Wild and P. Żółtowski, 28th Meeting of the International Society of Electrochemistry, Varna (1977), Extended Abstracts, Vol. 2, p. 202.

[127] N.S. Lidorenko and V.E. Dmitrenko, 28th Meeting of the International Society of Electrochemistry, Varna (1977), Extended Abstracts, vol. 2, p. 433.

[128] V. Obretenov, T. Kossev, V. Bostanov and E. Budevsky, *J. Electroanal. Chem.*, **159**, 257 (1983).

[129] J. Jowulski and L. Suski, *J. Appl. Electrochem.*, **14**, 135 (1984).

[130] Y. Geronov, F. Schwager and R.H. Muller, *J. Electrochem. Soc.*, **129**, 1422 (1982).

[131] J. Molenda, C. Delmas, P. Dordor and A. Stokłosa, Solid State Ionics, 12, 473 (1982).

[132] E. Budevski, Scientific Session of the CMEA Experts, Sofia (1980), report.

[133] R. Skowroński and J. Przyłuski, 6th Scientific Session of MR I-11 Problem, University of Łódź, Łódź (1985).

[134] E. Budevski, Conference on the multilateral co-operation programme of the Academies of Sciences of the socialist countries "Electrochemistry", Sofia (1984), report.

[135] E. Dutkiewicz, in: *The Electrochemistry of Interfaces in Ionogenic Media* (in Polish) (W. Libuś and E. Dutkiewicz, eds.), PWN, Warszawa (1984), p. 152; A. Kisza, *ibid.*, p. 99; T. Krogulec and Z. Galus, *ibid.*, p. 165; J. Sobkowski, *ibid.*, p. 205.

[136] I.A. Kedrynskii, V.E. Dmitrenko, Y.M. Povarov and I.I. Grudyanov, *Chemical Sources of Current with Lithium Electrodes* (in Russian), Izd. Krasnoyarskogo Universiteta, Krasnoyarsk (1983).

[137] Y. Matsuda, *Progr. Batt. Solar Cells*, **6**, 17 (1987).

[138] W.L. Bowden, L.H. Barnette and D.L. Muth, J. Electrochem. Soc., 136, 1614 (1989).

[139] J.A.R. Stiles, *J. Power Sources*, **26**, 233 (1989).

[140] J.M. Tarascon, T.P. Orlando and M.J. Neal, *J. Electrochem. Soc.*, **135**, 804 (1988).

[141] K.M. Abraham, D.M. Pasquariello and D.A. Schwartz, *J. Power Sources*, **26**, 217 (1989).

[142] F.C. Laman and K. Brandt, *J. Power Sources*, **24**, 195 (1988).

[143] P.H. Smith, S.D. James, K.M. O'Neil and M.H. Wilson, *J. Electrochem. Soc.*, **136**, 1625, 1631 (1989).

[144] M. Helbig, H. Hoerhold, D. Raabe, U. Scherf, F. Leonhardt and E. Triebel, Ger. (East) Pat. DDR 261027 (12 Oct. 1988).

[145] S. Szpak, P.A. Mosier-Boss, J.J. Smith and R.J. Nowak, *Proc. Electrochem. Soc.*, **1989**, 89-4, 142 (Chem. Abstr. 111: 10192).

[146] W.K. Behl and Der-Tau Chin, *J. Electrochem. Soc.*, **135**, 21 (1988).

[147] Ch.W. Walker, *J. Power Sources*, **25**, 13 (1989).

[148] F. Deligiannis, D.H. Shen and G. Halpert, *J. Power Sources*, **26**, 211 (1989) (Chem. Abstr. 111: 81321).

[149] D. Vallin and M. Broussely, *J. Power Sources*, **26**, 201 (1989) (Chem. Abstr. 111: 81320).

[150] A.J. Hills and N.A. Hampson, *J. Power Sources*, **24**, 253 (1988).

[151] M. Mizutani and Y. Okamura, *Progr. Batteries and Solar Cells*, **6**, 48 (1987).

[152] C.W. Walker, M. Binder, W.L. Wade and S. Gilman, *J. Electrochem. Soc.*, **132**, 1536 (1985).

[153] K.A. Klinedienst, *J. Electrochem. Soc.*, **131**, 492 (1984).

[154] J. Heydecke, *Progr. Batt. Solar Cells*, **6**, 93 (1987).

[155] N. Marincić, in: *Modern Aspects of Electrochemistry*, Vol. 15 (R.E. White, J.O'M. Bockris and B.F. Conway, eds.), 1982, p. 167.

[156] E.S. Takeuchi, C. Bessel and E. Eads, *J. Power Sources*, **24**, 229 (1988).

[157] A.N. Dey, H.C. Kuo, P. Piliero and M. Kallianidis, *J. Electrochem. Soc.*, **135**, 2115 (1988).

[158] R.J. Mammone and B. Binder, *J. Electrochem. Soc.*, **134**, 37 (1987).

[159] M.L. Gapikanth, *Bull. Electrochem.*, **4**, 443 (1988).

[160] H.K. Park, D.E. Stilwell, D.E. Bliss and M.H. Miles, *Proc. Electrochem. Soc.*, **1989**, 89-4 (Proc. Symp. Lithium Batt. 1988), p. 180.

[161] *Comprehensive Inorganic Chemistry*, Vol. 2, (J.C. Bailar, H.J. Emeléus, R. Nyholm and A.F. Trotman-Dickenson, eds.), Pergamon Press, Oxford (1973), p. 1517.

[162] L.W. Shacklette, T.R. Jow, M. Maxfield and R. Hatami, *Synth. Metals*, **28**, C 655 (1989).

[163] E. Plichta, S. Slane, M. Uchijama, M. Salomon, D. Hua, W.B. Ebner and H.W. Lin, *J. Electrochem. Soc.*, **136**, 1865 (1989).

[164] K. Wiesener, U. Eckholdt, J. Müller, W. Schneider and D. Rahner, *Bull. Electrochem.*, **5**, 23 (1989).

[165] J.R. Ackridge and H. Vourlis, *Progr. Batt. Solar Cells*, **6**, 81 (1987).

[166] T. Sotomura, T. Kambara and S. Sekido, *Progr. Batt. Solar Cells*, **6**, 25 (1987).

[167] A. Rogero, B. Scrosati, M. Andrei, S. Passerini and U. Pedretti, Eur. Pat. 312,160 (19 Apr. 1989).

[168] S. Kondo, T. Tonomura, T. Kanbara and K. Takada, Jap. Pat. 8921870 (25 Jan. 1989).

[169] G.A. Nazari, D.M. McArthur and J.F. O'Gara, *Polym. Prepr.*, **30**, 430 (1989).

[170] M. Alamgir, R.K. Reynolds and K.M. Abraham, *Proc. Electrochem. Soc.*, **1989**, 89-4, p. 321.

[171] R.G. Kelly and P.J. Morgan, *J. Electrochem. Soc.*, **134**, 25, 31 (1987).

[172] C. Sarrazin, *La Recherche*, No 229 (February 1991), p. 170.

[173] I.I. Astakhov *et al.*, *Electrochim. Acta*, **36**, 1127 (1991).

[174] Kokama Noboru, Jap. Pat. 03 31, 458 (1991), after *Chem. Abstr.*, **114**, 232088.

[175] P. Bro and S.C. Levy, *Quality and Reliability Methods for Primary Batteries*, Wiley, N.Y. (1990).

[176] L. Werblan, J. Izydorek and J. Lesiński, *Proceedings of the 5th International Meeting on Lithium Batteries, Beijing, China, May 27–June 1, 1990*, pp. 353–360.

[177] L. Werblan, A. Białkowska, J. Downarowicz, B. Polszakiewicz, B. Szczęśniak, R. Szłapka and I. Walas, 6th International Meeting on Lithium Batteries, May 10–15, 1992, University of Munster, Germany, Abstracts.

[178] J. Przyłuski and M. Zagórska, *Electrochemistry of Conductive Polymers*, International Autumn School, September 21–27, 1986, Kazimierz nad Wisłą, Poland, Warsaw University of Technology, pp. 105–118; *ibid.*, A. Proń, pp. 92–104.

[178a] K. Jackowska, J. Bukowska, *Electrochemistry of Condustive Polymers*, International Autumn School, Kazimierz nad Wisłą, Poland 1990, Abstracts; *Synthetic Metals*, **35**, (1990) 143.

[179] K.M. Abraham, *Proceedings of the 5th IMLB, Beijing, China, May 27–June 1, 1990*, pp. 13–27.

[180] Cha, Chuan-sin, *Proceedings of the 5th IMLB, Beijing, China, May 27–June 1, 1990*, pp. 5–9.

[181] Qiu Ruizeng and Chen Baozeng, *Proceedings of the 5th IMLB, Beijing, China, May 27–June 1, 1990*, pp. 268–271.

[182] E. Peled, R. Cohen, J. Kimel, Y. Lavi and A. Melman, *Proceedings of the 5th IMLB, Beijing, China, May 27–June 1, 1990*, p. 10.

[183] Y. Geronov, B. Puresheva, V. Moshtev, P. Zlatilova, B. Pistoia and M. Pasquali, *Proceedings of the 5th IMLB, Beijing, China, May 27–June 1, 1990*, pp. 199–206.

[184] Z. Takehara, K. Kanamura, N. Imanishi, S.Yonezawa, *Proceedings of the 5th IMLB, Beijing, China, May 27–June 1, 1990*, pp. 188–198.

[185] W. Greatbatch and C.F. Holmes, *Proceedings of the 5th IMLB, Beijing, China, May 27–June 1, 1990*, pp.71–84.

[186] D.P. Wilkinson, H. Blom and K. Brandt, *Proceedings of the 5th IMLB, Beijing, China, May 27–June 1, 1990*, pp.145–153.

[187] V.S. Kolosnitsyn, A.V. Popov and Yu.E. Nikitin, *Proceedings of the 5th IMLB, Beijing, China, May 27–June 1, 1990*, pp. 329–332.

[188] M. Garreau and J. Thevenin, *Proceedings of the 5th IMLB, Beijing, China, May 27–June 1, 1990*, pp. 396–398.

[189] Chen Rongjin, Li Jin'e and Qiu Erzhan, *Proceedings of the 5th IMLB, Beijing, China, May 27–June 1, 1990*, pp. 576–581.

[190] J.O. Besenhard, Th. Bergknecht, P. Castella and M. Hess, *Proceedings of the 5th IMLB, Beijing, China, May 27–June 1, 1990*, pp 142–144.

[191] F. Walsh, *Proc. 26th Intersoc. Energy Convers. Eng. Conf.*, **1991**, 477.

[192] W.P. Kilroy, *J. Electrochem. Soc.*, **138**, 2220 (1991).

[193] J.R. Dahn, U. v. Sacken, M.W. Juzkow and H. Al-Janaby, *J. Electrochem. Soc.*, **138**, 2207 (1991).

[194] S.D. Jones and J.R. Akridge, *Proc. Int. Symp. Ionic Mixed Cond. Ceram., Electrochem. Soc.*, **1991**, 91–112, p. 145.

[195] P.A. Nelson and T.D. Kaun, *Proc. 26th Intersoc. Energy Convers. Eng. Conf.*, **1991**, 423.

[196] V.J. Farozic and G.A. Prentice, *J. Appl. Electrochem.*, **21**, 767 (1991).

Index